MASS AND HEAT TRANSFER

This book allows instructors to teach a course on heat and mass transfer that will equip students with the pragmatic, applied skills required by the modern chemical industry. This new approach is a combined presentation of heat and mass transfer, maintaining mathematical rigor while keeping mathematical analysis to a minimum. This allows students to develop a strong conceptual understanding and teaches them how to become proficient in engineering analysis of mass contactors and heat exchangers and the transport theory used as a basis for determining how the critical coefficients depend on physical properties and fluid motions.

Students will first study the engineering analysis and design of equipment important in experiments and for the processing of material at the commercial scale. The second part of the book presents the fundamentals of transport phenomena relevant to these applications. A complete teaching package includes a comprehensive instructor's guide, exercises, design case studies, and project assignments.

T. W. Fraser Russell is the Allan P. Colburn Professor of Chemical Engineering at the University of Delaware. Professor Russell is a member of the National Academy of Engineering and a Fellow of the American Institute of Chemical Engineering (AIChE). He has been the recipient of several national honors, including the AIChE Chemical Engineering Practice Award.

Anne Skaja Robinson is an Associate Professor of Chemical Engineering at the University of Delaware and Director of the National Science Foundation (NSF) Integrative Graduate Education and Research Traineeship program in biotechnology. She has received several national awards, including the NSF Presidential Early Career Award for Scientists and Engineers (PECASE/Career).

Norman J. Wagner is the Alvin B. and Julia O. Stiles Professor and Chair of the Department of Chemical Engineering at the University of Delaware. His international teaching and research experience includes a Senior Fulbright Scholar Fellowship in Konstanz, Germany, and a sabbatical as a Guest Professor at ETH, Zurich, as well as at "La Sapienza," Rome, Italy.

Mass and Heat Transfer

ANALYSIS OF MASS CONTACTORS AND HEAT EXCHANGERS

T. W. FRASER RUSSELL
University of Delaware

ANNE SKAJA ROBINSON
University of Delaware

NORMAN J. WAGNER
University of Delaware

CAMBRIDGE
UNIVERSITY PRESS

CAMBRIDGE UNIVERSITY PRESS
Cambridge, New York, Melbourne, Madrid, Cape Town, Singapore, São Paulo, Delhi

Cambridge University Press
32 Avenue of the Americas, New York, NY 10013-2473, USA

www.cambridge.org
Information on this title: www.cambridge.org/9780521886703

First published 2008

A catalog record for this publication is available from the British Library.

Library of Congress Cataloging in Publication Data

Russell, T. W. F., 1934–
    Mass and heat transfer : analysis of mass contactors and heat exchangers / T.W.
Fraser Russell, Anne Skaja Robinson, Norman J. Wagner.
    p. cm. – (Cambridge series in chemical engineering)
Includes bibliographical references and index.
ISBN   978-0-521-88670-3 (hardback)
1. Heat exchangers. 2. Chemical engineering – Equipment and supplies.

I. Robinson, Anne Skaja, 1966– II. Wagner, Norman Joseph, 1962– III. Title.

    TP363.R87 2008
    621.402'2–dc22

                                                                    2007045343

ISBN   978-0-521-88670-3 hardback

This book is dedicated to our families:

Shirley, Bruce, Brian, Carey

Clifford, Katherine, Brenna

Sabine

Contents

PART II

Preface

Chemical engineers educated in the undergraduate programs of departments of chemical engineering have received an education that has been proven highly effective. Chemical engineering educational programs have accomplished this by managing to teach a methodology for solving a wide range of problems. They first did so by using case studies from the chemical process industries. They began case studies in the early part of the 20th century by considering the complete processes for the manufacture of certain chemicals and how they were designed, operated, and controlled. This approach was made much more effective when it was recognized that all chemical processes contained elements that had the same characteristics, and the education was then organized around various unit operations. Great progress was made during the 1940s and 1950s in experimental studies that quantified the analysis and design of heat exchangers and equilibrium stage operations such as distillation. The 1960s saw the introduction of reaction and reactor analysis into the curriculum, which emphasized the critical relationship between experiment and mathematical modeling and use of the verified models for practical design. We have built upon this approach, coupled with the tools of transport phenomena, to develop this text.

Our approach to teaching mass and heat transfer has the following goals:

1. Teach students a methodology for rational, engineering analysis of problems in mass and heat transport, i.e., to develop model equations to describe mass and heat transfer based on the relationship between experimental data and model.
2. Using these model equations, teach students to design and interpret laboratory experiments in mass and heat transfer and then to effectively translate this knowledge to the operation and design of mass and heat transfer equipment.
3. Develop the students' molecular understanding of the mechanisms of mass and heat transfer in fluids and solids and application in the estimation and correlation of mass and heat transfer coefficients.

To achieve these goals we use the following methods:

- Emphasize the critical role of experiment coupled with the development of an appropriate model.
- Focus attention on analysis and model development rather than on mathematical manipulation of equations. This is facilitated by organization of the analysis method into *levels*.

- Provide a rational framework for analyzing mass and heat transfer phenomena in fluids and the associated equipment based on a simple fluid mechanical model of the devices.
- Treat mass transfer on an equal level with heat transfer, and, wherever possible, provide a parallel development of mass and heat transfer phenomena.

The *levels of analysis* introduced in Chapter 1, Table 1.1, provide a guide to the rational analysis of engineering transport equipment and transport phenomena in increasing orders of complexity. The information obtainable from each level of analysis is delineated and the order of analysis preserved throughout the textbook.

We present the material in a manner also suitable for nonmajors. Students with a basic college-level understanding of thermodynamics, calculus, and reaction kinetics should be prepared to follow the presentation. By avoiding the more tedious and sophisticated analytical solution methods and relying more on simplified model equations and, where necessary, modern mathematical software packages, we strive to present the philosophy and methodology of engineering analysis of mass and heat transfer suitable for nonmajors as well. Note that a course in fluid mechanics is not a prerequisite for understanding most of the material presented in this book.

Engineering starts with careful analysis of experiment, which naturally inspires the inquiring mind to synthesis and design. Early emphasis on developing model equations and studying their behavior enables the instructor to involve students in problem-based learning exercises and transport-based design projects right from the beginning of the course. This and the ability to challenge students to apply their analysis skills and course knowledge to transport phenomena in the world around them, especially in emerging technologies in the nanosciences and environmental and biological sciences, result, in our experience, in an exciting and motivating classroom environment. We sincerely hope that you as reader will find this approach to transport phenomena to be as fresh and invigorating as we have.

> Get the habit of analysis—analysis will in time enable synthesis to become your habit of mind.
> — Frank Lloyd Wright

To the Student

This text is designed to teach you how to carry out quantitative analysis of physical phenomena important to chemical professionals. In the chemical engineering curriculum, this course is typically taught in the junior year. Students with adequate preparation in thermodynamics and reactor design should be successful at learning the material in this book. Students lacking a reactor design course, such as chemists and other professionals, will need to pay additional attention to the material in Chapter 2 and may need to carry out additional preparation by using the references contained in that chapter. This book uses the logic employed in the simple analysis of reacting systems for reactor design to develop the more complex analysis of mass and heat transfer systems.

Analysis is the process of developing a mathematical description (model) of a physical situation of interest, determining behavior of the model, comparing the behavior with data from experiment or other sources, and using the verified model for various practical purposes.

There are two parts in the analysis process that deserve special attention:

- developing the mathematical model, and
- comparing model behavior with data.

Our experience with teaching analysis for many years has shown that the model development step can be effectively taught by following well-developed logic. Just what constitutes agreement between model behavior and data is a much more complex matter and is part of the art of analysis. This is more difficult to learn and requires one to consider many different issues; it always depends on the reasons for doing the analysis. Time constraints have a significant impact on this decision, as do resources available. We will illustrate this aspect of analysis by examining chemical reactors, heat exchangers, and mass contactors, equipment of particular interest to chemical professionals.

Determining model behavior requires you to remember some calculus—how to solve algebraic equations and some simple differential equations. This step in analysis is often given too much emphasis because it is the easiest part of analysis to do and is the step for which students have the best background. Do not fall into the common trap of assuming that analysis is primarily concerned with determining model behavior—it is not! Analytical methods to solve algebraic or differential equations are most useful if the manipulations leading to solution give insights into the physical situation being examined. Tedious algebraic manipulations are not help-ful and seriously distract one from the real purpose of analysis. You should stop and ask questions of any instructor who performs a lot of algebra at the board without constantly referring back to what the manipulations mean in terms of the physical situation being studied. In this day and age, computer programs that solve sets of equations are so readily available that tedious algebra is not required.

Once you have mastered how to obtain the model equations, you need to devote your creative energies to deciding if behavior matches experiment. Just what consti-tutes a match is not trivial to determine.

The model development step is simplified by considering the level of complexity required to obtain useful practical results. We define six levels of complexity in this text:

The first level employs only the laws of conservation of mass and/or energy. Time is the only dependent variable in the differential equations considered in Level I analysis, but many problems of considerable significance assume steady state and eliminate time as a variable. In this case the model equations become algebraic.

The second level also employs these two conservation laws, but, in addition, phase, thermal, or chemical equilibrium is assumed. The model equations in a Level II analysis are algebraic because time is not an independent variable when equilibrium is assumed.

A Level III analysis requires a constitutive relationship to be employed. The six constitutive relations needed in studying reactors, heat exchangers, and mass contactors are shown in Tables 1.4 and 1.5. These relations have been veri-fied by various experiments that we will discuss in some detail. Level III analy-sis assumes simple fluid motions, either well mixed or plug flow. Completely stagnant fluids or solid phases can also be handled at this level of analysis. A background in fluid mechanics is not required. A Level III analysis allows one to complete equipment design at the laboratory, pilot, and commercial scales for most single-phase systems. The Level III model equations for well-mixed fluids contain time as the only independent variable if steady state is not assumed. Plug-flow fluid motions require one independent spatial variable in the steady state and time if the steady state cannot be assumed.

To deal effectively with multiphase systems, a Level IV analysis needs to be per-formed. The Level IV analysis also assumes simple fluid motions but requires application of the conservation laws of mass and energy coupled with consti-tutive relations for both phases.

A Level V analysis is restricted to single-phase systems but can employ all the conservation laws. It is the first level in which the law of conservation of momentum is used. In its most complicated form, the model equations of Level V can have time and all three spatial coordinates as independent vari-ables. A Level V analysis considering time and only one spatial direction will be sufficient for most problems we will analyze in this book.

Multiphase systems with complex fluid motion require a Level VI analysis, which we will not consider in this text.

There are two parts to this book. An introduction to the material and method of approach is followed by chapters on chemical reactor analysis (Chapter 2), heat exchanger analysis (Chapter 3), and mass contactor analysis (Chapter 4). These chapters have been developed to highlight the similarities in the analysis methods and in the process equipment used. By using experimentally determined values of the rate constant (k), the heat transfer coefficient (U), the mass transfer coefficient (K_m), and the interfacial area (a), you will be able to solve problems in mass and heat transfer and develop operating and design criteria.

Part II features additional chapters that focus on the microscopic analysis of control volumes to estimate U or K_m for a broad range of systems. Correlations for K_m and U are developed that facilitate the design of equipment.

Chapter 7 provides methods for calculating the area for mass transfer in a variety of mass contacting equipment. Chapter 8 illustrates the technically feasible design procedure through case studies of common mass contactors and heat exchangers.

On successful completion of a course using this textbook, you should understand the basic physical principles underlying mass and heat transfer and be able to apply those principles to analyze existing equipment and design and analyze laboratory experiments to obtain data and parameters.

Finally, you should be capable of performing technically feasible designs of mass contactors and heat exchangers, as well as reading the technical literature so as to continue your education and professional development in this field.

Acknowledgments

The preparation of this text has benefited from significant contributions from numerous Teaching Fellows, teaching assistants, undergraduate students, and colleagues in the Department of Chemical Engineering at the University of Delaware. In particular, we wish to acknowledge the Teaching Fellow program in the Chemical Engineering Department at Delaware, which provides a fellowship semester to a senior graduate student who wishes an internship in university education methods and theory. This competitive program has been in existence since 1992 and has supported 24 student Teaching Fellows (1992–2007). To date, more than 10 former Fellows have become faculty members at a number of institutions.

The Teaching Fellows work closely with faculty in lesson planning, classroom delivery, and discussion of classroom performance. All in-class teaching by Fellows is monitored by the faculty mentors. Regular lively discussions over course content and teaching methods have proven to be infectious, such that many more graduate students and faculty also benefit from active discussions about educational methods and theory. This in fact may well be one of the most important aspects of our program.

This textbook represents a course organization that is fundamentally different from all other courses and textbooks on mass and heat transfer. Our approach builds upon the principles of analysis developed in the text by Russell and Denn, *Introduction to Chemical Engineering Analysis*. The authors, our Teaching Fellows, and teaching assistants evolved the present text through much spirited debate. Our first Teaching Fellow to work on the material in this text was *Will Medlin*, who taught with *Anne Skaja Robinson* and *T. W. Fraser Russell*. Will's enthusiastic acceptance of a different approach to teaching mass and heat transfer and his lively debates with other graduate students helped simplify and categorize the types of fluid motions required for modeling mass and heat transfer equipment and the level analysis that is the hallmark of the modeling approach used here. *Jonathan Romero*, our next Teaching Fellow, helped organize the transfer coefficient correlations. Will's office mate was *Suljo Linic*, an undergraduate physics major who came to Delaware to do graduate work in chemical engineering. This serendipity produced some very lively debates, which often migrated to a faculty office. Typical of such discussions was the following:

> "Why can't we solve all problems in mass and heat transfer with Fourier's or Fick's law and the appropriate set of differential equations?"

Answered by:

> "In the first place, they are not laws but constitutive equations, which themselves have only been verified for solid control volumes or liquid control volumes with no fluid motion . . ."

Suljo was awarded the Teaching Fellow position the next year, with *Norman J. Wagner* and *I. W. Fraser Russell* coteaching the course. His willingness to question the fundamental principles of our analysis disciplined all of us in systematically using level analysis to approach problems in heat transfer. *Wim Thielemans*, the fourth Fellow, helped redraft the chapter on heat transfer based on undergraduate student comments and solved a number of models numerically to illustrate model behavior. *Mark Snyder* accepted the Teaching Fellow position in our fifth year and made significant contributions by classifying mass transfer unit operations equipment using our simplified fluid mechanics analysis. *Yakov Lapitsky*, *Jennifer O'Donnell*, and *Michelle O'Malley*, our sixth, seventh, and eighth Teaching Fellows contributed to numerous examples and tested material in class.

We have also been blessed by an enthusiastic cadre of Delaware undergraduates, who have both challenged us to become better educators and, in some special cases, have made significant contributions to the course content through original research projects. In particular, we would like to thank *Patrick Schilling*, who contributed to the organization and numerical examples found in Chapter 3. Patrick's interest in the topic expanded over the following summer, when he performed original research under our direction on predicting interfacial areas in fluid–fluid systems for use in mass transfer operations, which are summarized in Chapter 7. Our undergraduate classes in the junior-level course in heat and mass transfer have helped clarify and correct errors in the manuscript that we used in class. Other undergraduates and alumni made significant contributions: *Matt Mische* (heat exchanger design), *Steven Scully* (manuscript review), *Brian J. Russell* (index), and *Josh Selekman* (graphs). Any remaining errors are the authors' responsibility.

We also wish to thank the numerous graduate students who contributed to this manuscript and the course through dedicated service as teaching assistants. *Brian Lefebvre, Kevin Hermanson, Nicole Richardson, Yakov Lapitsky, Amit Kumar, Matt Helgeson*, and *Rebecca Brummitt* all made significant contributions to the materials found herein. Brian, Kevin, and Yakov were inspired to become the Teaching Fellows in various courses after their positive experience in the mass and heat transfer course. *Damien Thévénin* took great care in preparing some of the figures in this text, and *Claudio Gelmi* helped analyze the fermentor data.

Multiple authors lead to lively and spirited debates, which results in chaos concerning a written manuscript. The authors are indebted to the organizational and secretarial skills of *Lorraine Holton*, who typed the final manuscript, and *Carrie Qualls*, who prepared the figures to Cambridge University Press standards.

Fraser Russell would like to recognize the influence on this text of his years at Delaware when he collaborated closely with *M. M. Denn*, who taught him how to effectively interpret mathematical models.

We also benefited from the support and encouragement of the Department of Chemical Engineering, our colleagues, and our families, to whom we are most grateful.

We greatly appreciated the effective efforts of *Michelle Carey*, Cambridge University Press, and *Katie Greczylo*, Aptara, Inc., in the production of this book.

Instructors' and Readers' Guide

This book is designed to teach students how to become proficient in engineering analysis by studying mass and heat transfer, transport phenomena critical to chemical engineers and other chemical professionals. It is organized differently than traditional courses in mass and heat transfer in that more emphasis is placed on mass transfer and the importance of systematic analysis. The course in mass and heat transfer in the chemical engineering curriculum is typically taught in the junior year and is a prerequisite for the design course in the senior year and, in some curricula, also a prerequisite for a course in equilibrium stage design. An examination of most mass and heat transfer courses shows that the majority of the time is devoted to heat transfer and, in particular, conductive heat transfer in solids. This often leads to overemphasis of mathematical manipulation and solution of ordinary and partial differential equations at the expense of engineering analysis, which should stress the development of the model equations and study of model behavior. It has been the experience of the authors that the "traditional" approach to teaching undergraduate transport phenomena frequently neglects the more difficult problem of mass transfer, despite its being an area that is critical to chemical professionals.

At the University of Delaware, chemical engineering students take this course in mass and heat transfer the spring semester of their junior year, after having courses in thermodynamics, kinetics and reactor design, and fluid mechanics. The students' analytical skills developed through analysis of problems in kinetics and reactor design provide a basis for building an engineering methodology for the analysis of problems in mass and heat transfer. This text is presented in two parts, as illustrated in Figure I. Part I of this text, shown on the figure as "Equipment-Scale Fluid Motion," consists of Chapters 1–4. Part II of the text is represented by the other two elements in the figure, titled "Transport Phenomena Fluid Motion" (Chapters 5 and 6) and "Microscale Fluid Motion" (Chapter 7). Chapter 8 draws on Parts I and II to illustrate the design of mass contactors and heat exchangers.

Part I of this text is devoted to the analysis of reactors, heat exchangers, and mass contactors in which the fluid motion can be characterized as well mixed or plug flow. Table I indicates how Chapters 2, 3, and 4 are structured and details the fluid motions in each of these pieces of equipment. Such fluid motions are a very good approximation of what is achieved pragmatically and in those situations in which the fluid motion is more complex. The Table I analysis provides useful limits on performance. The model equations developed in Part I are essential for the analysis of

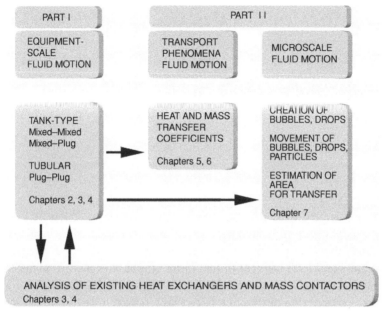

Figure I. Analysis of existing heat exchangers and mass contactors.

existing equipment and for the design of new equipment. Experiments performed in existing equipment, particularly at the laboratory scale, determine reaction-rate constants, heat transfer coefficients, mass transfer coefficients, and interfacial area and are necessary to complete the correlations developed in Part II. Carefully planned experiments are also critical to improving operation or control of existing laboratory-, pilot-, or commercial-scale equipment.

Another way to characterize our approach to organizing the analysis of equipment and transport problems is shown in Table II (see p. xxiv). This is presented to give guidance to the emphasis instructors might like to place on the way they teach from

Table I. Equipment fluid motion classification

Reactors: Single phase	Reactors: Two phase	Heat exchangers	Mass contactors
Single control volume	Two control volumes	Two control volumes	Two control volumes
Tank type	Tank type	Tank type	Tank type
Mixed–mixed	*Mixed–mixed*	*Mixed–mixed*	*Mixed–mixed*
• Batch	• Batch	• Batch	• Batch
• Semibatch	• Semibatch	• Semibatch	• Semibatch
• Continuous	• Continuous	• Continuous	• Continuous
	Mixed–plug	*Mixed–plug*	*Mixed–plug*
	• Semibatch	• Semibatch	• Semibatch
	• Continuous	• Continuous	• Continuous
Tubular	Tubular	Tubular	Tubular
Plug flow	*Plug flow*	*Plug–plug flow*	*Plug–plug flow*
	• Cocurrent	• Cocurrent	• Cocurrent
		• Countercurrent	• Countercurrent

this text. Level I and Level II analyses are discussed in the first sections of Chapters 2, 3 and 4. Chapters 2 and 3 require a Level III analysis. Chapter 4 demonstrates the importance of a Level IV analysis. Part II continues with Level I, II, and III analyses in Chapter 5 but introduces two new constitutive equations, shown in Table 1.4. Chapter 6 requires a Level V analysis to develop relationships for mass and heat transfer coefficients. This text does not deal with any Level VI issues except in a minor way in Chapter 7, which provides methods for estimating interfacial areas in mass contactors. In teaching the material in this text it is crucial that students understand the critical role of experiment in verifying the constitutive equations for rate of reaction, rate of heat transfer, and rate of mass transfer summarized in Table 1.5. It is these constitutive equations that are used in Chapters 2, 3, and 4 in the model equations for the fluid motions, as outlined in Table I. The most critical elements in Part I of this text are therefore

- 2.1 The Batch Reactor
- 2.2 Reaction Rate and Determination by Experiment
- 3.1 Batch Heat Exchangers
- 3.2 Rate of Heat Transfer and Determination by Experiment
- 4.1 Batch Mass Contactors
- 4.2 Rate of Mass Transfer and Determination by Experiment

The students in our course at the University of Delaware have taken a course in chemical engineering kinetics, so we expect students to know how to obtain reaction-rate expressions and how to use the verified rate expression in the design of continuous tank-type and tubular reactors. Of course, some review is always necessary because it is important for students to realize that we build carefully on the reaction analysis to study mass and heat transfer. We try to limit this review to one to two class periods with appropriate homework.

In teaching Chapter 3 on heat transfer we believe that one should cover, in addition to Sections 3.1 and 3.2, the following sections, which demonstrate the utility of the constitutive equation for heat transfer:

- 3.3.2.1 Semibatch Heat Exchanger, Mixed–Mixed Fluid Motions
- 3.4 Tubular Heat Exchangers

We often add another heat exchanger analysis, such as Subsection 3.3.3, so we have model equations that we can compare with the mass contactor analysis. We normally devote between 6 and 8 class hours to heat exchanger analysis of existing equipment for which the heat transfer coefficient U is known. Prediction of U is covered in Chapters 5 and 6.

Our major emphasis in the course we teach is Chapter 4, and we believe that it deserves between 9 and 12 hours of class time. The model equations are developed for the two control volumes as for heat exchangers so one can draw comparisons that are useful to cement the students' understanding of the modeling process. The major differences between heat exchanger analysis and mass contactor analysis are the equilibrium issues, the approach to equilibrium conclusions, and the issues raised by direct contact of the two phases. In addition to the mass transfer coefficient K_m,

Table II. Level definitions

	Level I	Level II	Level III	Level IV	Level V	Level VI
Outcome	Overall mass and energy balances	Allows number of equilibrium stages to be determined *but does not allow stage design to be achieved*	Equipment design at the laboratory, pilot, and commercial *scale* can be achieved		Molecular conduction and diffusion analysis are quantified; mass and heat transfer coefficient correlations are developed	
Conservation of mass	Required	Required	Required	Required	Required	Required
Conservation of energy	Required	Required	Required	Required	Required	Required
Conservation of momentum	Not required	Not required	Not required	Not required	Required	Required
Equilibrium constitutive relations	Not required	Required	May be required	May be required	May be required	May be required
Rate equations	Not required	Not required	Required	Required	Not required	Not required
Conduction–diffusion– viscosity relations	Not required	Not required	Not required	Not required	Required	Required

there is the interfacial area a to be considered. Methods to estimate K_m are covered in Chapters 5 and 6. Procedures for estimating a are given in Chapter 7.

Chapter 5 is devoted to experimental justification of the two constitutive relations commonly referred to as Fourier's and Fick's "Laws." This development, in Sections 5.1 and 5.2, thus parallels our discussions in Chapters 2, 3, and 4 that provide experimental evidence for the constitutive relations for rate of reaction, rate of heat transfer, and rate of mass transfer. The derivations for the overall coefficients, U and K_m, in terms of individual resistances can be skipped if time is short, but the resulting expressions are essential. The material on membrane diffusion may be of interest in some situations.

Chapter 6 also contains more material than one can reasonably cover in a typical 40 hours of class time, so choices have to be made depending on the emphasis one desires. It is probably necessary to cover most of Sections 6.2, 6.3, and 6.4, but one needs to avoid long lectures in which there is excessive algebraic manipulation—it is the resulting correlations that are critical. These are summarized in Section 6.4.

In Chapter 7 we treat the challenging problem of estimating interfacial areas in both tank-type and tubular mass contactors. This is an area of active research today, but we have tried to present the current state of the art so this critical parameter for rational scale-up and design can be estimated.

Part II concludes with Chapter 8, which presents designs that can be completed once the mathematical models from Part I are available and methods for estimating U, K_m, and a are available from Part II. This is illustrated in Figure II. These design

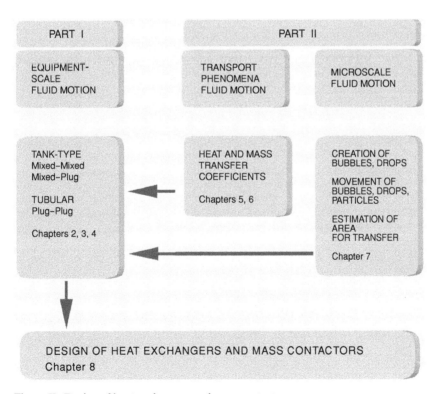

Figure II. Design of heat exchangers and mass contactors.

case studies evolved from in-class problem-based learning exercises as well as from group semester project assignments and can be used as bases for such activities.

There is a good deal more material in this text than one can reasonably cover in 40 hours of class time. We have endeavored to produce a text that gives the instructor and student maximum flexibility without sacrificing the logic of sound engineering analysis.

This book is not a reference book, nor is it an exhaustive compendium of phenomena, knowledge, and solved problems in mass and heat transfer. Suitable references are provided in each chapter for further study and for aid in the analysis of phenomena not treated herein in depth. As a first course in mass and heat transfer, this book is limited in scope and content by design. As an instructor, we hope you can build upon this book and tailor your lectures to incorporate your own expertise and experiences within this framework to enrich the course for your students.

MASS AND HEAT TRANSFER

PART I

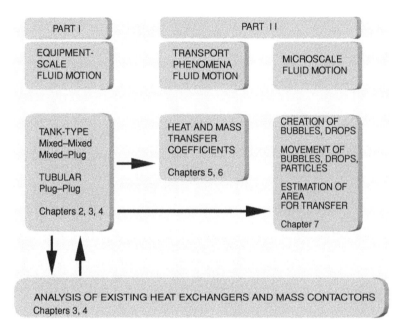

PART I

EQUIPMENT-
SCALE
FLUID MOTION

PART II

TRANSPORT
PHENOMENA
FLUID MOTION

MICROSCALE
FLUID MOTION

TANK-TYPE
Mixed–Mixed
Mixed–Plug

TUBULAR
Plug–Plug

Chapters 2, 3, 4

HEAT AND MASS
TRANSFER
COEFFICIENTS

Chapters 5, 6

CREATION OF
BUBBLES, DROPS

MOVEMENT OF
BUBBLES, DROPS,
PARTICLES

ESTIMATION OF
AREA
FOR TRANSFER

Chapter 7

ANALYSIS OF EXISTING HEAT EXCHANGERS AND MASS CONTACTORS
Chapters 3, 4

1   Introduction

All physical situations of interest to engineers and scientists are complex enough that a mathematical model of some sort is essential to describe them in sufficient detail for useful analysis and interpretation. Mathematical expressions provide a common language so different disciplines can communicate among each other more effectively. Models are very critical to chemical engineers, chemists, biochemists, and other chemical professionals because most situations of interest are molecular in nature and take place in equipment that does not allow for direct observation. Experiments are needed to extract fundamental knowledge and to obtain critical information for the design and operation of equipment. To do this effectively, one must be able to quantitatively analyze mass, energy, and momentum transfer (transport phenomena) at some level of complexity. In this text we define six levels of complexity, which characterize the level of detail needed in model development. The various levels are summarized in Table 1.1.

Level I, Conservation of Mass and/or Energy. At this level of analysis the control volume is considered a black box. A control volume is some region of space, often a piece of equipment, that is designated for "accounting" purposes in analysis. Only the laws of conservation of mass and/or energy are applied to yield the model equations; there is no consideration of molecular or transport phenomena within the control volume. It is a valuable approach for the analysis of existing manmade or natural systems and is widely employed. The mathematical expressions needed to describe Level I problems are algebraic or simple first-order differential equations with time as the only independent variable:

A calculation of the flows and stream compositions to and from a continuously operating distillation column illustrates this level of analysis. A sketch of the system is shown as Figure 1.1. A typical column has internals or some type of trays, a condenser, a reboiler, and pumps for circulation of liquids. The mass flow rate to the column, F, must be equal to the mass flow rate of the distillate, D, plus the mass flow rate of the bottoms product, B,

$$F = D + B.$$

This simple mass balance always holds and is independent of the column diameter, type, or number of trays or the design of the condenser or the reboiler. We must know

Table 1.1. Level definitions

	Level I	Level II	Level III	Level IV	Level V	Level VI
Outcome	Overall mass and energy balances	Allows number of equilibrum stages to be determined but does not allow stage design to be achieved	Equipment design at the laboratory, pilot and commercial scale can be achieved		Molecular conduction and diffusion analysis are quantified; mass and heat transfer coefficient correlations are developed	
Conservation of mass	Required	Required	Required	Required	Required	Required
Conservation of energy	Required	Required	Required	Required	Required	Required
Conservation of momentum	Not required	Not required	Not required	Not required	Required	Required
Equilibrium constitutive relations	Not required	Required	May be required	May be required	May be required	May be required
Constitutive relations						
Rate equations	Not required	Not required	Required	Required	Not required	Not required
Conduction–diffusion–viscosity relations	Not required	Not required	Not required	Not required	Required	Required

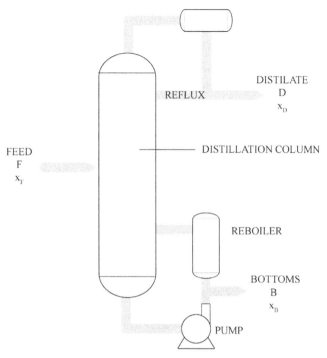

Figure 1.1. Distillation column.

two of the quantities in this equation before we can calculate the third. Such values can come from a distillation column already operating or we may specify the required quantities if considering a process design.

We can also write component mass balances for the column, and these are subsequently presented for a two-component system. The mass fraction of one component in the feed stream to the column is x_F, and the total amount of this component entering the column is Fx_F; similarly, the amounts leaving in the distillate and the column bottoms can be represented by Dx_D, and Bx_B. The component balance relation becomes

$$Fx_F = Dx_D + Bx_B.$$

These two simple equations are the first step in any analysis of a distillation column. Many other useful relations can be readily derived by selecting portions of the column as a control volume. A very clear discussion of this type of Level I analysis is presented in Chapter 18 of Unit Operations of Chemical Engineering by McCabe et al. (1993).

Level II, Conservation of Mass and/or Energy and Assumption of Equilibrium. At this level, transport phenomena are not considered. The analysis of molecular phenomena is simplified by assuming that chemical, thermal, or phase equilibrium is achieved in the control volume of interest. Chemical equilibrium is characterized by K_{eq}, a quantity that can be obtained from tabulated values of the free energy for many reactions. Thermal equilibrium is achieved in devices that exchange heat when all temperatures are the same (for a closed two-volume control system $T_{1\infty} = T_{2\infty}$). The constitutive relationships describing phase equilibria can be complex, but decades of research have produced many useful constitutive relationships for phase equilibria. The simplest for a liquid–liquid system is Nernst's "Law," which relates species concentration in one liquid to that in another by a constant called a distribution coefficient. Henry's "Law" is another simple example relating concentration of a dilute species in a liquid phase to the concentration of the same species in a vapor phase by a single constant. Assuming phase equilibria has proven most valuable in determining the number of theoretical stages to accomplish some stated goal of mass transfer between phases, but does not allow the stage design to be specified. When equilibrium is assumed for reacting systems, the concentrations of product and reactant can be determined but not the volume of the reactor. Almost all situations of interest at Level II require only algebraic equations.

In our example of the distillation column, detailed computer procedures that can handle complex equilibrium relations are available to determine the number of theoretical stages. To do so, mass and energy balances need to be derived by first selecting as a control volume an individual stage (tray) in the column. The set of such algebraic equations can become quite complex and requires a computer for solution. Widely used programs to do this are contained in the software package from Aspen Technology (see reference at the end of this chapter). However, even with a numerical solution, such an analysis is not able to specify the tray design or the number of actual trays needed.

Level III, Conservation of Mass and/or Energy and Use of Constitutive Relationships. This is the first level at which the rate of transport of mass and energy is considered. Momentum transfer is simplified by assuming simple single-phase fluid motions or no fluid motions within the chosen control volume. It is the level of analysis that is almost always employed for laboratory-scale experiments and is one in which pragmatic equipment design can be achieved. Only two types of limiting fluid motion are considered:

- well-mixed fluid motion,
- plug-flow fluid motion.

Well-mixed fluid motion within a control volume is easy to achieve with gases and low-viscosity liquids and almost always occurs in small-scale batch laboratory experimental apparatus. It is also relatively easy to achieve in pilot- or commercial-scale equipment. A very complete analysis of mixing and mixing equipment is presented in the *Handbook of Industrial Mixing* (Paul et al., 2004). By "well mixed," we mean that one can assume there is no spatial variation in the measured variable within the control volume. This is easy to visualize with a batch system because we are not introducing any fluid into the vessel as the process takes place. However, there is a conceptual and sometimes a pragmatic difficulty when one considers a semibatch (sometimes referred to as a fed-batch) in which one fluid is introduced to the system over the time the process is taking place or a continuous-flow system in which fluids both enter and leave the vessel. The well-mixed assumption requires that any fluid introduced immediately reach the average property of the bulk fluid in the vessel.

Of course, this is not exactly what occurs in a real system. Close to the point of introduction, the fluid in the vessel will have properties that are some intermediate value between that of the bulk fluid and that of the incoming fluid. In the majority of situations this will not materially affect the model equation behavior that is developed with the well-mixed assumption. There are a few cases with either chemical or biochemical reactors in which special attention must be paid to the fluid introduction. These are discussed in Chapters 2, 4, and 7. The well-mixed assumption also requires that any fluid leaving the vessel have the same properties as those of the well-mixed fluid in the vessel. This is almost always the case in well-designed vessels.

Plug-flow motion is the other extreme in fluid motion behavior, and the analysis assumes that changes occur in one spatial direction only. This is frequently a good assumption if one is concerned with chemical reaction, mass and/or heat transfer in pipes. A Level III analysis is particularly valuable for the study of experiments to investigate molecular phenomena. Many simple chemistry experiments are carried out in well-mixed batch apparatus.

The models in a Level III analysis of a fluid system are first-order ordinary differential equations with either time or one spatial dimension as the independent variable.

In a solid or for the case in which there is no fluid motion, one can model using time and all three spatial coordinates if required.

In our distillation column example, a Level III analysis is not useful because we are dealing with two phases, a liquid and a vapor. When more than one phase is present, we need a Level IV analysis.

A Level III problem is illustrated in the discussion following the definition of levels.

Level IV, Level III Equivalent for Multiple Phases. In problems in which there is more than one phase, the fluid motions are often extremely complex and difficult to quantify. However, many significant problems can be solved by assuming simple two-phase fluid motions. The following fluid motion categories have proven useful:

- both phases well mixed,
- both phases in plug flow,
- one phase well mixed, one phase in plug flow.

Assuming these simplified fluid motions allows an analysis of experiment and pragmatic equipment design to be achieved.

Level IV analysis of gas–liquid systems is illustrated in a series of papers that have been widely employed for analysis of experiment and equipment design (Cichy et al., 1969; Schaftlein and Russell, 1968).

Level V, Complex Analysis of Single-Phase Transport Phenomena. This level considers the analysis of transport phenomena by considering time and, if required, all three spatial variables in the study of single-phase fluid motions. It is the first level in which we may need detailed fluid mechanics. Most problems of interest to us in this text will involve time and one spatial direction and are discussed in Part II of the text.

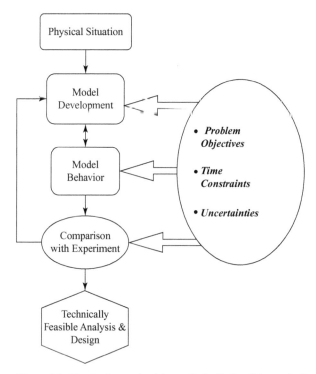

Figure 1.2. The logic required for technically feasible analysis and design.

Level VI, Complex Analysis of Multiple-Phase Transport Phenomena. The analysis of Level VI is extended to multiphase systems and is an area of active research today. We will not consider problems at this level.

In Part I of this text we consider those physical situations that we can model at Levels I through IV, with most of our analysis concentrating on Level III. In Part II we solve some Level V problems to gain insight into the small-scale fluid motions affecting heat and mass transfer.

Figure 1.2, modified from that presented in *Introduction to Chemical Engineering Analysis* by Russell and Denn (1972), identifies the critical issues in the analysis of mass and heat transfer problems that we discuss in this text. The logic diagram and the definition of levels guide us to the proper choice of model complexity.

Defining problem objectives and uncertainties, and identifying time constraints, are of critical importance but are often ignored. The objectives can have a great impact on the complexity of the model, which in turn affects time constraints on any analysis. One should always strive to have the simplest model that will meet the problem objectives (which almost always include severe time constraints and uncertainties). The problem complexity levels defined in Table 1.1 are critical in this evaluation. There are many reasons for this, which we will illustrate as we proceed to develop our approach to heat and mass transfer.

Models of the physical situations encountered in heat and mass transfer are almost always a set of algebraic or differential equations. A straightforward application of the laws of conservation of mass, energy, and momentum allows one to derive the model equations if one attains a certain level of skill in selecting the right constitutive equation to use in the basic conservation law equations. Comparison of model behavior with data from laboratory experiments or other sources requires that model behavior be determined. One can do this by analytical procedures or by numerically solving equations using any one of a number of numerical software packages. It is often this part of analysis that is given the most attention because determining behavior follows well-established rules and is thus easier to teach even though it can be extremely tedious. Overconcentration on model behavior can take away from the time available to deal with the more critical issues of analysis, such as model development and evaluation of model uncertainty.

In our view, the most critical and indeed the most interesting part of analysis is deciding when a satisfactory agreement has been reached when one compares model behavior with experimental reality. *Model evaluation and validation is a nontrivial task and is the art of any analysis process.* Note that it is dependent on the problem objectives. It is a matter we will return to many times as we develop a logical approach to mass and heat transfer.

A verified model allows us to plan additional experiments if needed or to use the equations for the design, operation, and control of laboratory- pilot- or commercial-scale equipment.

We can illustrate the analysis process most effectively with a simple example using a reaction and reactor problem. A single-phase liquid reacting system is the simplest Level III problem chemical engineers and chemical professionals encounter, and its analysis effectively illustrates the crucial link among experiment, modeling, and design. We first must derive the model equations.

Figure 1.3, modified from Figure 2.1 presented in *Introduction to Chemical Engineering Analysis* by Russell and Denn (1972), is a well-tested guide needed to obtain the model equations. We review in this book that part of model development critical to mass and heat transfer. A variety of physical situations that require analysis are discussed. In academic environments professors provide prose descriptions of the physical situation that students are then expected to use for model development. For any situation of reasonable complexity it is difficult to provide a completely adequate prose description, and this can lead to frustrations. In what is often called the "real world," the physical situation is defined by dialogue with others and direct observation.

EXAMPLE 1.1. In our example reaction and reactor problem, the physical situation that must be modeled first is a batch reactor. We assume that our experiments will be carried out in a well-mixed 1-L glass flask that is kept at the temperature of interest by immersion in a temperature-controlled water bath. Figure 1.4 shows the same steps as in Figure 1.3 for this physical situation. For this example problem, a photo of this experimental apparatus is shown in Figure 1.5. We consider that a liquid-phase reaction is taking place in which a compound "A" is converted to a product "D." Previous experiment has shown that the chemical equation is

$$A \rightarrow D.$$

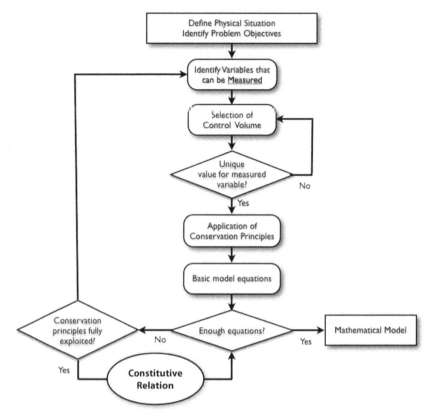

Figure 1.3. Model development logic.

One may need one or more of the basic variables of mass, energy, and momentum to begin model development. These quantities are conserved within a given control volume. The word statement that allows us to begin an equation formulation is presented as Figure 1.6 [Russell and Denn (1972)]. For most situations encountered in mass and heat transfer we will make more frequent use of the conservation of mass and energy than of momentum. When it is not possible to measure mass or energy directly, we must express them with dependent variables that can be measured.

Our conserved variable in the batch reacting system is mass. Because the total mass in the flask is constant we are concerned with the mass of the various species, which also satisfy the word statement of the conservation laws.

The second block on the logic diagrams, Figures 1.3 and 1.4, selection of variables that can be measured, is very important in developing our quantitative understanding of any physical situation. It is easiest to visualize experiments for measuring mass. For example, if we are concerned with the flow of a liquid from a cylindrical tank mounted on a scale, m (mass) can be directly measured. If a scale were not available we could determine mass of liquid by measuring the height of liquid h, the liquid density ρ, (available in physical property tables), and the measured area of the tank A, thus expressing the mass of liquid in the tank by the combination ρAh. Temperature T and concentration of a species C are two of the most common measured state variables

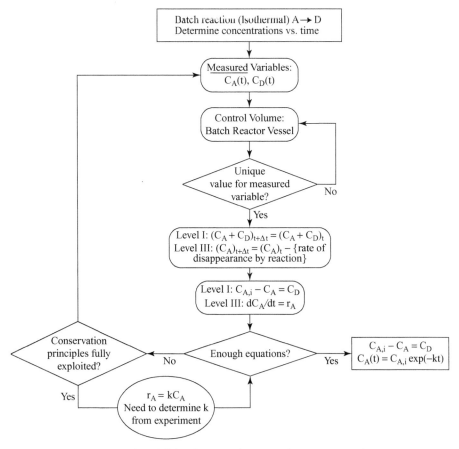

Figure 1.4. Application of model development (Figure 1.3) to batch reaction example.

used in mass and heat transfer problems, but there are many other variables that may be needed. Some require a specially designed experiment, and some are obtained from physical property tables in handbooks or from software programs (experiments done by others and tabulated). The experiment to obtain any variable should be clearly understood; an ability to do so is critical for effective interpretation of the model equations.

For the batch reacting system we assume that component A can be measured by some analytical technique, which gives its concentration in mass per volume; here, we use the symbol C_A to denote this concentration in g-mole per liter. Thus the mass of A will be $VC_A\,MW_A$, where V is the volume of liquid in the flask measured in liters and MW_A is the molecular weight of A in grams per mole.

To make use of the word statement in Figure 1.6, a control volume must be defined. In a great many problems of interest the control volume can be taken as the piece of equipment in which the operation of interest is occurring. When this is done it is assumed that there is a unique value for the measured variables that is the same everywhere within the piece of equipment at any time (note that application of the word statement always produces an equation with time as an independent variable). This is often called a lumped analysis because it is assumed that there is no

Figure 1.5. Laboratory-scale experiment (photo by Katherine S. Robinson).

spatial variation in the measured variables. Level III problems in which we assume a well-mixed control volume are in this category. This assumption of complete mixing within the control volume is widely used and of considerable pragmatic significance, particularly in the analysis of batch laboratory-scale experiments and the design of pilot- and commercial-scale equipment.

Our choice of control volume (third block in Figure 1.4) is the liquid in the flask. The concentration of A, C_A, will be the same at any place a sample is taken from the liquid in the flask provided the liquid is well mixed.

Selection of a control volume within the piece of equipment leads in its most complex application to model equations with measured variable dependence on all three spatial directions. This produces equations with time and all three spatial coordinates as independent variables. Such equations are derived in all of the many books on transport phenomena and are presented in Table 1.2 for momentum, energy, and a component mass balance. The uppercase D in these equations is shorthand to represent the substantial time derivative, meaning that this derivative is calculated only for fixed initial coordinates (Hildebrand, 1976), a quantity that we will discuss later. The gradient ∇ in the equations is shorthand to account for variations in all the spatial directions and is employed to save writing equations for each spatial coordinate (as will be discussed in Chapter 6).

The equations are rarely used with more than time and one spatial direction as dependent variables because of the difficulty of obtaining experimental results for any systems other than solids, stagnant fluids, or fluids in laminar motion.

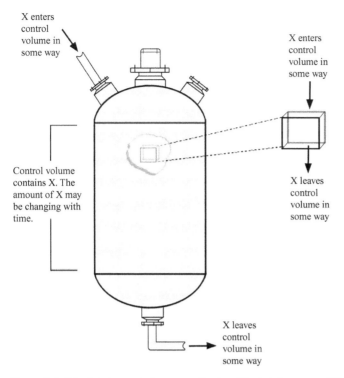

Figure 1.6. Word statement of conservation law. The total amount of X (which may be mass, momentum, or energy) contained within the control volume at time t + Δt must be equal to the total amount of X contained within the control volume at time t, plus the total amount of X that has appeared in the control volume in the time interval Δt by all processes, less the total amount of X that has disappeared from the control volume in the time interval Δt by all processes.

Table 1.2. Momentum, energy, and component mass balances

Momentum equation for a Newtonian fluid:

$$\rho\frac{D\vec{v}}{Dt} = -\nabla P + \mu\nabla^2\vec{v} + \rho\vec{g}$$

Energy equation for constant ρ, μ, k:

$$\rho\hat{C}_p\frac{DT}{Dt} = k\nabla^2 T + \mu\Phi_v$$

Conservation of species for constant D_{AB} and multiple reaction (j):

$$\frac{DC_A}{Dt} = D_{AB}\nabla^2 C_A + \sum_j r_{Aj}$$

EXAMPLE 1.1 CONTINUED. We can perform a Level I analysis for our reacting system using the word statement (Figure 1.6). Because mass is conserved in the control volume,

$$V\left(C_A MW_A + C_D MW_D\right)_{t+\Delta t} = V\left(C_A MW_A + C_D MW_D\right)_t,$$

where MW_A and MW_D, are the molecular weight of A and D, respectively.

Designating C_A and C_D at time $t = 0$ with the subscript i to indicate the initial values gives

$$(C_A - C_{Ai})\, MW_A = (C_{Di} - C_D)\, MW_D.$$

Because A re-forms into D, $MW_A = MW_D$:

$$C_A - C_{Ai} = C_{Di} - C_D.$$

We can carry out the Level I analysis for reacting systems by using molar concentrations if we use information given in the chemical equation, in this case A→D. Further, we consider a case in which there is no product D at time zero so $C_{Di} = 0$.

Next, we analyze the rate of conversion of A to D by performing a Level III analysis on the batch reactor. The word statement for our reacting system is

(VC_A) at $t + \Delta t = (VC_A)$ at $t +$ amount of A that enters the control volume by reaction during $\Delta t -$ amount of A that leaves the control volume by reaction.

Because the chemical equation shows that A is only consumed by reaction, we can simplify the word statement:

(VC_A) at $t + \Delta t = (VC_A)$ at $t -$ amount of A that leaves the control volume by reaction during Δt.

The first problem is always to remove the prose statements from the word statement and replace them with relations containing measured variables. The loop in Figures 1.3 and 1.4, which counts equations and unknowns, is needed to ensure that all the conservation laws for mass, energy, and momentum have been exploited and appropriate constitutive relations found. Except for very simple situations, virtually all problems require at least one constitutive equation. Finding the proper constitutive equation is a nontrivial task often involving well-planned experimental and trial-and-error procedures.

For our reacting system we can postulate that the rate at which A leaves the control volume by conversion to D may be represented by the constitutive relation kC_A. This is a quantity with dimensions of mass per time per unit volume and is often designated by the symbol r_A. The numerical value of the specific reaction-rate constant k (with units of reciprocal time) and the functional form we have just assumed must be verified by experiment.

We obtain the model equation for the batch reactor by substituting kC_A into the word statement:

$$VC_{A_{t+\Delta t}} = VC_{A_t} - kC_A V \Delta t.$$

All terms are divided by $V\Delta t$, and the limit is taken as Δt approaches 0 to yield

$$\frac{dC_A}{dt} = -kC_A. \tag{1.1}$$

We can also obtain this first-order differential equation directly from the species conservation equation in Table 1.2 by simplifying for a batch, homogeneous reactor with the constitutive relation $r_A = -kC_A$. The equation is separable, i.e., we can divide both sides by C_A and then integrate to yield

$$\ln \frac{C_A}{C_{Ai}} = -kt. \tag{1.2}$$

Table 1.3. *Simplified model equations for batch heat exchangers and mass transfer contactors*

Batch reactor $(A + B \rightarrow D)$ (irreversible, first order in A and B):

$$\frac{dC_A}{dt} = -kC_A C_B$$

$$\frac{dC_B}{dt} = -kC_A C_B$$

$$\frac{dC_D}{dt} = kC_A C_B$$

Batch heat exchanger (average thermal properties):

$$\rho_1 V_1 \hat{C}p_1 \frac{dT_1}{dt} = -Ua\left(T_1 - T_2\right)$$

$$\rho_2 V_2 \hat{C}p_2 \frac{dT_2}{dt} = Ua\left(T_1 - T_2\right)$$

Batch mass contactor (dilute in A):

$$\frac{V^I dC_A^I}{dt} = -K_m a\left(C_A^I - MC_A^{II}\right)$$

$$\frac{V^{II} dC_A^{II}}{dt} = K_m a\left(C_A^I - MC_A^{II}\right)$$

Typical equationsfor the Level III lumped analysis of heat exchangers and mass contactors are shown in Table 1.3. The equations presented in Table 1.3 represent the energy and mass balances for the batch systems shown in Figure 1.7 and are obtained by use of the word statement in Figure 1.6. We present them here without derivation so the model development logic can be illustrated. We will derive them in Chapters 3 and 4.

All Level III problems in heat and mass transfer require a constitutive relation to complete them, and the equations in Tables 1.4 and 1.5 contain those constitutive relations most commonly employed. Table 1.5 also shows a constitutive equation for rate of reaction for comparison purposes.

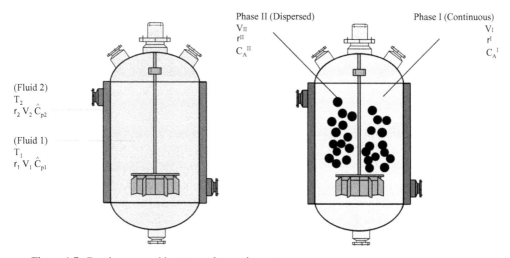

Figure 1.7. Batch mass and heat transfer equipment.

Table 1.4. Constitutive equations for mass, heat,
and momentum transfer

Fick's "Law":	$J_{A,x} = -D_{AB}\dfrac{dC_A}{dx}$
Fourier's "Law":	$\dfrac{Q_x}{n} = -k\dfrac{dT}{dy}$
Newton's "Law" of viscosity:	$\tau_{xy} = -\mu\dfrac{dV_y}{dx}$

Table 1.5. Constitutive equations for rate

Rate of reaction $A + B \to D$ (first order in A and B):	$r = kC_AC_B$
Rate of heat transfer:	$Q = Ua(T_1 - T_2)$
Rate of mass transfer:	$ar_A = K_ma(C_A^I - MC_A^{II})$

The constitutive equations in Table 1.4 for the flux of heat transfer and mass transfer were obtained in the 19th century by experiments carried out by Fourier and Fick, respectively. These are discussed in detail in Chapter 5. The constitutive equations in Table 1.5 for heat transfer have been verified by many experiments. It is possible to estimate U for almost all cases of interest by making use of experimental data and judicious use of the model equations in Table 1.2, simplified so time and one spatial direction are specified as independent variables. The constitutive equation for rate of mass transfer is not so readily verified because experiments are much more difficult to carry out. The main reason for this is the difficulty of designing and carrying out experiments, which will allow the interfacial area a and the mass transfer coefficient, K_m, to be estimated independently.

Equation (1.1) is a model of much simpler form than those shown in Tables 1.2 and 1.3, but it serves to illustrate the analysis process shown in Figure 1.2 and provides a guide to the rational analysis of the more experimentally and mathematically complex problems in mass and heat transfer.

EXAMPLE 1.1 CONTINUED. If our problem objective is to find a value for the rate constant k, we need experimental data, and these are presented in Figure 1.8. The figure is plotted as the natural logarithm of the concentration versus time as indicated by Eq. (1.2). Rearranging this equation yields an exponentially decaying concentration with time:

$$C_A = C_{Ai}e^{-kt}. \tag{1.3}$$

The slope of the line in Figure 1.8 is the rate constant k and for this example it has a value of 0.031 min^{-1}. Comparison of the model behavior with the simple experiment has allowed k to be estimated. If our problem objective is to publish a short research note, we can consider that the agreement is satisfactory and the analysis process complete. If we wish to use the experimental results to design a process for the sale of product D, more analysis needs to be done. This often requires a pilot-scale operation to check and expand on the laboratory experiment and to provide product for testing.

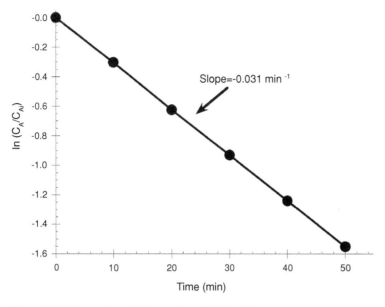

Figure 1.8. Comparison of data and model behavior.

Our problem objective (Figure 1.2) now becomes the design of a pilot-scale reactor to make 100 000 g-mol of D per year for product testing. This is a production rate of 0.198 g-mol/min. A continuous-flow reactor to accomplish this is sketched in Figure 1.9. The raw material A flows to the reactor at a rate q (in liters per minute) and is available at a concentration of $C_{AF} = 0.2$ g-mol/liter. The component balance model equations for this reactor operating at steady-state conditions (no changes with time) are

$$0 = qC_{AF} - qC_A - kC_A V, \tag{1.4}$$

$$0 = -qC_D + kC_A V. \tag{1.5}$$

We can obtain a Level I result by adding Eqs. (1.4) and (1.5) to yield

$$C_{AF} - C_A = C_D.$$

We can obtain this result without any knowledge of the constitutive relation $r_A = kC_A$.

To carry out a Level II analysis, we would need enough additional data at longer times to establish a rate relationship for the reverse reaction D → A, if any reversibility is suspected. Otherwise, at long time all A will be converted.

A technically feasible design is any combination of flow rate q and reactor liquid volume V that will produce 0.198 g-mol/min of D. The utility of the technically feasible design in teaching is discussed in a paper by Russell and Orbey (1992). In this book we make extensive use of the technically feasible design concept.

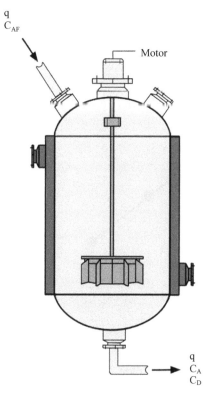

Figure 1.9. Continuous-flow well-mixed reactor.

The quantity qC_D is equal to 0.198 g-mol/min, and Eq. (1.5) can be readily rearranged to obtain an expression for V:

$$V = qC_D/kC_A.$$

Because we have found $k = 0.031$ min^{-1} from a batch experiment, the well-mixed tank reactor volume V can be obtained for any value of C_A less than the raw material feed stream concentration, $C_{AF} = 0.2$ g-mol/L. This assumes that the estimation of k from the batch experiment adequately represents the molecular phenomena and is not equipment dependent.

This will not be an optimal design, but we can always obtain critical insights into the effect of the reactor design on the downstream process design by examining technically feasible designs. An optimal design for this simple system, which considers income from sales and operating and capital costs is presented in Introduction to Chemical Engineering Analysis (Russell and Denn, 1972).

If half the A is reacted to D, $C_A = 0.1$ g-mol/L. and the reactor liquid volume to accomplish this is V = 63.8 L. This requires a flow rate of 1.98 L/min and produces a product D concentration of 0.1 g-mol/min [$C_D = C_{AF} - C_A$, obtained by adding Eqs. (1.4) and (1.5)]. A higher conversion of the raw material A to product D will require larger reactor volumes and for $C_A = 0.01$ g-mol/L, V is 639 L, q = 1.04 L/min, and $C_D = 0.19$ g-mol/L. The capital and operating costs of any separation unit downstream of the reactor will increase with increasing concentrations of A in the reactor *product stream, so a balance between reactor and separation cost must be sought in any commercial-scale design. This is not so important for the pilot-scale process.*

In the chapters of Part I, we first examine in detail the analysis of single-phase reacting systems. These are important problems for chemical professionals. The analysis can be carried out with a minimum of mathematical manipulation, which allows us to illustrate the essential relationship among experiment, modeling, and use of verified models to plan experiments or to design pilot- or commercial-scale equipment. This review is followed by a parallel development of the analysis and technically feasible design of heat exchangers (Chapter 3) and mass contactors (Chapter 4). Part II of the book discusses important issues of conduction, diffusion, and convective transport that enable estimating the phenomenological transport coefficients appearing in the rate laws, as well as completing the technically feasible design of heat and mass contactors.

REFERENCES

Aspen Technology, Inc. *Software: Aspen Engineering Suite*, 10 Canal Park, Cambridge, MA, 06141.

Cichy PT, Ultman JS, Russell TWF. Two-phase reactor design tubular reactors— Reactor model development. I&EC 1969; 61(8): 6.

Hildebrand FB. *Advanced Calculus for Applications* (2nd ed.). Englewood Cliffs, NJ: Prentice-Hall, 1976.

McCabe WL, Smith JC, Harriott P. *Unit Operations of Chemical Engineering* (5th ed.). New York: McGraw-Hill, Chemical Engineering Series, 1993.

Paul EL, Atiemo-Obeng VA, Kresta SM, eds. *Handbook of Industrial Mixing: Science and Practice*. New York: Wiley-Interscience, 2004,

Russell TWF, Denn MM. *Introduction to Chemical Engineering Analysis*. New York: Wiley, 1972.

Russell TWF, Orbey N. The technically feasible design. CEE 1993; Summer: 166–9.

Schaftlein RW, Russell TWF. Two-phase reactor design, tank-type reactors. I&EC 1968; 60: 12.

2 Chemical Reactor Analysis

A substantial portion of this chapter is taken from *Introduction to Chemical Engineering Analysis* by Russell and Denn (1972) and is used with permission.

This short chapter is a review of the analysis of simple reacting systems for chemical engineers. It is also designed to teach the fundamentals of analysis to other chemical professionals.

Reactor analysis is the most straightforward issue that chemical professionals encounter because rates of reaction can be obtained experimentally. The analysis of experimental data for reacting systems with mathematical models and the subsequent use of the verified model equations for design provide a template for the analysis of mass and heat transfer. We begin with simple reacting systems because the laboratory-scale experiments enable determination of reaction-rate constants that, with reasonable assumptions, can be used in model equations for design and operation of pilot- or commercial-scale reactors.

The same general principles apply to the design of mass contactors and heat exchangers, although it is more difficult to get the necessary values for mass and heat transfer coefficients from experiment. As we will see, mass transfer analysis is further complicated by the need to determine interfacial areas.

All chemical reaction and reactor analysis begins with experiment. Most experiments are carried out in batch equipment in a laboratory, and efforts are made to ensure the vessel is well mixed. In batch experiments, the concentrations of critical reacting species and products are measured over a given time period, sometimes until equilibrium is reached.

Large-scale equipment that mirrors these well-mixed batch systems are tank-type reactors. In a batch reactor, there is no flow to or from the reactor. Raw materials are added to the reactor at time $t = 0$, and the mixture is kept in the reactor until the desired conversion to products is obtained. In a semibatch reactor, one reactant is added initially, and other reacting materials are added to the reactor over some period of time. Products are collected when the desired conversion is obtained. Here the volume of material in the reactor varies with time, resulting in changes in the concentration of both reactants and products. Continuous-flow reactors are operated such that the flow into the reactor equals the flow leaving the reactor. Conversion of reactant to product is controlled by the average amount of time the material resides in the reactor, the residence time.

Table 2.1. *Equipment classifications considered in Chapter 2*

Reactors: Single phase	Reactors: Two phase	Heat exchangers	Mass contactors
Single control volume	Two control volumes	Two control volumes	Two control volumes
Tank type	Tank type	Tank type	Tank type
Mixed	*Mixed–mixed*	*Mixed–mixed*	*Mixed–mixed*
• **Batch**	• Batch	• Batch	• Batch
• **Semibatch**	• Semibatch	• Semibatch	• Semibatch
• **Continuous**	• Continuous	• Continuous	• Continuous
	Mixed–plug	*Mixed–plug*	*Mixed–plug*
	• Semibatch	• Semibatch	• Semibatch
	• Continuous	• Continuous	• Continuous
Tubular	Tubular	Tubular	Tubular
Plug flow	*Plug flow*	*Plug–plug flow*	*Plug–plug flow*
	• Cocurrent	• Cocurrent	• Cocurrent
		• Counter current	• Counter current

In some cases, experiments are carried out in tubular reactors. In tubular reactors there is a steady movement of material from one end of the reactor to the other without external mixing. The reactor in this case takes the form of a pipe or tube, and the fluid motion in this system is typically nonmixing plug flow. When several reactions occur simultaneously, and product distribution is critical, tubular reactors often yield the best performance. For example, ethylene oxide and its derivatives are typically produced in a tubular reacting system because, for a given input ratio of reactants, more of the desired monoproduct is produced.

A Level I analysis for a reactor provides species concentration relationships between raw materials and finished product. Level II analysis gives a limit on the concentration of product that can be produced. Because, for practical purposes, many reactions are assumed irreversible, a Level II analysis may not be required. A Level III analysis is required for analyzing experimental data and for designing pilot- and commercial-scale equipment. The fluid motions most often encountered are shown in Table 2.1, where we have included the fluid motions in heat exchanger and mass contactors for comparison.

The analysis of single-phase reactors requires us to consider only a single control volume. Two-phase reactors, heat exchangers, and mass contactors all require model equations for two control volumes.

We begin our review in this chapter by analyzing a well-mixed batch reactor and some typical reactions that illustrate how the batch reactor is modeled and how the model behavior is compared with experimental data. Then we develop model equations for semibatch and continuous-flow tank reactors and finally tubular reactors.

2.1 The Batch Reactor

A reactor that is operated by charging reactants and then letting a chemical reaction take place over some time period, usually an hour or less, is called a batch

Figure 2.1. Typical laboratory reactors (photo by Katherine S. Robinson).

reactor. Extremely fast chemical reactions that are of the order of seconds or portions of a second are an exception. For such fast reactions, the experiments are much
more complicated and require one to study microscale mixing. Batch reactors are
used in the laboratory and pilot plants and for commercial-scale production of specialty chemicals. Before we can consider the design and operation of pilot-scale or
commercial-scale reactors, we need to examine the very important use of the batch
reactor as an experimental apparatus.

A typical laboratory experiment is often done at pressures close to atmospheric
in a piece of simple glassware, such as a beaker or flask. If high pressures are necessary, the laboratory equipment can become much more complex, requiring different
materials of construction and inclusion of the reactor in specially designed enclo-
sures. Some typical laboratory reactors are shown in Figure 2.1. Such a vessel is
usually mixed with a magnetic or pneumatic stirrer and may be placed in a water or
oil bath or on a heating mantle to maintain constant temperature.

To illustrate the development of a mathematical model, we consider the following
problem. A and B are charged into the vessel at time zero and a chemical reaction
occurs to form product D. We will use shorthand notation to describe the chemical

reaction of 1 mol of A and 1 mol of B reacting to make n mol of D. D may decompose to form A and B but in many cases this reverse reaction may not be significant:

$$A + B \rightarrow nD.$$

Figure 1.3 provides the logic to obtain a mathematical model. The problem we consider is to predict the concentrations of A, B, and D (C_A, C_B, and C_D, in moles per liter, respectively) with time for an isothermal, well-mixed batch reactor. The variables we wish to measure are then C_A, C_B, and C_D, and the initial concentrations C_{Ai}, C_{Bi}, and C_{Di}. The piece of laboratory apparatus in which the reaction takes place is the control volume. The fluid is well mixed so that the concentrations within the volume vary only as functions of time and not spatial position. This means that one can sample from any place in the vessel to determine reactant or product concentration as a function of time. A well-mixed condition is readily achieved in laboratory-scale vessels containing low-viscosity liquids or gases.

A Level I analysis yields

$$(C_A MW_A + C_B MW_B + C_D MW_D)_{t+\Delta t}$$
$$= (C_A MW_A + C_B MW_B + C_D MW_D)_t,$$

which, when evaluated at some time t and the initial time yields

$$MW_A (C_A - C_{Ai}) + MW_B (C_B - C_{Bi}) + MW_D (C_D - C_{Di}) = 0. \qquad (2.1.1)$$

This overall mass balance contains the concentrations of all three components at any time. The Level I analysis procedes for each component as follows:

$$(C_A V)_{t+\Delta t} = (C_A V)_t + \begin{pmatrix} \text{moles of A that appear in the control} \\ \text{volume by all processes in } \Delta t \end{pmatrix}$$
$$- \begin{pmatrix} \text{moles of A that disappear from the} \\ \text{control volume by all processes in } \Delta t \end{pmatrix}. \qquad (2.1.2)$$

Equation (2.1.2) is a mass balance in which we have cancelled the molecular weight of A from each term. $(C_A V)_t$ represents the total moles of A, measured in moles per liter, at time t, and $(C_A V)_{t + \Delta t}$ represents the total moles of A at a future time increment Δt.

Species A can appear or disappear from the control volume only as a result of chemical reaction because there is no flow into or out of the batch reactor. To obtain the model equations for the appearance and disappearance we need to express the word statements in Eq. (2.1.2) in terms of the *variables* that can be obtained by experiment. The first step is to define the *rate of reaction*:

r_{A+} as the rate at which A appears in the control volume in moles per volume per time,

r_{A-} as the rate at which A disappears from the control volume in moles per volume per time.

The rate of reaction is written on a volume basis because the moles of any species formed or depleted will be proportional to the fluid volume in the reactor. Note that the symbols r_{A+} and r_{A-} are just shorthand for the word statements.

$$\left(\begin{array}{c} \text{moles of A that appear in the control} \\ \text{volume by all processes in } \Delta t \end{array}\right) = r_{A+}V\Delta t,$$

$$\left(\begin{array}{c} \text{moles of A that disappear from the} \\ \text{control volume by all processes in} \Delta t \end{array}\right) = r_{A-}V\Delta t.$$

The species mole balance equation for species A becomes

$$(C_A V)_{t+\Delta t} - (C_A V)_t = (r_{A+} - r_{A-}) V\Delta t.$$

We obtain the differential form by taking the limit as the time step (Δt) approaches zero:

$$\lim_{\Delta t \to 0} \left(\frac{(C_A V)_{t+\Delta t} - (C_A V)_t}{\Delta t} \right) = (r_{A+} - r_{A-}) V$$

$$\frac{dC_A V}{dt} = (r_{A+} - r_{A-}) V \tag{2.1.3}$$

In a similar manner we can obtain the basic model equations for species B and D by using the known reaction chemistry. From the chemical equation, we know that every time a mole of A reacts one mole of B must also react. Therefore,

$$r_{A+} - r_{A-} = r_{B+} - r_{B-}. \tag{2.1.4}$$

Furthermore, when a mole of A and B react, n moles of D are formed. Thus,

$$nr_{A-} = r_{D+},$$

$$\frac{dC_B V}{dt} = (r_{B+} - r_{B-}) V = (r_{A+} - r_{A-}) V = \frac{dC_A V}{dt},$$

$$r \equiv r_{A-} - r_{A+} = \frac{r_{D+} - r_{D-}}{n}. \tag{2.1.5}$$

Typically, only a single rate is needed to describe the reaction, sometimes called the *intrinsic rate of reaction*, denoted by r:

$$\frac{dC_D V}{dt} = nrV. \tag{2.1.6}$$

Using shorthand symbols for rate of reaction allows us to write the basic model equations in terms of symbols, but in order to obtain a useful mathematical model we must, as directed by Figure 1.3, find a constitutive equation that will relate rate of reaction to those variables that we can measure by experiment, the concentrations of A, B, and D.

Before doing this we can gain some useful information by continuing our Level I analysis. The species mole balance for species B, obtained with the same procedure as that for Eq. (2.1.3), is

$$\frac{dC_B V}{dt} = (r_{B+} - r_{B-}) V = (r_{A+} - r_{A-}) V, \tag{2.1.7}$$

where the relationship between rates, Eq. (2.1.4), is employed. Comparison with Eq. (2.1.3) shows that

$$\frac{dC_B V}{dt} = \frac{dC_A V}{dt},$$

which on integration yields

$$C_A(t) - C_{Ai} = C_B(t) - C_{Bi}. \tag{2.1.8}$$

Here, C_{Ai} and C_{Bi} are the initial concentrations of A and B, respectively, at time t equal to zero. Equation (2.1.8) states that the number of moles of A that have reacted equals the number of moles of B that have reacted. Similarly, Equations (2.1.3) and (2.1.6) taken together are

$$\frac{dC_D}{dt} = -n\frac{dC_A}{dt},$$

which yields, on integration,

$$C_D(t) - C_{Di} = n\left[C_{Ai} - C_A(t)\right], \tag{2.1.9}$$

where C_{Di} is the initial concentration of D, and is often equal to zero. Thus the concentrations of B and D can be expressed in terms of the concentration of A, so it is necessary to know the concentration of only one species to define the system at any time.

A Level II analysis to determine if chemical equilibrium has been achieved will give the maximum concentration of D that can be achieved for given values of initial A and B concentrations.

2.1.1 Chemical Equilibrium

When a chemical reaction is carried out until the concentrations of the components A, B, and D no longer vary with time, the reaction is considered to be at chemical equilibrium. For the bulk solution, we can measure the concentrations of the components at equilibrium, C_{Ae}, C_{Be}, and C_{De}. If $n = 2$ for the reaction we are considering, the equilibrium constant is then defined as

$$K_e \equiv \frac{\text{products}}{\text{reactants}} = \frac{C_{De}^2}{C_{Ae}C_{Be}}. \tag{2.1.10}$$

Many of these experiments have been carried out for established chemical reactions. If the experiments reach chemical equilibrium, the value of the equilibrium constant K_e, defined as the ratio of the concentration of the products to the reactants at equilibrium, can be obtained. K_e data can often be determined from tabulated standard Gibbs free-energy values, a procedure discussed in any thermodynamics text (Sandler, 2006). This value sets the limits on the final concentrations of each species in chemical reacting systems.

Before verifying the rate expressions by obtaining concentration versus time data, we examine our reaction $A + B \leftrightarrows 2D$ with a Level II analysis and assume that chemical equilibrium has been reached. When chemical equilibrium is reached, Eq. (2.1.8) becomes

$$C_{Ae} - C_{Ai} = C_{Be} - C_{Bi}, \tag{2.1.11}$$

and Eq. (2.1.9) becomes

$$C_{De} - C_{Di} = 2(C_{Ai} - C_{Ae}).\qquad (2.1.12)$$

If we consider a simplified case, in which $C_{Ai} = C_{Bi}$ and $C_{Di} = 0$, then we can solve for C_{De} in terms of only C_{Ai}, n, and K_e:

$$C_{De} = n\left(C_{Ai} - \sqrt{\frac{C_{De}^n}{K_e}}\right).\qquad (2.1.13)$$

This simplifies to the following expression for $n = 2$:

$$C_{De} = \frac{2C_{Ai}}{1 + \frac{2}{\sqrt{K_e}}}.\qquad (2.1.14)$$

Equation (2.1.14) gives the maximum value of D that can be obtained in any reactor.

To continue the model development, we see that we have three unknowns, $C_A(t)$, $C_B(t)$, and $C_D(t)$, and only two relationships, (2.1.8) and (2.1.9). The final required equation is found from a Level III analysis and requires further consideration of the constitutive equation, as discussed in the next section.

2.2   Reaction Rate and Determination by Experiment

2.2.1 Rate Expression

Thus far, we have used Level I and Level II analyses to determine the behavior of a batch reacting system. The Level I analysis enabled determination of the concentration of one of the reactants if the other two were known. The Level II analysis showed that, when the system reaches equilibrium, the value of the concentrations depends on only the stoichiometry and the equilibrium constant. This may be useful in determining the limiting behavior of any reactor. However, it does not yield the time-dependent behavior of the reaction, which is still necessary for sizing a batch reactor that does not reach equilibrium conditions.

A Level III analysis is needed to fully describe the chemical reaction and requires us to plan an experiment or to analyze experimental results that someone else has obtained in order to determine the rate of reaction. We must postulate a relationship, solve the basic model equations, and then compare the model prediction with data.

The *reaction rates* r_{A-}, r_{B-}, and r_{D+} must be related to the concentrations C_A, C_B, and C_D. This constitutive equation may be deduced in part by theoretical considerations, but it ultimately requires experimentation for a complete determination.

It is clear that the reaction cannot proceed at all if either A or B is missing, so r must be a function of C_A and C_B and go to zero if either C_A or C_B vanishes. Furthermore, we know from elementary chemistry that the chemical reaction is the result of a collision between one molecule of A and one molecule of B. Clearly, the more molecules of a substance in a given volume, the more likely a collision will occur. Thus the rate at which the reaction proceeds should increase with increased C_A or C_B. The simplest functional form, which embodies both of these features, is

$$r_{A-} = kC_A C_B,\qquad (2.2.1)$$

where k is a constant determined experimentally. We can, however, postulate many other forms that meet our simple requirements of vanishing when C_A or C_B does and of increasing when C_A or C_B increases. Some examples are

$$r_{A-} = kC_A^n C_B^m,$$

$$r_{A-} = \frac{kC_A^2 C_B^3}{1 + \beta C_B}.$$

There are many expressions that meet the simple requirements, so we need additional information to determine the correct constitutive relation between r_{A-} and C_A, C_B. Typically data of concentration versus time are fit to the model to determine the reaction order and the kinetic constant. However, for the remaining analysis, we assume that Eq. (2.2.1) fits the reacting system of interest.

If the reaction is reversible then a rate expression is required for the disappearance of D. A possible expression for r_{D-} is

$$r_{D-} = k_r C_D^2, \tag{2.2.2}$$

where $r_{D-} = nr_{A+} = nr_{B+}$. Eq. (2.2.2) also must be verified by experiment, and other possibilities exist.

Using rate expressions (2.2.1) and (2.2.2), we can substitute into Eq. (2.1.3) derived from the Level I analysis to obtain system equations that include the rate expression:

$$\frac{dC_A}{dt} = \left(k_r C_D^2 - kC_A C_B\right). \tag{2.2.3}$$

V, the volume of the reactor, will be constant in the isothermal batch system and has been canceled from both sides of the equation. For gas-phase reactions or strongly nonideal mixtures with an increase or decrease of density with time, an overall mass balance is required for completing the model. The more complicated model behavior is discussed in Russell and Denn (1972).

To determine the full concentration–time behavior of the system, we need only to solve Eq. (2.2.3) by integration. To illustrate the procedure with a minimum of algebraic complexity, we first assume that $k_r C_D^2$ is very small compared with $kC_A C_B$, i.e., the reaction is essentially irreversible over the time of interest:

$$\frac{dC_A}{dt} = -kC_A C_B.$$

Using Equation (2.1.8) to relate C_A to C_B, we obtain

$$\frac{dC_A}{dt} = -kC_A \left(C_A + C_{Bi} - C_{Ai}\right). \tag{2.2.4}$$

It is convenient to define the constant $M = C_{Bi} - C_{Ai}$. Equation (2.2.4) then becomes

$$\frac{dC_A}{dt} = -kC_A \left(C_A + M\right). \tag{2.2.5}$$

A solution to Equation (2.2.5) can be obtained directly, although it is necessary to consider separately the two cases $M = 0$ and $M \neq 0$. In the former case, corresponding to equal starting concentrations of reactants A and B, the solution is obtained by the sequence of steps shown in the left-hand column of Table 2.2, giving

Table 2.2. *Integration of the equation for* C_A *in a batch reactor with second-order reaction* $A + B \rightarrow nD$ *for equal* $(M = 0)$ *and unequal* $(M \neq 0)$ *starting concentrations*

$M = 0$	$M \neq 0$
$\dfrac{dC_A}{dt} = -kC_A^2$	$\dfrac{dC_A}{dt} = -kC_A(M + C_A)$
$\dfrac{dC_A}{C_A^2} = -kdt$	$\dfrac{dC_A}{C_A(M + C_A)} = -kdt$
$\displaystyle\int_{C_{Ai}}^{C_A(t)} \dfrac{dC}{C^2} = -k\int_0^t d\tau$	$\displaystyle\int_{C_{Ai}}^{C_A(t)} \dfrac{dC}{C(M + C)} = -k\int_0^t d\tau$
$-\dfrac{1}{C_A(t)} + \dfrac{1}{C_{Ai}} = -kt$	$\dfrac{1}{M}\ln\dfrac{C_A(t)}{C_{Ai}} - \dfrac{1}{M}\ln\left(\dfrac{C_A(t) + M}{C_{Ai} + M}\right) = -kt$

$$M = 0: \quad \frac{1}{C_A(t)} = \frac{1}{C_{Ai}} + kt. \tag{2.2.6}$$

For $M \neq 0$ the solution steps are shown in the right-hand column of Table 2.2, giving

$$M \neq 0: \quad \ln\left[\left(\frac{C_{Ai}}{C_A(t)}\right)\left(\frac{M + C_A(t)}{M + C_{Ai}}\right)\right] = Mkt, \tag{2.2.7}$$

or, equivalently,

$$M \neq 0: \quad \ln\frac{C_B(t)}{C_A(t)} + \ln\frac{C_{Bi}}{C_{Ai}} = Mkt. \tag{2.2.8}$$

Thus, if the rate is truly represented by Eq. (2.2.1) then, according to Eq. (2.2.6), for equal starting concentrations a plot of $1/C_A$ versus t will yield a straight line with slope k. Similarly, for unequal initial concentrations, Eq. (2.2.8) indicates that a plot of a more complicated function of C_A versus t will give a straight line with slope Mk.

This completes the successful development of a model equation for the batch reactor. Model validation, however, requires comparison with experiment. This is discussed in the following examples.

EXAMPLE 2.1. The data in Table 2.3 were obtained for the reaction of sulfuric acid with diethyl sulfate, in aqueous solution at 22.9 °C. The initial concentrations of H_2SO_4 and $(C_2H_5)_2SO_4$ were each 5.5 g-mol/L. The reaction can be represented as follows:

$$H_2SO_4 + (C_2H_5)_2SO_4 \rightarrow 2C_2H_5SO_4H.$$

SOLUTION. In our symbolic nomenclature, H_2SO_4 is designated as A, $(C_2H_5)_2SO_4$ is designated as B, n is 2, and $C_2H_5SO_4H$ is designated as D.

As required by Eq. (2.2.6), the data are plotted in Figure 2.2 as $1/C_A$ versus t. There is still some experimental scatter, but a straight line can easily be drawn through the data, confirming the adequacy of the rate expression, Eq. (2.2.1), for the data shown. The parameter k is determined from the slope in Figure 2.2 as

$$k = 6.05 \times 10^{-4} \text{ L/g-mol min.}$$

Table 2.3. Concentration of H_2SO_4 vs. time for the reaction of sulfuric acid with diethyl sulfate in aqueous solution at 22.9 °C (Hellin and Jungers, 1957)

Time (minutes) t	Concentration of H_2SO_4 (g-mol/L) C_A
0	5.50
41	4.91
48	4.81
55	4.69
75	4.38
96	4.12
127	3.84
162	3.59
180	3.44
194	3.34
212	3.27
267	3.07
318	2.92
379	2.84
410	2.79
∞	2.60

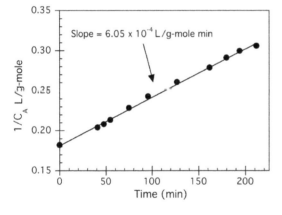

Figure 2.2. Computation of the second-order rate constant for the reaction between sulfuric acid and diethyl sulfate in aqueous solution.

EXAMPLE 2.2 FERMENTOR ANALYSIS. The complexity of microbial growth has often led to lumped models for growth in bioreactors, where substrates and microorganisms are introduced at some time, and products are collected at some later time, and the behavior of the reaction is simply based on the components. We will take this example through the book and examine how well the simple models we can develop work for this complex system.

Taking an experimental study (Tobajas et al., 2003), we examine the growth of *Candida utilis* in a Biostat UD 30 internal-loop airlift bioreactor, as shown in Figure 2.3. This is a semibatch mixed–plug reactor, but, as we will show, the pertinent model equations are the same as those for a batch reactor because the concentration of oxygen in the flowing gas stream is essentially constant.

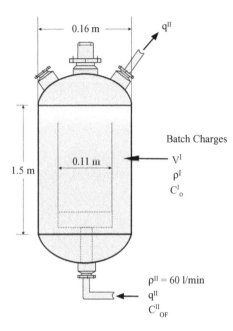

0.16 m

q^{II}

1.5 m

0.11 m

Batch Charges

V^I

ρ^I

C^I_o

$\rho^{II} = 60$ l/min

q^{II}

C^{II}_{OF}

Figure 2.3. Sketch of semibatch bioreactor, based on the Biostat UD 30 Airlift Bioreactor (B Braun, Germany). Vessel is not drawn to scale.

The working volume of this fermentor is 33 L, and the fermentor is initially charged with media containing glucose, salts, and yeast extract to provide a proper environment for growth of the microorganisms and maintained at a pH of 5.0 and a temperature of 30 °C throughout the growth. During growth, air is sparged through the culture at sufficient levels to provide the required oxygen for metabolism.

In this chapter, we focus on the consumption of the major limiting substrate, glucose, by the cells, and the subsequent growth of the cells (biomass) as a result of this. These values are typically easy to measure – carbohydrates can be determined by a spectroscopic method or with a glucose probe. Biomass can be determined by a direct cell count by microscopy or by an indirect measurement of cell mass (wet or dry cell mass) or optical density and conversion by use of a standard curve developed in prior studies. Similar to chemical reacting systems, the critical aspect is to decide what we want or need to measure and to develop the characteristic equations.

SOLUTION. Because they can be measured, we consider the concentrations of the cell mass C_A and the substrate glucose C_B. The Level III component mass balances for the semibatch experiment in the fermentor are

$$\frac{dC_A}{dt} = +r_{A+},$$ (2.2.9)

$$\frac{dC_B}{dt} = -r_{B-}.$$ (2.2.10)

The simplest constitutive equation we can assume is the following relationship:

$$r_{A+} = +k_1 C_A C_B,$$ (2.2.11)

where r_A is the rate at which the substrate disappears from the control volume and the rate at which cell mass appears. Note that bioengineers often define k_1 as μ, the intrinsic growth rate. It is often better to fit the data with two separate rates, in which the rate of substrate depletion is then

$$r_{B-} = +k_2 C_A C_B. \tag{2.2.12}$$

This difference in k_1 and k_2 is a natural consequence of using r_A, r_B defined in terms of mass concentration.

We have not included the component mass balance equations for the oxygen in the flowing gas phase or in the liquid phase. The impact of oxygen balances is discussed in Chapter 8.

Thus the model equations for the batch system become

$$\frac{dC_A}{dt} = +k_1 C_A C_B, \tag{2.2.13}$$

$$\frac{dC_B}{dt} = -k_2 C_A C_B. \tag{2.2.14}$$

To determine whether this model is adequate, we compare it with the experimental data reported by Tobajas et al. (2003).

An analytical solution of the model equations, with $M = C_{Bi} - C_{Ai} \neq 0$, according to Table 2.2, is possible. However, solution of the model equations is also straightforward with Matlab or another program, by minimizing the error between model and experimental data. By setting this constraint as a fixed minimum, or demanding a certain number of iterations, we can then find that $k_1 = 0.165$ L/g h and $k_2 = 0.096$ L/g h, giving the resulting plots for the model fits shown in Figure 2.4.

Using the simplest possible constitutive equation fits the data reasonably well, but not perfectly, particularly at late times, because the growth depends on the substrate in a more complex way. One possibility is the effect of oxygen depletion, necessitating an oxygen balance equation, which is discussed in Chapter 8. A curve fit of the data also does not necessarily give insight into the mechanism of growth. When the cells consume nutrients, they use them for other metabolic functions besides growth. This is particularly true in early and late growth stages. These inefficiencies in growth

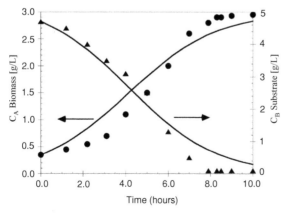

Figure 2.4. Comparison of C_A and C_B, concentrations of biomass and substrate determined experimentally (symbols) by Tobajas et al. (2003), with those determined by a model fit (curves) to Eqs. (2.2.13) and (2.2.14).

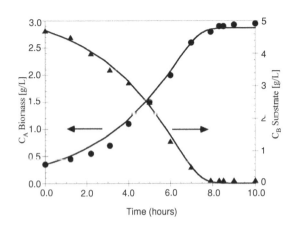

Figure 2.5. Comparison of C_A and C_B, concentrations of biomass and substrate, with model predictions obtained with Monod-type constitutive relationship and Eqs. (2.2.15) and (2.2.16).

are well established and can be accounted for in many ways. One way to do this is to limit the examination of the fit of the experimental behavior during the cell's maximal growth stage, when nutrients are in excess, and cells are growing at their maximal rate, often called "exponential growth."

To try to fit all of the experimental data, we utilize a constitutive equation that is typically used to describe the inefficiencies in substrate consumption during early and late stage growth, a Monod-type relationship. Thus the model equations for cell growth and substrate consumption become

$$\frac{dC_A}{dt} = +k_1 C_A \frac{C_B}{K_B + C_B}, \qquad (2.2.15)$$

$$\frac{dC_B}{dt} = -k_2 C_A \frac{C_B}{K_B + C_B}, \qquad (2.2.16)$$

where K_B is essentially another fit parameter. Figure 2.5 shows the result of the model fit for these equations, where $k_1 = 0.34$ L/g h, $k_2 = 0.64$ L/g h, and $K_B = 0.48$ g/L. As we see from the figure, the data over all the time are well fit by the new model. This model is sufficient for this organism to describe the growth of the cell and consumption of substrate in batch, and could likely be used to scale up the system to larger growth. However, if the organism were grown under different conditions, an alternative fit (alternative rate constants) would likely be needed.

2.2.2 Approach to Equilibrium

The data for the conversion of sulfuric acid in Table 2.3 show that the concentration approaches an equilibrium value at long times, which was not considered in Example 2.1. Hence the reaction is reversible, and the reverse reaction must be considered to model the full-time behavior. Given that the forward reaction rate has been established from the short-time data, the equilibrium result can be used with the full model, Equation (2.2.3), to determine the reverse reaction rate, k_r.

Because $C_{Ai} = C_{Bi}$ and $C_{Di} = 0$, the Level I analysis yields [Eq. (2.1.7)]

$$C_D(t) = 2[C_{Ai} - C_A(t)].$$

The Level II analysis is described by the equilibrium relationship given in Eq. (2.1.12) and the Level I balances:

$$K_e = \frac{k}{k_r} = \frac{C_{De}^2}{C_{Ae}C_{Be}} = \frac{4(C_{Ai} - C_{Ae})^2}{C_{Ae}^2}.$$

Given the initial and final data in Table 2.2, calculation shows that $K_e = 1.25$ and hence

$$k_r = 7.56 \times 10^{-4} \text{ L/g-mol min}.$$

The Level III analysis given in Eq. (2.2.3) can be simplified and solved with the Level I balances previously discussed, as

$$\frac{dC_A}{dt} = k\left[\frac{1}{K_e}(C_{Ai} - C_A)^2 - C_A^2\right]. \qquad (2.2.17)$$

This can be separated and integrated:

$$\int_{C_{Ai}}^{C_A} \frac{K_e dC}{(C_{Ai} - C)^2 - K_e C^2} = \int_0^t k d\tau.$$

The solution is

$$\ln\left[\frac{C_A(1 - \sqrt{K_e}) - C_{Ai}}{C_A(-1 - \sqrt{K_e}) + C_{Ai}}\right] = \frac{2C_{Ai}k}{\sqrt{K_e}}t. \qquad (2.2.18)$$

The experimental data in Table 2.3 can be plotted according to this equation to yield a straight line, which validates the model.

The right-hand side of Eq. (2.2.18) defines a characteristic time for a process as $\sqrt{K_e}/(2C_{Ai}k)$, which has the units of time. The time required for reaching equilibrium is approximately given by $2C_{Ai}kt_{eq}/\sqrt{K_e} \approx 3$. This time is sufficient to achieve ~95% of the final, total conversion. For the example under consideration,

$$t_{eq} \approx \frac{3\sqrt{K_e}}{2C_{Ai}k} = \frac{3\sqrt{1.25}}{(2 \times 5.50 \text{ g-mol/L})(6.05 \times 10^{-4} \text{ L/g-mol min})} = 504 \text{ min}.$$

This calculation is consistent with the data in Table 2.3. Such estimations of the time required for reaching equilibrium are valuable in analyzing an operating reactor as well as in designing new processes.

2.3 Tank-Type Reactors

In the previous section we defined the rate of reaction, and we discussed reaction-rate expressions, their role in the model development, and the batch reactor experiments needed to obtain constitutive relations between rate and concentrations for certain simple reactions. We are now prepared to consider the final step in the analysis process — using tested mathematical descriptions for design and predictive purposes.

Reactors can be classified into two broad categories, based on simplified assumptions regarding the fluid motion: *tank type and tubular*. We derived the general mathematical description for a tank-type batch reactor in Section 2.1 for the single reaction A + B → nD, the resulting model equations being Eqs. (2.1.3)–(2.1.6). The model equations for other tank-type reactors are developed in this section.

A tank-type reactor is a vessel in which the fluid is assumed to be well mixed. In its most common form at the pilot or commercial scale, it frequently has a height-to-diameter ratio between about 1:1 and 3:1. We further classify tank-type reactors by the way in which raw materials (reactants) are introduced into the reactor and by the way in which product and unreacted raw material are removed from the reactor.

A semibatch or fed-batch reactor is one in which one of the reactants is fed into the reactor over the time the reaction is taking place. The volume of fluid thus changes with time. The well-mixed assumption holds provided the means of introduction of the reactant is well designed for effective dispersion into the mixed vessel and the time for substantial reaction is long compared with the mixing time. This is the case in the vast majority of reacting systems, and we will assume that these reactors are well mixed for purposes of model development.

A significant problem can arise if the reaction is very fast or, in the case of biochemical reactions, the incoming material is toxic to the microorganisms at its inlet concentrations. Such issues require more complex modeling and different equipment.

A continuous-flow tank-type reactor has one or more reactants continuously introduced to the reactor and products continuously removed. The well-mixed assumption requires that the product concentration be the same as that in the vessel and that the incoming reactants quickly reach the concentration of reactants in the reactor. This is a sound assumption for the majority of reacting systems, although it is conceptually difficult to visualize just how the reactants reach the vessel concentration quickly. They do, of course, take some time to mix, but if this is small compared with the time for reaction, it does not materially affect the concentrations used for modeling purposes.

2.3.1 Batch Reactors

A batch reactor is one in which the raw materials are brought together at time t = 0 and then allowed to react until the desired conversion of reactants is achieved. There is no flow into or out of the reactor. Some examples of laboratory reactors are shown in Figure 2.1. Pilot and commercial scale batch reactors are similar to the vessels shown in Figures 2.3 and 2.6.

2.3.2 Semibatch Reactors

One operates a semibatch reactor by placing one reactant in the reactor and initiating the reaction by adding another reactant or reactants over a period of time. Such an operation is often carried out to achieve better temperature control with exothermic reactors. There is no flow out of the reactor; therefore the volume of fluid in the semibatch reactor varies with time. Species concentrations change with time because of reaction, increase in fluid volume, and flow into the reactor. A typical semibatch reactor is shown in Figure 2.6.

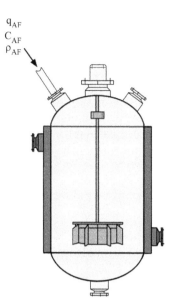

q_{AF}
C_{AF}
ρ_{AF}

Figure 2.6. Semibatch reactor.

These model equations describe the situation in which an initial volume V_i of reactant B of concentration C_{Bi} is put into the reactor. At time $t = 0$ the reaction $A + B \rightarrow nD$ is initiated by the introduction of a stream containing A at a concentration C_{AF} and at a volumetric flow rate q_{AF}, where ρ_{AF} is the density of the stream. A mass balance on the reactor yields

$$\frac{d\rho V}{dt} = \rho_{AF} q_{AF}. \tag{2.3.1}$$

A species mass balance yields

$$\frac{dVC_A}{dt} = q_{AF} C_{AF} - rV, \tag{2.3.2}$$

$$\frac{dVC_B}{dt} = -rV, \tag{2.3.3}$$

$$\frac{dVC_D}{dt} = nrV, \tag{2.3.4}$$

$$r = r_{A-} = r_{B-} = \frac{r_{D+}}{n}.$$

Equation (2.3.1) can be simplified if we assume that density is linear in concentration and that the rate of change of density with concentration is proportional to molecular weight of each species in the reactor (Russell and Denn, 1979):

$$\frac{dV}{dt} = q_{AF}.$$

In both batch and semibatch reactors the concentration of any species varies as a function of time. We have seen in previous sections how the behavior of the model equation is determined for batch systems. We now briefly consider the same problem for semibatch reactors. To initiate our discussion we again consider the following reaction and rate expression:

$$r_{A-} = r_{B-} = \frac{1}{n}r_{D+} \equiv r = kC_AC_B.$$

In a batch reactor we fill the reactor with the required amounts of A and B and allow the reaction to proceed to some desired degree of conversion of one of the raw materials. If the initial molar concentrations of A and B are equal and the reactor is reacting isothermally, we find that the concentration of A at any time can be expressed by Eq. (2.2.6):

$$M = 0: \quad \frac{1}{C_A(t)} = \frac{1}{C_{Ai}} + kt.$$

We obtained Eq. (2.2.6) by integrating the component mass balance equation for A, after we used the chemical equation and the component mass balance for A and B to relate C_A and C_B [Eq. (2.1.8)]:

$$C_A(t) - C_{Ai} = C_B(t) - C_{Bi}.$$

In a semibatch reactor the model behavior is not so readily obtained. If all of material B is put into the reactor and if the stream containing A is fed at a constant rate q_{AF} to the reactor over a period of time, model equations (2.3.1)–(2.3.4) apply.

For constant density, Eq. (2.3.1) can be integrated directly to yield V as a function of time;

$$V = V_i + q_{AF}t. \tag{2.3.5}$$

Equations (2.3.2), (2.3.3), and (2.3.4), written with the assumed form for the reaction rate, become

$$\frac{dVC_A}{dt} = q_{AF}C_{AF} - kC_AC_BV, \tag{2.3.6}$$

$$\frac{dVC_B}{dt} = -kC_AC_BV, \tag{2.3.7}$$

$$\frac{dVC_D}{dt} = nkC_AC_BV. \tag{2.3.8}$$

There are two major differences between these equations and the corresponding equations for a batch system. V is not a constant and must be retained inside the derivative on the left-hand side of the equations. The component mass balance equation for A, (2.3.6), has an additional term accounting for the convective flow of A into the system.

Combining Eqs. (2.3.2) and (2.3.3) or (2.3.6) and (2.3.7) leads to the relation

$$\frac{dVC_A}{dt} = \frac{dVC_B}{dt} + q_{AF}C_{AF}$$

that can be integrated for constant $q_{AF}C_{AF}$ to yield

$$VC_A - V_iC_{Ai} = VC_B - V_iC_{Bi} + q_{AF}C_{AF}t. \tag{2.3.9}$$

Here, V_i is the volume at $t = 0$. Notice the contrast with the equations for the batch reactor. Similarly, from Eqs. (2.3.3) and (2.3.4) or (2.3.7) and (2.3.8),

$$V_i C_{Di} - V C_D = n V C_B - n V_i C_{Bi}. \tag{2.3.10}$$

The system behavior can then be expressed in terms of a single equation for C_B by use of Eqs. (2.3.9) and (2.3.7):

$$\frac{dV C_B}{dt} = -k C_B (V C_B + M + q_{AF} C_{AF} t), \tag{2.3.11}$$

where V is defined by Eq. (2.3.5) and $M = V_i(C_{Ai} - C_{Bi})$.

We can solve Eq. (2.3.11) using a change of variables solution, by a substitution $V C_B = 1/y$. We obtain the equation

$$\frac{dy}{dt} = \frac{k(M + q_{AF} C_{AF} t) y}{V_i + q_{AF} t} + \frac{k}{V_i + q_{AF} t}.$$

This is a linear first-order equation. An analytical solution is obtained that, following some tedious algebra, can be expressed as

$$\frac{V C_B}{V_i C_{Bi}} = \left(\frac{V_i}{V}\right)^{\alpha-1} e^{-k C_{AF} t} \left[1 + \frac{k C_{Bi} V_i}{q_{AF}} \int_0^{(V/V_i)-1} \frac{e^{-(k C_{AF} V_i / q_{AF})\xi}}{(1 + \xi)^\alpha} d\xi \right]^{-1},$$

$$\alpha = \frac{k}{q_{AF}} (M - V_i C_{AF}) + 1. \tag{2.3.12}$$

Equation (2.3.12) expresses the ratio of moles of B that are unreacted to the initial number of moles. We shall not discuss in detail the behavior of this equation, but note that the conversion–time relation depends on only a number of parameters with units of inverse time, $k C_{AF}$, $k C_{Bi}$, $k C_{Ai}$ (probably zero), and q/V_i. We can therefore design the system for given production requirements.

Semibatch reactors are used commercially when there is some particular advantage to be gained by adding one reactant continuously to the other. This can occur when there are several reactions taking place, and the distribution of products can be regulated by adding reactant over a period of time or when there are severe thermal effects that can be regulated by controlled addition of a reactant. The latter case requires application of the principle of energy conservation, but the analysis leading to Eq. (2.3.12) forms a part of the total set of model equations.

2.3.3 Continuous Flow

A continuous-flow stirred tank reactor (CFSTR) is one in which reactants are continuously fed into the reactor and product and unreacted raw material continuously removed from the reactor. It is the only kind of tank-type reactor that can be used in a continuous commercial-scale chemical processing unit, and it is commonly employed to carry out reactions that produce millions of kilograms of product per year. A typical system is shown in Figure 2.7. The continuous-flow tank reactor is operated so that the flow of reactants to the reactor (q) is equal to the flow of product and unreacted raw material from the reactor, yielding a constant volume (V) in the reactor. Conversion of reactant to product is controlled by the time that the fluid remains

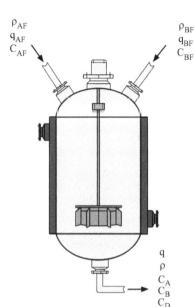

ρ_{AF}
q_{AF}
C_{AF}

ρ_{BF}
q_{BF}
C_{BF}

q
ρ
C_A
C_B
C_D

Figure 2.7. Continuous-flow stirred tank reactor (CFSTR).

in the reactor, V/q. For a *constant-density system*, the basic model equations for a reaction $A + B \rightarrow nD$ are as follows.

A Level I balance relating the concentrations of A and B in the exit and entrance flows is derived following the procedure employed in the previous cases, namely, species mass balances are derived and combined as

$$\frac{dVC_A}{dt} = q_{AF}C_{AF} - qC_A - (r_{A-} - r_{A+})V,$$

$$\frac{dVC_B}{dt} = q_{BF}C_{BF} - qC_B - (r_{B-} - r_{B+})V,$$

$$(r_{A-} - r_{A+}) = (r_{B-} - r_{B+}).$$

Therefore

$$\frac{dVC_A}{dt} - \frac{dVC_B}{dt} = q_{AF}C_{AF} - q_{BF}C_{BF}$$
$$- q(C_A - C_B).$$

Further, assuming *steady-state* operation and recognizing from the overall mass balance that, for ideal mixing,

$$q = q_{AF} + q_{BF},$$

we obtain

$$C_A = C_B + \frac{q_{AF}C_{AF}}{q_{AF} + q_{BF}} - \frac{q_{BF}C_{BF}}{q_{AF} + q_{BF}}, \tag{2.3.13}$$

$$C_D = -nC_B + \frac{nq_{BF}C_{BF}}{q_{AF} + q_{BF}}. \tag{2.3.14}$$

If we assume that the reaction is essentially irreversible, then a Level II analysis is not necessary. The Level III analysis proceeds with the total mass balance,

$$\frac{dV}{dt} = q_{AF} + q_{BF} - q, \tag{2.3.15}$$

and species mass balances,

$$\frac{dVC_A}{dt} = q_{AF}C_{AF} - qC_A - rV, \tag{2.3.16}$$

$$\frac{dVC_B}{dt} = q_{BF}C_{BF} - qC_B - rV, \tag{2.3.17}$$

$$\frac{dVC_D}{dt} = -qC_D + nrV. \tag{2.3.18}$$

When the reactor is operating at steady state, the time derivatives are all equal to 0, and for ideal mixing, $q_{AF} + q_{BF} = q$. We need the complete model if we are concerned with transient behavior, as we are during start-up and shutdown of the process or when considering control of the reactor. The design of the reactor, however, is carried out for steady-state operation. At steady state, flows to and from the reactor are equal and there are no changes in level or concentration with time. Thus all time derivatives are zero and Eqs. (2.3.15)–(2.3.18) simplify to

$$0 = q_{AF} + q_{BF} - q, \tag{2.3.19}$$

$$0 = q_{AF}C_{AF} - qC_A - rV, \tag{2.3.20}$$

$$0 = q_{BF}C_{BF} - qC_B - rV, \tag{2.3.21}$$

$$0 = -qC_D + nrV. \tag{2.3.22}$$

For simplicity we assume a specific form for the rate expression:

$$r = kC_AC_B. \tag{2.3.23}$$

Substituting the rate into Eqs. (2.3.20)–(2.3.22) and solving for q from Eq. (2.3.19), we obtain the final working equations:

$$0 = q_{AF}C_{AF} - (q_{AF} + q_{BF})C_A - kC_AC_BV, \tag{2.3.24}$$

$$0 = q_{BF}C_{BF} - (q_{AF} + q_{BF})C_B - kC_AC_BV, \tag{2.3.25}$$

$$0 = (q_{AF} + q_{BF})C_D + nkC_AC_BV. \tag{2.3.26}$$

Assuming n and k are known from batch experiments, these three equations relate eight unknowns: C_{AF}, C_{BF}, q_{AF}, q_{BF}, V, C_A, C_B, and C_D. Thus, five values must be specified by other means, with the remaining three quantities calculated from Eqs. (2.3.24)–(2.3.26). The particular five quantities that are specified will depend on the problem to be solved. We might, for example, be given the reactor volume, flow rates, and feed composition and be asked to calculate the composition of the exit stream. Or we might be given feed and effluent compositions and be asked to calculate reactor volume and flow rates. In the several examples that follow we use the numbers for the sulfuric acid–diethyl sulfate reaction from Example 2.1. Thus, n = 2 and, assuming the reaction to be irreversible, $k = 6.05 \times 10^{-4}$ L/g-mol min.

EXAMPLE 2.3. Given $V = 25.4$ L, $q_{AF} = q_{BF} = 0.1$ L/min, and $C_{AF} = C_{BF} = 11$ g-mol/L, determine C_D.

SOLUTION. We can then calculate

$$\text{reactor holding time: } \Theta = \frac{V}{q_{AF} + q_{BF}} = \frac{25.4}{0.1 + 0.1} = 127 \text{ min,}$$

$$\text{Equation (2.3.13): } C_A = C_B + \frac{0.1 \times 11}{0.1 + 0.1} - \frac{0.1 \times 11}{0.1 + 0.1} = C_B,$$

$$\text{Equation (2.3.14): } C_D = -2C_B + \frac{2 \times 0.1 \times 11}{0.1 + 0.1} = 11 - 2C_B.$$

Thus Eq. (2.3.24) becomes

$$0 = 0.1 \times 11 \, (0.1 + 0.1) \, C_A - (6.05 \times 10^{-4}) C_A^2 \, (25.4).$$

The quadratic has a solution

$$C_A = C_B = 4.15 \text{ g-mol/L,}$$

$$C_D = 2.70,$$

where the negative root for the quadratic has been discarded.

EXAMPLE 2.4. Given $V = 25.4$ L, $q_{AF} = q_{BF} = 0.2$ L/min, and $C_{AF} = C_{BF} = 11$ g-mol/L, determine C_D.

SOLUTION. This example is the same as the previous one except that the flow rate is doubled. Then, as before,

$$\Theta = \frac{V}{q_{AF} + q_{BF}} = 63.5 \text{ min,}$$

$$C_A = C_B,$$

$$C_D = 11 - 2C_B.$$

The quadratic for C_A, Eq. (2.3.20), is now

$$0 = 0.2 \times 11 - (0.2 + 0.2) \, C_A - (6.05 \times 10^{-4}) C_A^2 (25.4),$$

$$C_A = C_B = 4.70 \text{ g-mol/L,}$$

$$C_D = 1.60.$$

This calculation can be easily repeated for various flow rates. Table 2.4 shows how C_A is affected by changing the flow $q_{AF} + q_{BF}$ for $q_{AF} = q_{BF}$, $C_{AF} = C_{BF} = 11$, $V = 25.4$. As we would expect, the amount of sulfuric acid converted to product decreases as the reactor holding time Θ decreases. Examination of the equations shows that the same effect is obtained if the flow rates are held constant and the feed concentrations of A and B are decreased.

Table 2.4. *Concentration in CFSTR for changing flow rate*

$q_{AF} + q_{BF}$ (L/min)	$\Theta = V/q$ (min)	C_A (g-mol/L)	C_D (g-mol/L)
0.2	127	4.15	2.70
0.3	84.7	4.45	2.10
0.4	63.5	4.70	1.60
0.5	50.9	4.80	1.40
0.6	42.8	4.90	1.20
0.8	31.8	5.00	1.00
1.0	25.4	5.05	0.90

EXAMPLE 2.5. Given $V = 25.4$ L, $q_{AF} = q_{BF}$, $C_{AF} = 11.0$ g-mol/L, $C_{BF} = 5.5$ g-mol/L, and $C_A = 4.0$ g-mol/L, determine C_D and the flow rates.

Here we have specified the desired conversion of sulfuric acid (A) and the feed concentrations. We desire the product concentration C_D and the flow rates. Notice that $q_{AF} = q_{BF}$ is only one relation, as we have not specified the value.

SOLUTION. We first note that C_B and C_D can be computed from Eqs. (2.3.13) and (2.3.14) without any knowledge of the total flow as long as the ratio q_{AF}/q_{BF} is known:

Equation (2.3.13):
$$C_B = C_A + \frac{C_{BF}}{1 + q_{AF}/q_{BF}} - \frac{C_{AF}}{1 + q_{BF}/q_{AF}}$$
$$= 1.25 \text{ g-mol/L},$$

Equation (2.3.14):
$$C_D = -2C_B + \frac{2C_{BF}}{1 + q_{AF}/q_{BF}} = 3.0.$$

The flow rate, $q_{AF} + q_{BF}$, is then computed easily from Eq. (2.3.26),

$$q_{AF} + q_{BF} = \frac{nkC_AC_BV}{C_D} = 0.0512 \text{ L/min},$$

$$q_{AF} = q_{BF} = 0.0256 \text{ L/min}.$$

The holding time is

$$\Theta = \frac{V}{q_{AF} + q_{BF}} = \frac{25.4}{0.0512} = 496 \text{ min}.$$

As the examples illustrate, we may solve steady-state CFSTR problems by manipulation of the set of algebraic equations that make up the mathematical description. For the system used in the examples the specification of five quantities allows the remaining three to be calculated. We cannot, of course, specify any five quantities arbitrarily, because both the physical situation and the model equations require that each component balance must be satisfied. If, for example, we have the situation in which $q_{AF} = q_{BF}$ and C_A, C_{AF}, and V are specified, we are not free to specify C_D. This quantity is known from Eq. (2.3.26) once the other variables are selected. To calculate q_{AF} or q_{BF} we must have an additional piece of information, which in this case is the value of C_{BF} or C_B.

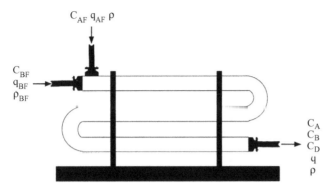

Figure 2.8. Tubular reactor.

2.4   Tubular Reactors

The continuous reacting systems discussed so far are for well-stirred tank-type systems. In the tubular reactor, there is a steady movement of material from one end of the reactor to the other with no attempt to induce mixing. Instead, the fluid motion is generally considered one of plug flow (Figure 2.8) and such devices are known as plug-flow reactors (PFRs). At any axial position z, a cross section of the tube is perfectly mixed, but no mixing takes place between adjacent cross sections. For this kind of fluid behavior to occur, the reactor takes the form of a tube or pipe. In situations in which plug-flow motion may not be a good assumption, the performance predicted with the assumption can often allow a sound design decision to be made.

We consider the same reaction we used for the tank-type reactor:

$$A + B \rightarrow nD.$$

If we begin our model development for the tubular reactor by selecting the reactor itself as a control volume, as we did for the tank-type reactors, we find that there is not a unique value for the measured quantities C_A, C_B, and C_D. These quantities will vary in the axial direction (with distance down the reactor) and, in accordance with Figure 1.3, we must select a new control volume. If we select a small incremental length of the reactor with thickness Δz, then the thin cylinder with volume $\pi D^2 \Delta z/4$ shown in Figure 2.9 can be tried as the simplest control volume. Within the incremental slice of thickness Δz, C_A, C_B, and C_D are considered to have unique values if we sample at any radial or circumferential position and find no change.

Selection of this control volume is equivalent to assuming that the fluid travels through the reactor in plug or piston flow and that the concentration of any species

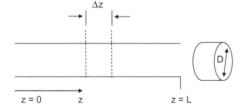

Figure 2.9. Control volume selection for a tubular reactor. Concentrations are approximately constant over the small distance Δz.

at a particular axial position is uniform throughout the cylindrical cross section. This uniformity in the radial and circumferential directions is achieved in many situations by random small-scale fluid motions that tend to produce a well-mixed fluid within the control volume, $\pi D^2 \Delta z/4$. The assumption that characterizing variables are essentially constant over the length Δz is consistent with the usual approximations made in applying the calculus and will be valid in the limit as $\Delta z \to 0$, an operation that we will perform subsequently.

   We begin with a Level I analysis, employing only the law of conservation of mass to the control volume $\pi D^2 \Delta z/4$ in the same manner as we have done with tank-type systems. We need to recognize when we set up the model that conditions at point z in the reactor are different from those at point $z + \Delta z$. The total mass in the control volume is $\rho \pi D^2 \Delta z/4$, and the overall mass balance can be written as follows:

$$\begin{pmatrix} \text{mass in control} \\ \text{volume at time} \\ t + \Delta t \end{pmatrix} = \begin{pmatrix} \text{mass in control} \\ \text{volume at time} \\ t \end{pmatrix} + \begin{pmatrix} \text{mass entered} \\ \text{during } \Delta t \text{ at} \\ \text{position z} \end{pmatrix} - \begin{pmatrix} \text{mass left} \\ \text{during } \Delta t \text{ at} \\ \text{position } z + \Delta z \end{pmatrix},$$

$$\rho_{(t+\Delta t)} \pi D^2 \Delta z/4 = \rho_{(t)} \pi D^2 \Delta z/4 + \rho_{(z)} q_{(z)} \Delta t - \rho_{(z+\Delta z)} q_{(z+\Delta z)} \Delta t. \tag{2.4.1}$$

As before we are using mean chemical property values and () denotes a functional dependence. There are now two independent variables: time t and position in the reactor z. If we divide Eq. (2.4.1) by $\Delta z \Delta t$ and rearrange slightly, we obtain

$$\frac{\rho_{(t+\Delta t)} \pi D^2/4 - \rho_{(t)} \pi D^2/4}{\Delta t} = - \frac{\rho_{(z+\Delta z)} q_{(z+\Delta z)} - \rho_{(z)} q_{(z)}}{\Delta z}.$$

   The quantity on the left is the difference approximation to a time derivative in which the quantities are evaluated at a mean position in the control volume. In the limit as $\Delta t \to 0$, this becomes the partial time derivative at a fixed position. Similarly, the quantity on the right is the difference approximation to a derivative with respect to z. In the limit as $\Delta z \to 0$ this becomes the partial spatial derivative, differentiation with respect to z at fixed t. Thus, taking the two limits ($\Delta t \to 0$, $\Delta z \to 0$), we obtain

$$\frac{\partial \rho \pi D^2/4}{\partial t} = - \frac{\partial \rho q}{\partial z}. \tag{2.4.2}$$

Using the definition of the plug-flow velocity,

$$v = 4q/(\pi D^2),$$

and assuming that the cross section is constant, so that D is independent of z, we obtain

$$\frac{\partial \rho}{\partial t} + \frac{\partial \rho v}{\partial z} = 0. \tag{2.4.3}$$

This is often referred to as the continuity equation and is most frequently encountered in studies of fluid mechanics. At steady state, ρ does not vary with time and Eq. (2.4.3) reduces to

$$\frac{\partial \rho v}{\partial z} = 0. \tag{2.4.4}$$

Equation (2.4.4) states that the quantity ρv is constant and independent of z. Furthermore, if ρ is constant, as it would be for a liquid system, we see that v, the liquid velocity, is a constant. None of these conclusions is very surprising, however, and we will see that the overall mass balance is not of primary importance in the analysis of this reacting system.

The component mass balances are readily developed for the control volume, and we will go through the derivation in detail for component A. The word statement of the conservation law is expressed in symbols as follows:

$$C_{A(t+\Delta t)}\pi D^2 \Delta z/4 = C_{A(t)}\pi D^2 \Delta z/4 + q_{(z)}C_{A(z)}\Delta t$$

$$-q_{(z+\Delta z)}C_{A(z+\Delta z)}\Delta t$$

$$-r_{A-}\pi D^2 \Delta z\Delta t/4. \tag{2.4.5}$$

Dividing each term by $\Delta z\Delta t$ yields

$$\frac{C_{A(t+\Delta t)}\pi D^2/4 - C_{A(t)}\pi D^2/4}{\Delta t}$$

$$= -\frac{q_{(z+\Delta z)}C_{A(z+\Delta z)} - q_{(z)}C_{A(z)}}{\Delta z} - r_{A-}\frac{\pi D^2}{4}.$$

Taking the limit as Δz and Δt approach zero we then obtain

$$\frac{\pi D^2}{4}\frac{\partial C_A}{\partial t} = -\frac{\partial q C_A}{\partial z} - r_{A-}\frac{\pi D^2}{4}. \tag{2.4.6}$$

We are generally concerned with a constant volumetric flow. In that case, the linear velocity in the tube is also constant. Rearranging Eq. (2.4.6) and using the definition of fluid velocity, we obtain

$$\frac{\partial C_A}{\partial t} = -v\frac{\partial C_A}{\partial z} - r_{A-}. \tag{2.4.7}$$

In a similar manner we can derive the component balances for substances B and D:

$$\frac{\partial C_B}{\partial t} = -v\frac{\partial C_B}{\partial z} - r_{B-}, \tag{2.4.8}$$

$$\frac{\partial C_D}{\partial t} = -v\frac{\partial C_D}{\partial z} + r_{D+}. \tag{2.4.9}$$

To examine the design problem we need to consider the steady-state situation, for which all time derivatives are zero. Equations (2.4.7)–(2.4.9) then reduce to

$$v\frac{\partial C_A}{\partial z} = -r_{A-}, \tag{2.4.10}$$

$$v\frac{\partial C_B}{\partial z} = -r_{B-}, \tag{2.4.11}$$

$$v\frac{\partial C_D}{\partial z} = +r_{D+}. \tag{2.4.12}$$

These equations are similar in form to those for the tank system [Eqs. (2.1.3)–(2.1.6)], except that, with the tank system, the dependent variable is time, rather than distance along the reactor. Despite the mathematical equivalence to the batch reactor, there is an essential difference from an engineering viewpoint; namely, that the tubular reactor can be employed as a continuous processing unit. The important comparison in terms of reactor efficiency and design is with the CFSTR, and we can see the differences by considering the following examples.

EXAMPLE 2.6. Using the information given in Example 2.3, we can determine the concentrations of A, B, and D in the exit stream of a tubular reactor of volume 25.4 L. All the inlet concentrations are as specified in Example 2.3 and the flow is the sum $q_{AF} + q_{BF}$, as given in the example.

SOLUTION. The model equations for the tubular reactor become the same as those for the batch reactor, with $\tau = z/v$ substituted for t:

$$\frac{dC_A}{d\tau} = -kC_AC_B,$$

$$\frac{dC_B}{d\tau} = -kC_AC_B,$$

$$\frac{dC_D}{d\tau} = 2kC_AC_B.$$

As was shown in previous sections the following relationships are obtained for the species concentrations:

$$C_A - C_{AF} = C_B - C_{BF},$$

$$C_{AF} - C_A = \frac{1}{2}C_D.$$

(Here we are denoting the concentration at $\tau = 0$ as C_{AF}, rather than C_{Ai}, as for the batch system.) Because C_{AF} and C_{BF} are equal in this example, $C_A = C_B$ and C_A is readily obtained by substitution into the first equation:

$$\frac{dC_A}{d\tau} = -kC_A^2.$$

Integration of this equation yields exactly the same result as that given by the equation in Table 2.2, with τ substituted for t:

$$\frac{1}{C_A} = \frac{1}{C_{AF}} + k\tau.$$

Here C_{AF} represents the inlet concentration of A in the combined stream containing both A and B. To maintain the same molar feed rate as in Example 2.3, the feed concentrations must be one-half those in the individual feed streams, so

$$C_{AF} = \frac{11}{2} = 5.5 \text{ g-mol/L},$$

$$C_{BF} = \frac{11}{2} = 5.5 \text{ g-mol/L}.$$

The exit of the reactor is at $\tau = \Theta = \pi D^2 L/4q = V/q$. For $q = 0.2$, $\Theta = 127$. Thus, C_A, the concentration of A in the exit stream from the tubular reactor, is

$$\frac{1}{C_A} = \frac{1}{5.5} + 6.05 \times 10^{-4} \times 127,$$

$$C_A = 3.84 \text{ g-mol/L}.$$

Table 2.5. *Concentration in a tubular reactor compared with that of a CFSTR with the same residence time*

q	$\Theta = V/q$	CFSTR			Tubular		
(L/min)	(min)	C_A	C_D	qC_D	C_A	C_D	qC_D
0.2	127	4.15	2.70	0.54	3.84	3.32	0.664
0.3	84.7	4.45	2.10	0.63	4.29	2.42	0.726
0.4	63.5	4.70	1.60	0.64	4.54	1.92	0.768
0.5	50.9	4.80	1.40	0.70	4.70	1.60	0.800
0.6	42.8	4.90	1.20	0.72	4.81	1.38	0.828
0.8	31.8	5.00	1.00	0.80	4.97	1.06	0.848
1.0	25.4	5.09	0.88	0.82	5.07	0.86	0.860

C_D, the product concentration, is easily computed as

$$C_D = 2[5.5 - 3.84]$$
$$= 3.32 \text{ g-mol/L.}$$

We see that, for the same residence time, $\Theta = 127$ min, we obtain a higher conversion of A in the tubular reactor than in the CFSTR. A series of calculations can be performed to yield Table 2.5, which compares the tubular reactor results with the CFSTR results contained in Table 2.4. It is clear that the tubular configuration is more efficient in that a smaller reactor is always required than that necessary to produce the same conversion in a CFSTR. The fact that a smaller reactor is required does not necessarily mean that it will be less expensive to construct, and in fact there are many situations in which a CFSTR is less expensive.

EXAMPLE 2.7. For a reaction A → D studied in a CFSTR, find the conversion and production rate in a tubular reactor with the same volume and flow rate.

SOLUTION. The conversion of A is described by the equations

$$\frac{dC_A}{d\tau} = -kC_A,$$
$$C_A(\tau) = C_{AF}e^{-k\tau},$$
$$C_D(\tau) = C_{AF} - C_A(\tau).$$

At the reactor exit, $\tau = \Theta$, we have, using $C_{AF} = 0.2$ g-mol/L, $k = 5 \times 10^{-3}$ min^{-1}:

$$C_A = 0.2e^{-0.005\Theta},$$
$$C_D = 0.2(1 - e^{-0.005\Theta}).$$

The results are shown in Table 2.6 for values of V and q that produce D at a rate of 50 g-mol/min. Analogous to the previous example, the production rate in the tubular reactor exceeds the value in the CFSTR, indicating that a smaller reactor volume is required.

Table 2.6. *Volume and conversion in a tubular reactor for a given production requirement compared with those of a CFSTR*

q	V	Θ	CFSTR			Tubular		
(L/min)	(L)	(min)	C_A	C_D	qC_D	C_A	C_D	qC_D
250	∞	∞	0	0.2	50.0	0	0.2	50.0
300	303 000	1000	0.033	0.167	50.0	0.00135	0.1987	59.6
400	133 000	333	0.075	0.125	50.0	0.0378	0.1622	64.9
500	100 000	200	0.10	0.10	50.0	0.0736	0.1264	63.2
800	72 600	90.7	0.1375	0.0625	50.0	0.1271	0.0729	58.3
1000	66 600	65	0.15	0.05	50.0	0.1445	0.0555	55.5
2000	58 100	27	0.175	0.025	50.0	0.1747	0.0253	50.6
4000	53 200	13.3	0.1875	0.0125	50.0	0.1871	0.0129	51.6

EXAMPLE 2.8. Consider the irreversible reaction $A + B \rightarrow nD$. Show that for a given conversion the residence time of a tubular reactor will always be smaller than that of a CFSTR.

SOLUTION. The tubular and CFSTR equations can be written as follows:

PFR: CFSTR:

$$\frac{dC_A}{dt} = -r,$$
$$0 = \frac{1}{\Theta_{CFSTR}}[C_{AF} - C_A] - r,$$

$$\frac{dC_A}{r} - dt,$$

$$-\int_{C_{AF}}^{C_A} \frac{dC}{r(C)} = \Theta_{PFR},$$

$$\Theta_{PFR} = \int_{C_A}^{C_{AF}} \frac{dC}{r(C)}. \qquad \Theta_{CFSTR} = \frac{C_{AF} - C_A}{r(C_A)}.$$

The development assumes that the chemical equation has been used to find r in terms of C_A. Now, consider r as an increasing function of C_A (e.g., $r = kC_A^n$). Thus, $1/r(C)$ is a decreasing function of C as C varies from C_A to C_{AF} ($C_A < C_{AF}$). Figure 2.10 shows a typical plot of $1/r(C)$ versus C, $C_A \leq C_{AF}$, 0. The integral of $1/r$ is the area under that curve. On the other hand, Θ is the area of the rectangle shown in the figure with base $C_{AF} - C_A$ and height $1/r(C_A)$. Thus, $\Theta_{CFSTR} > \Theta_{PFR}$.

2.5 Reactor Energy Balance

In this chapter, we developed mathematical models for tank and tubular reactors that react isothermally in the liquid phase. We are now in a position to examine nonisothermal operation of the batch reactor by adding an energy balance to the mass balances and reaction rate we developed previously by using a Level III analysis.

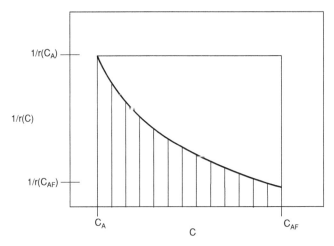

Figure 2.10. Reciprocal of rate vs. concentration to determine residence time.

For the reaction $A + B \leftrightarrow nD$ the equations of conservation of mass and energy are given by

$$\frac{dC_A V}{dt} = -rV, \tag{2.5.1}$$

$$\frac{dC_B V}{dt} = -rV, \tag{2.5.2}$$

$$\frac{dC_D V}{dt} = +nrV, \tag{2.5.3}$$

$$\rho V \hat{C}_p \frac{dT}{dt} = (-\Delta \underline{H}_{rxn}) rV + Q. \tag{2.5.4}$$

[The derivation of Eq. (2.5.4) is given in Appendix A of Chapter 3.] The energy balance includes the heat supplied or removed from the reactor (Q) as well as the *heat of reaction* ($\Delta \underline{H}_{rxn}$). For given concentrations and temperature at time zero and for a specified rate constitutive equation, these equations can be integrated to give the concentrations and temperature as functions of time. In fact, the equations simplify considerably in seeking a solution. Only one concentration equation is really required, as C_B and C_D can be related to C_A and the initial values of the concentration of A, B, and D. Furthermore, the concentration and temperature are related, so that ultimately only a single equation needs to be integrated.

To illustrate this point without algebraic complexity, we consider the slightly simpler case of an adiabatic reactor (Q = 0) with an irreversible reaction whose rate expression is

$$r = k_0 e^{-E_A/RT} C_A C_B,$$

and if $C_{Ai} = C_{Bi}$ then $C_A = C_B$ at all times. We take the density as linear in concentrations and the rate of change of density with concentration proportional to the molecular weight of each component in the reactor. Thus the volume is constant and Eqs. (2.5.1) and (2.5.4) simplify to

$$\frac{dC_A}{dt} = -k_0 e^{-E_A/RT} C_A^2, \tag{2.5.5}$$

$$\frac{dT}{dt} = \frac{(-\Delta\underline{H}_{rxn})}{\rho\hat{C}_p} k_0 e^{-E_A/RT} C_A^2, \tag{2.5.6}$$

$$C_A = C_{Ai}, T = T_i \quad \text{at } t = 0.$$

Assume further that $\Delta\underline{H}_{rxn}$ and $\rho\hat{C}_p$ are constants. (Neither assumption is necessary but the algebra is less complicated and any error introduced is usually negligible.) Then Eqs. (2.5.5) and (2.5.6) combine to

$$\frac{dT}{dt} = -\frac{(-\Delta\underline{H}_{rxn})}{\rho\hat{C}_p} \frac{dC_A}{dt},$$

$$T - T_i = -\frac{(-\Delta\underline{H}_{rxn})}{\rho\hat{C}_p} (C_A - C_{Ai}). \tag{2.5.7}$$

Combining Equations (2.5.5) and (2.5.7) entirely in terms of the concentration, we obtain

$$\frac{dC_A}{dt} = -k_0 C_A^2 \exp\left(-\frac{E_A}{R\left[T_i + \frac{-\Delta\underline{H}_{rxn}}{\rho\hat{C}_p}(C_{Ai} - C_A)\right]}\right). \tag{2.5.8}$$

Equation (2.5.8) is separable. The solution can be expressed in terms of an integral:

$$\int_{C_{Ai}}^{C_A} \frac{dC}{C^2 \exp\left(-\frac{E_A}{R\left[T_i + \frac{-\Delta\underline{H}_{rxn}}{\rho\hat{C}_p}(C_{Ai}-C)\right]}\right)} - -k_0 t, \tag{2.5.9}$$

which must be evaluated numerically.

EXAMPLE 2.9. The reaction A + B → products has the following physical and chemical properties:

$$\Delta\underline{H}_{rxn} = -21\,111 \text{ cal/g-mol of A,}$$
$$E_A = 18\,500 \text{ cal/g-mol,}$$
$$k_i = 6.2 \times 10^4 \text{ m}^3/\text{g-mol min,}$$
$$\rho = 1000 \text{ kg/m}^3,$$
$$\hat{C}_p = 1 \text{ cal/g }^\circ\text{C}$$

Initial conditions are $C_{Ai} = C_{Bi} = 8020$ g-mol/m³, $T_i = 93.3\,^\circ\text{C}$. Estimate the temperature and concentration variation with time.

SOLUTION. According to Eq. (2.5.7) the temperature will rise monotonically from T_i to some limiting value as C_A varies from C_{Ai} to zero. It is convenient to calculate the limiting temperature directly as a check on the subsequent calculations. Setting

$C_A = 0$ in Eq. (2.5.7), we obtain

$$t \to \infty : T = T_i + \frac{(-\Delta \underline{H}_{rxn}) C_{Ai}}{\rho \hat{C}_p}$$

$$= 93.3 + \frac{(21\,111\,\text{cal/g-mol})(8020\,\text{g-mol/m}^3)}{(10^3\,\text{kg/m}^3)(1\,\text{cal/g}\,^\circ\text{C})(10^3\,\text{g/kg})}$$

$$= 262.6\,^\circ\text{C}$$

The activation energy is given in cgs units, for which the gas constant is $R = 1.99$ cal/g-mol K. For consistency, the temperature must be expressed in degrees Kelvin, using Eq. (2.5.7) as follows:

$$T = T_i + \frac{(-\Delta \underline{H}_{rxn})}{\rho \hat{C}_p} (C_{Ai} - C_A)$$

$$= 93.3 + 169.3 - 0.021 C_A + 273\,\text{K}$$

$$= (535.6 - 0.021 C_A)\,\text{K}.$$

Equation (2.5.9) can then be written as

$$t = -\frac{1}{k_0} \int_{C_{Ai}}^{C_A} \frac{dC}{C^2 \exp(-E_A/RT)}$$

$$= -\frac{1}{6.2 \times 10^4\,\text{m}^3/\text{g-mol min}} \int_{8020}^{C_A} \frac{dC}{C^2 \exp[-18\,500/1.99(535.6 - 0.021 C)]}$$

$$= -1.6 \times 10^{-5} \int_{8020}^{C_A} \frac{dC}{C^2 \exp[-9296.5/(535.6 - 0.021 C)]}\,\text{min}$$

The integral cannot be readily evaluated in terms of known functions, but for each value of C_A the integration can be carried out numerically. Table 2.7 contains the computed concentration–time relations and the corresponding temperatures. As seen, for this exothermic reaction ($\Delta H_{rxn} < 0$) the reactor temperature rises significantly. A similar analysis can be performed to determine Q, the heat necessary to be removed, if isothermal operation were desired, or necessitated because of product degradation, safety, or unwanted side reactions.

Table 2.7. Concentration and temperature vs. time for the reaction A + B → products in an adiabatic batch reactor

C_A (g-mol/L)	T ($^\circ$C)	t (min)
8.0	93	0
6.4	127	19
4.8	161	24
3.2	194	26
1.6	228	27
0	262	∞

This concludes our overview of batch, semibatch, and CFSTRs and PFRs. The principles of the level approach to develop model equations for analysis and design have been illustrated. These principles and procedures are now applied to the analysis of heat exchange equipment (Chapter 3) and mass contactors (Chapter 4).

REFERENCES

Hellin M, Jungers JC. Absorption de l'éthylène par l'acide sulfurique. Bull. Soc. Chim. Fr. 1957; 2: 386–400.

Russell TWF, Denn MM. *Introduction to Chemical Engineering Analysis.* New York: Wiley, 1972.

Sandler SI. *Chemical and Engineering Thermodynamics* (4th ed.). New York: Wiley, 2006.

Tobajas M, García-Calvo E, Wu X, Merchuk J. A simple mathematical model of the process of *Candida utilis* growth in a bioreactor. World J. Microbi. Biotechnol. 2003; 19: 391–8.

PROBLEMS

2.1 Suppose that the irreversible reaction $2A + B \rightarrow nD$ taking place in a batch reactor has a rate constitutive equation $r_A = kC_A^2 C_B$. Denote the initial concentration of A and B as C_{Ai} and C_{Bi}, and suppose that there is initially no D in the reactor. Determine the model behavior for the case when $C_{Ai} = C_{Bi}$ and when $C_{Ai} \neq C_{Bi}$. How would you use experimental data to determine k?

2.2 The data in Table P2.1 for the reaction between sulfuric acid and diethyl sulfate are represented by the rate expression $r = kC_A C_B$. Because of the scatter in experimental data, it may often be difficult or impossible to distinguish between models. Using the data in Table P2.1, check the use of the constitutive equation $r = k'C_A$. Determine the rate constants for this reaction.

Table P2.1. Concentration of H_2SO_4 vs. time for the reaction of sulfuric acid with diethyl sulfate in aqueous solution at 22.9 °C (Hellin and Jungers, 1957)

Time (min) t	Concentration of H_2SO_4 (g-mol/L) C_A
0	5.50
41	4.91
48	4.81
55	4.69
75	4.38
96	4.12
127	3.84
162	3.59
180	3.44
194	3.34

2.3 The reaction $CH_3COCH_3 + HCN \leftrightarrow (CH_3)_2CCNOH$ was studied in 0.0441 N acetic acid, 0.049 sodium acetate by Svirebly and Roth (J. Am. Chem. Soc. 1953;75:3106). They report the data given in Table P2.2 from a batch experiment.

Find a constitutive equation for the rate of $(CH_3)_2CCNOH$ production. Determine the rate constants for this reaction.

Table P2.2. *Concentration of acetic acid and sodium acetate vs. time*

t (min)	HCN (mol/L)	CH_3COCH_3 (mol/L)
0	0.0758	0.1164
4.37	0.0748	
73	0.0710	
170	0.0655	
265	0.0610	
347	0.0584	
434	0.0557	
∞	0.0366	0.0772

2.4 For the reaction sequence

$$A \to R, \quad r_A = k_1 C_A,$$
$$R \to S, \quad r_S = k_2 C_R,$$

carried out in a well-mixed batch reactor with no R or S initially present, the concentration of R will go through a maximum as a function of time. Derive the model equations for this situation and solve them for $C_R(t)$. Compute the time at which the maximum occurs, t_M, and the value $C_R(t_M)$. [It will be necessary to solve an equation of the form $dx/dt + kx = b(t)$.]

2.5 For the reaction sequence

$$A \to R, \quad r_A = k_1 C_A,$$
$$R \to S, \quad r_S = k_2 C_R,$$

carried out in an isothermal well-mixed batch reactor with no R or S initially present, the concentration of R will go through a maximum as a function of time. Both reaction rates follow the Arrhenius temperature dependence:

$$k_i = k_{i,0}\, e^{\frac{-E_a}{RT}}.$$

The activation energies of the two sequential reactions are 15 and 25 kcal/mol, respectively. It is also known that, at 25 °C, k_1 is 0.2 min^{-1} and k_2 is 0.3 min^{-1}. Assuming that the initial C_A is 1.0 M, compute:

a. C_R versus time for temperatures of 0, 25, 50, 100, and 200 °C.
b. How the time at which the *maximum* C_R is achieved varies with temperature.
c. How the maximum possible C_R depends on reactor temperature.

2.6 The decomposition of dibromosuccinic (DBS) acid has the rate expression

$$r_{DBS^-} = 0.031\ min^{-1}\ c_{DBS}, \quad DBS \to B + C.$$

a. Find the time required in an isothermal batch reactor for reducing the concentration of DBS from 5 to 1.5 g/L.
b. Find the size of an isothermal CFSTR to achieve the same conversion of DBS if the steady-state flow rate is 50 L/min.
c. What residence time is required?

2.7 Two identical CFSTRs are to be connected so that the effluent from one is the feed to the other. What size should each be if the same conversion of DBS acid is to be achieved as that specified in Problem 2.6? If these two reactors were operated in parallel with each receiving half the total raw material, what exit concentration of DBS acid would be achieved in each?

2.8 There is a landfill by a river, leaking toxic Agent X into the river so that the concentration of Agent X is 0.01 M. The river 220 ft. wide and has an average depth of 13 ft. Luckily, Agent X is biodegradable and its degradation follows the kinetic equation

$$\frac{dC_X}{dt} = \frac{-k\,C_X}{K + C_X}.$$

The Department of Natural Resources determines the kinetic constants k and K to be 7.8×10^{-5} M s^{-1} and 8.3×10^{-2} M, respectively, and determines that it is safe to consume fish only where the concentration of X is lower than 1×10^{-4} M. If the river flows at a flow rate of 51 000 ft^3/s, how far downstream from the landfill should the fish be safe to eat? You may assume that the fish in question are very territorial, and do not move any appreciable distance up and down the river. *Hint: You may assume that the river is approximately plug flow.*

2.9 It has been found that the reaction A $\rightarrow$ D is first order and has a rate constant of 0.005 min^{-1}. The reagent A is provided at a concentration of 0.2 M. Provide a technically feasible design for a reactor that would generate D at a rate of 50 mol/min. Your design should specify the reactor type and the reactor volume that will achieve the desired result.

2.10 It has been found that the reaction 2A $\rightarrow$ D is second order, and has a rate constant k of 0.003 M^{-1} min^{-1}. The reagent is provided at a concentration C_A of 0.3 M. Provide a technically feasible design for a reactor that would generate D at a rate of 50 mol/min. Your design should specify the reactor type and the reactor volume that will achieve the desired result.

2.11 You have been charged with the task of converting an environmentally unfriendly plant liquid purge stream of component A at a concentration of C_{Af} in part to a benign liquid product D according to EPA regulations. This conversion can be completed with the simple reaction

$$A \rightarrow D$$

You have commissioned a small batch reactor to analyze the kinetics of this reaction with additional on-line analytical equipment capable of measuring the concentration of A throughout the experiment. You charge the batch reactor with a volume V (L) of component A at a concentration of C_{Ao} (g-mol/L) and collect time-dependent data for this system. These data are subsequently plotted in several forms.

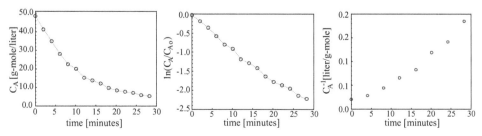

Figure P2.1.

Ultimately, you must use these data to characterize the kinetics describing the reaction.

a. **Level I analysis:** First, derive a mathematical model describing the batch system pictured here and the time-dependent concentrations of species A [C_A (g-mol/L)] and D [C_D (g-mol/L)] for the general reaction rate r [(g-mol/Ls)], and system volume V (L).

b. **Level II analysis:** Explain why a Level II analysis is not needed for this system.

c. **Level III analysis:**

   i. Propose a constitutive equation to describe the relation between the reaction rate (r) and the measured concentration of A in the system (i.e., a reaction-rate expression in which r = k ...).

   ii. Use this constitutive equation to solve the mathematical model for the batch reactor for the time-dependent concentration of A(C_A).

   iii. Examine the functional form of the solution. Which preceding plot (a, b, or c) would you use to validate the proposed constitutive relation? Why? Determine the value of k describing the reaction using data from the appropriate plot.

What assumptions are made in the kinetic model and its solution? How could you check to see if these assumptions are met in practice?

2.12 In Problem 2.11 you extracted the necessary information to describe the kinetics of a simple reaction. You must now complete the design of a CFSTR in order to convert A to product D.

a. **Level I and Level III analyses:** Derive the mathematical model describing the CFSTR system using the constitutive relation proposed in Problem 2.3.

b. The EPA has requested that the plant reduce its emission of A by 95%. Determine the volume V of the CFSTR and associated reactor residence time required to do this, given that the feed stream is a plant purge stream of 50 g-mol of A/L (C_{Af}) at a volumetric flow rate of $q_f = 100$ L/min.

3 Heat Exchanger Analysis

Heat exchanger analysis follows the same procedures as chemical reactor analysis with the added complication that we need to deal with two control volumes separated by a barrier of area a. As shown in Table 3.1 we need to consider both tank type and tubular systems and allow for mixed–mixed, mixed–plug, or plug–plug fluid motions.

A heat exchanger is any device in which energy in the form of heat is transferred. The types of heat exchangers in which we will be most interested are devices in which heat is transferred from a fluid at one temperature to another at a different temperature. This is most commonly done by the confinement of both fluids in some geometry in which they are separated by a conductive material. In such devices the area available for heat transfer is set by the device type and size, and is often the object of a design calculation. If the two fluids are immiscible it is possible to exchange heat by direct contact of one fluid with the other, but such direct contact heat exchange is somewhat unusual. It is, however, the configuration most commonly employed to transfer mass between two fluids, and as such, will be treated in detail in Chapter 4. Note that the area for heat (and mass) transfer is much more difficult to measure or calculate in direct fluid–fluid contacting.

A simple heat exchanger, easily constructed, is one in which the fluids are pumped through two pipes, one inside the other. This double pipe exchanger is shown in Figure 3.9 in Section 3.4. The fluids can flow in the same direction (cocurrent) or in opposite directions (countercurrent). Heat is transferred between fluids because of temperature differences. One fluid is heated and one fluid is cooled. There may or may not be a phase change in one or the other of the fluids. As we will see, there are more complex geometries employed to transfer heat between two fluids. As shown in Table 3.1 the fluid motions within each fluid volume of any exchanger configuration can be classified as one of the following:

- mixed–mixed,
- mixed–plug,
- plug–plug.

For instance, a reactor with a jacket for heating or cooling the reactor contents can be considered as well mixed, a condition easily achieved with low-viscosity liquids and gases. The reactor jacket needs to be well mixed for effective heat exchange, and thus the system can be modeled as mixed–mixed. The same reactor with an internal coil can be modeled as mixed for the reactor contents and plug for the fluid pumped

Table 3.1. Equipment classifications considered in Chapter 3

Reactors: Single phase	Reactors: Two phase	**Heat exchangers**	Mass contactors
Single control volume	Two control volumes	**Two control volumes**	Two control volumes
Tank type	Tank type	**Tank type**	Tank type
Mixed	*Mixed–mixed*	***Mixed–mixed***	*Mixed mixed*
• Batch	• Batch	• **Batch**	• Batch
• Semibatch	• Semibatch	• **Semibatch**	• Semibatch
• Continuous	• Continuous	• **Continuous**	• Continuous
	Mixed–plug	***Mixed–plug***	*Mixed–plug*
	• Semibatch	• **Semibatch**	• Semibatch
	• Continuous	• **Continuous**	• Continuous
Tubular	Tubular	**Tubular**	Tubular
Plug flow	*Plug flow*	***Plug–plug flow***	*Plug–plug flow*
	• Cocurrent	• **Cocurrent**	• Cocurrent
		• **Countercurrent**	• Countercurrent

through the coil. The double-pipe exchanger (see Figure 3.9) can be modeled as a plug–plug system.

The analysis proceeds just as for chemical reactors, but now requires two control volumes, one for each fluid involved in the heat exchange. To develop the model equations describing the rate of heat transfer, we use the laws of conservation of energy and mass with the appropriate constitutive equations.

As shown in the analysis of reactors in Chapter 2, we need to consider both tank-type and tubular systems. Tank-type systems can be operated in a batch, semibatch, or continuous mode, but all tubular systems are operated continuously.

By the end of the chapter you should be able to derive the model equations for the types of heat exchangers just defined, and you should understand what experiments are necessary to verify the validity of the model equations as well as extract process parameters. You should be able to use the model equations to design simple exchangers given the values of the heat transfer coefficients from experiment. The second part of this text will show how the heat transfer coefficients can be obtained from correlations developed partly from analysis of smaller control volumes within the equipment and partly from years of experiments under various conditions.

3.1   Batch Heat Exchangers

Consider two liquids at different temperatures, each contained in an enclosed, well-stirred chamber. There is no flow of mass into or between the tanks. The chamber walls are in contact with one another in such a way that heat is readily transferred between the two. The two-tank system is insulated (i.e., there is no heat flow to the surroundings). This batch heat exchanger is not commonly used, but it allows us to illustrate important aspects of analysis for the mixed–mixed fluid motion situation.

The analysis to follow is valid for any two control volumes in which heat can be exchanged. An example of a typical industrial batch heat exchanger is the jacketed

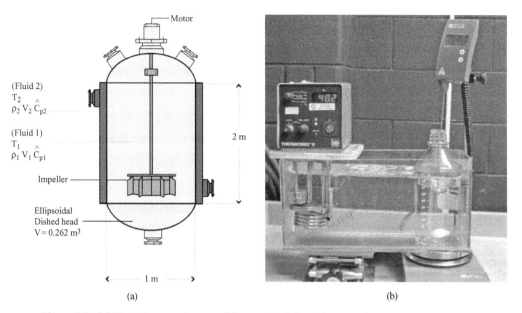

Figure 3.1. (a) Batch heat exchanger; (b) water-bath batch heat exchanger.

dished head vessel shown in Figure 3.1(a). The jacket in this vessel is mixed by the fluid motion induced by the proper design of inlet nozzle. The main part of the vessel is mixed by a mechanical mixer. An example of a laboratory batch heat exchanger is shown in Figure 3.1(b). The control volumes in the laboratory apparatus are mixed with either a mechanical agitator or a magnetic stirrer.

3.1.1 Level I Analysis

We begin with a level I analysis employing only the law of conservation of energy. A derivation of an energy conservation equation is outlined in the appendix to this chapter. Several simplifications can be made for this case. Because the system is closed to mass flows and insulated, energy cannot be added or removed. The kinetic and potential energy in each container will not vary with time. The shaft work imparted to the system by the impellers is generally negligible. Finally, the pressure–volume (PV) work can be combined with the internal-energy term to give the enthalpy of the system.

Using the word statement of the conservation of energy (Figure 1.6) applied to a control volume consisting of both compartments yields the following result:

$$(H_1 + H_2)_{t+\Delta t} = (H_1 + H_2)_t,$$
$$\frac{d(H_1 + H_2)}{dt} = 0. \tag{3.1.1}$$

Integration of this expression gives

$$(H_1 - H_{1i}) = -(H_2 - H_{2i}). \tag{3.1.2}$$

The subscripts "1" and "2" denote the "process" and "utility" fluids, respectively, terminology that is used throughout the chapter. The "i" subscript denotes the initial condition at time t = 0. A common situation often encountered requires either cooling or heating of a quantity of fluid to a specified temperature.

Equation (3.1.2) as written is not immediately useful. It is difficult to compare model behavior with experiment, as quantities like enthalpy cannot be measured directly. As indicated in the model development logic diagram, Figure 1.3, it is necessary to quantify a system in terms of variables such as temperature, which are easily obtained in an experiment. For fluids that remain as a single phase, the enthalpy H (in kilojoules) can be defined in terms of the density ρ (kilograms per cubic meter), volume V (cubic meters), heat capacity $\hat{C}_p$ (kilojoules per kilogram times degrees Kelvin), and temperature T (degrees Kelvin), as discussed in the appendix to this chapter. Using average values of the fluid properties across the range of temperatures of interest in Eq. (3.1.1) produces a Level I relationship between the temperatures in both tanks valid for all times:

$$\rho_1 V_1 \hat{C}_{p1} (T_1 - T_{1i}) = -\rho_2 V_2 \hat{C}_{p2}(T_2 - T_{2i}). \tag{3.1.3}$$

If the fluid in each control volume is the same, Eq. (3.1.3) simplifies to

$$T_2 = T_{2i} - \frac{V_1}{V_2}(T_1 - T_{1i}). \tag{3.1.4}$$

T_{1i} and T_{2i} are the initial temperatures of each fluid at the time an operation begins. Density and heat capacities are tabulated in a number of readily available sources. Some selected values for ρ and $\hat{C}_p$ are shown in Table 5.1 in Chapter 5.

The level I analysis, Equation (3.1.3), provides a relationship between the temperatures in the two tanks at all times. This Level I analysis is somewhat simpler to carry out than the Level I analysis for reactors because we do not have to account for species appearance or disappearance within the control volume that is due to reactions. We must, however, recognize that the second law of thermodynamics constrains the values of T_1 and T_2 such that cold fluids in contact with warm fluids will heat and vice versa.

3.1.2 Level II Thermal Equilibrium

We need to perform a Level II analysis to see if there are limits on the value of T_1 or T_2.

Two substances initially at different temperatures eventually reach the same temperature when in thermal contact. In this case, the condition for thermal equilibrium between two systems is $T_{1,eq} = T_{2,eq} = T_\infty$.

Substituting T_∞ for all instances of T_1 and T_2 in (3.1.3), we can solve for the equilibrium temperature that the batch system will ultimately achieve:

$$T_\infty = \frac{\rho_1 V_1 \hat{C}_{p1} T_{1i} + \rho_2 V_2 \hat{C}_{p2} T_{2i}}{\rho_1 V_1 \hat{C}_{p1} + \rho_2 V_2 \hat{C}_{p2}}. \tag{3.1.5}$$

It can be seen that the equilibrium temperature is just a weighted average of the initial temperatures by their respective thermal capacities. In a batch system, the thermal capacity is $\rho V \hat{C}_p$.

We can compare this result with our intuition by looking at simpler cases. For instance, if we consider the equilibrium result when both chambers contain the same fluid, both the density and specific heat capacity cancel from Equation (3.1.5):

$$T_\infty = \frac{V_1 T_{1i} + V_2 T_{2i}}{V_1 + V_2}. \tag{3.1.6}$$

This is just a volume-weighted average of the initial temperatures. This equation tells us that the final temperature will be closer to the initial temperature of the larger-volume chamber, which is the expected result. Finally, if the volumes are the same, the equilibrium temperature is a simple average of the two initial temperatures, or

$$T_\infty = 0.5(T_{1i} + T_{2i}). \tag{3.1.7}$$

EXAMPLE 3.1. A fine chemical operation requires that 1400 L (1.4 m³) of a waterlike material be cooled from $T_{1i} = 80\,°C$ to $T_1 = 50\,°C$ before it can be further processed. There are two jacketed dished head pressure vessels available [Figure 3.1(a)]. The process fluid can be cooled by placing it in one of the vessels with a charge of chilled water in the jacket available at $T_{2i} = 5\,°C$.

Vessel A has a jacket volume of 1.4 m³ and Vessel B has a jacket volume of 0.5 m³. We need to determine which vessel or vessels can be used and what final utility water temperature T_2 will be reached. The warm utility water must be returned to the chilled water unit. Both liquids can be assumed to have the density and heat capacity of water.

SOLUTION. As a first critical step, we need to find the final equilibrium temperature that can be achieved in both vessels. This is a Level II analysis.

For Vessel A, the simplified form Eq. (3.1.7) applies:

$$\begin{aligned} T_\infty &= \frac{T_{1i} + T_{2i}}{2} \\ &= \frac{80 + 5}{2} \\ &= 42.5\,°C. \end{aligned}$$

Because this temperature is less than the $T_1 = 50\,°C$ required, Vessel A can be used to carry out the required cooling process.

For Vessel B, we need to use Eq. (3.1.6):

$$\begin{aligned} T_\infty &= \frac{V_1 T_{1i} + V_2 T_{2i}}{V_1 + V_2} \\ &= \frac{1.4(80) + 0.5(5)}{1.9} \\ &= 60.3\,°C. \end{aligned}$$

This calculation shows that, using Vessel B, we can cool the process fluid to only 60.3 °C, some 10 °C above that which we need to achieve.

We can gain additional insight into this simple problem by introducing the concept of a heat load, Q_{Load}. This is a quantity widely used in analyzing heat exchangers.

The heat load Q_{Load} is defined as the total heat to be transferred between the two control volumes for a batch exchanger. Q_{Load} is a quantity of some considerable pragmatic importance. For the batch system in which there is no phase change, Q_{Load} is obtained from Eq. (3.1.3):

$$Q_{Load} = \rho_1 V_1 \hat{C}_{p1}(T_1 - T_{1i}) = -\rho_2 V_2 \hat{C}_{p2}(T_2 - T_{2i}). \tag{3.1.8}$$

We need to calculate Q_{Load} (in kilojoules) for almost all heat exchanger analysis. Example 3.1, subsequently continued, illustrates this for the batch system.

Specifying the heat load, the process initial and final temperatures, and the process fluid properties allow one to request a heat exchanger design from a firm who designs, builds, and sells exchangers, a procedure known as "going out on performance."

EXAMPLE 3.1 CONTINUED. Using Eq. (3.1.8) we can compute the process heat load for Vessel A:

$$\rho_1 V_1 \hat{C}_{p1}(T_1 - T_{1i}) = (1000 \text{ kg/m}^3)(1.4 \text{ m}^3)(4.186 \text{ kJ/kg K})(50 - 80 \text{ K})$$
$$= -176 \times 10^3 \text{ kJ}.$$

This is negative because we are cooling the process stream.

Because there are 3.6×10^6 J to a kilowatt hour (kW h), this heat load is 48.8 kW h and must be supplied by the cooling fluid in the jacket.

3.2   Rate of Heat Transfer and Determination by Experiment

As with the reacting system, it is necessary to establish a constitutive equation relating the rate of heat transfer Q to other system variables, specifically the temperatures T_1 and T_2 (Figure 1.3). Once such an expression is verified, it is important to be able to extract relevant parameters from an experiment.

We need to take each chamber and the fluid within as an independent control volume and apply the word statement of the conservation of energy. The following equations are obtained:

$$\frac{dH_1}{dt} = Q_1, \tag{3.2.1}$$

$$\frac{dH_2}{dt} = Q_2. \tag{3.2.2}$$

These equations are similar in form to that of Eqs. (2.1.3) and (2.1.5) for the batch chemical reactor. Heat flow rates into each tank are represented by Q (in kilojoules per second or kilowatts).

By comparing overall energy balance (3.1.1) with the sum of (3.2.1) and (3.2.2), we see that the heat flows must be identical in magnitude, but opposite in sign, or

$$Q_1 = -Q_2. \tag{3.2.3}$$

Both heat flows can be replaced with the common symbol Q. The common convention for a single control volume is to define Q to be positive if heat is added to the system. Because our system consists of two control volumes that are insulated from the environment and no heat is generated, added, or removed from the closed isolated system, heat flowing into one control volume must have come from the other. Although it does not matter which control volume (1 or 2) we consider, we must define a control volume into which heat (Q) flows when Q is positive. As a convention, we define Q to be positive as a flow of heat from control volume 1 into volume 2 and use this choice for all model developments in this chapter.

The equations are then

$$\frac{dH_1}{dt} = -Q, \tag{3.2.4}$$

$$\frac{dH_2}{dt} = Q. \tag{3.2.5}$$

Provided the specific heat capacity does not change appreciably over the course of an operation for either substance, the enthalpy can be rewritten as

$$\frac{dH}{dt} = m\hat{C}_p \frac{dT}{dt} = \rho V \hat{C}_p \frac{dT}{dt},$$

where m is the mass of the fluid and ρ is the density (in kilograms per cubic meter). If the heat capacities are constant over the range in temperature expected or if an average value of the heat capacity of the fluid is used, the governing system equations become

$$\rho_1 V_1 \hat{C}_{p1} \frac{dT_1}{dt} = -Q, \tag{3.2.6}$$

$$\rho_2 V_2 \hat{C}_{p2} \frac{dT_2}{dt} = Q. \tag{3.2.7}$$

Addition of these equations and integration yields Eq. (3.1.3), as should be the case. Note that integration of either equation yields the heat load, Q_{Load}:

$$Q_{Load} = \int_0^t Q\, dt,$$

$$\text{at } t = 0, \quad T_1 = T_{1i},$$

$$t = t, \quad T_1 = T_1$$

$$Q_{Load} = \int_0^t \rho_1 V_1 \hat{C}_{p1} \frac{dT_1}{dt} dt$$

$$= \int_{T_{1i}}^{T_1} \rho_1 V_1 \hat{C}_{p1}\, dT_1$$

$$= \rho_1 V_1 \hat{C}_{p1}(T_1 - T_{1i}).$$

3.2.1 Rate Expression

We have made use of Level I and Level II analyses to determine the behavior of a batch heat exchanger. The Level I analysis enables determination of the temperature of the utility (or process) fluid at any time, but only if the temperature of the other fluid is first known. This is most useful in the analysis of existing equipment. The level I analysis also yields the heat load for a desired change in the process fluid.

Neither the Level I nor the Level II analysis gives any information about the temperatures of the two substances at a specific time. (The Level II analysis gives the final equilibrium temperature, which both fluids will achieve if given infinite time. This is essential for determining the limiting behavior of a system.)

For time-dependent descriptions of the temperature, a Level III analysis is needed. Such an analysis requires a constitutive relation for the heat flow Q in terms of system variables such as T_1 and T_2 (Figure 1.3). From experience and experiment, we know that

- Q is directly proportional to the area a (in square meters), through which heat is being transferred,
- Q is proportional to the magnitude of the temperature difference between two substances, in this case $T_1 - T_2$.

A simple constitutive equation that incorporates these properties is

$$Q = Ua(T_1 - T_2). \tag{3.2.8}$$

Here, U refers to the system's overall heat transfer coefficient (in kilowatts per square meter times degrees Kelvin), which is always positive.

Equations (3.2.4) and (3.2.5) state that a positive heat flow is out of control volume 1 and into control volume 2. Heat flows from hot to cold, and so if $T_1 > T_2$, Q is positive and Equation (3.2.4) shows that the enthalpy (i.e., temperature) in control volume 1 decreases, whereas the enthalpy (i.e., temperature) in control volume 2 increases. This analysis of the signs of the heat flows and temperature changes in the model equations is a critical component of model development.

Values of the heat transfer coefficient measured for some batch systems are given in Table 3.2. A more complete listing can be found in the seventh edition of the *Chemical Engineers' Handbook* (1997).

In an experiment with an existing exchanger, the quantities Q, a, and $(T_1 - T_2)$ can all be measured directly or indirectly. Equation (3.2.8) can be used to determine the heat transfer coefficient from such an experiment. In this chapter, U will typically be specified. In Part II of this book, we show how this parameter is dependent on the fluid properties, the wall between the fluid through which heat is transferred, and the fluid motion within the control volumes.

Table 3.2. Jacketed vessel U values (*Chemical Engineers' Handbook*, 1997, Table 11-7)

Process fluid in vessel	Wall material	Jacket fluid	U (kW/m^2 K)
Brine	Stainless steel	Water	0.23–1.625
Brine	Glass lined	Water	0.17–0.45
Heat transfer oil	Stainless steel	Water	0.28–1.14
Brine	Stainless steel	Light oil	0.20–0.74

Table 3.3. *System properties: Batch heat transfer experiment*

Property	Symbol	Fluid 1 (H_2O)	Fluid 2 (Engine oil)
Initial temperature (K)	T_i	300	430
Density (kg/m³)	ρ	1000	825
Volume (m³)	V	0.1	0.2
Specific heat capacity (kJ/kg K)	$\hat{C}_p$	4.18	2.30
Thermal contact area (m²)	a	0.5	0.5

With the constitutive relation just developed, we can proceed with the Level III analysis to validate the constitutive relation for heat transfer [Eq. (3.2.8)] and to determine the overall heat transfer coefficient for the system. A Level III analysis allows us to determine U for existing exchangers and to size heat exchanger equipment for a given Q_{Load}.

The system equations derived from the Level I analysis in Section 3.2, Eqs. (3.2.6) and (3.2.7), describe the batch heat exchanger performance. When the rate expression is substituted into these equations, we obtain

$$\rho_1 V_1 \hat{C}_{p1} \frac{dT_1}{dt} = -Ua(T_1 - T_2), \tag{3.2.9}$$

$$\rho_2 V_2 \hat{C}_{p2} \frac{dT_2}{dt} = Ua(T_1 - T_2). \tag{3.2.10}$$

Table 1.3 shows these equations and also those for the batch reactor and mass contactor to illustrate the similarities in the mathematical models. These are sufficient for describing a batch heat exchanger system and, although coupled, can be solved to obtain temperatures as functions of time.

An experimental system to determine U consists of two well-mixed compartments with the area between them of 0.5 m². The complete apparatus is well insulated so that there is no heat loss to the outside. The fluids used and their properties are shown in Table 3.3. The experimental data are shown in Figure 3.2.

We can express T_2 in terms of T_1 to uncouple the batch system equations. From Eq. (3.1.3), we find

$$T_2 = T_{2i} - \frac{1}{\Theta}(T_1 - T_{1i}), \tag{3.2.11}$$

where Θ is defined as the thermal capacity ratio of the fluids: $\Theta = [(\rho_2 V_2 \hat{C}_{p2})/(\rho_1 V_1 \hat{C}_{p1})]$.

Substitution of (3.2.11) into (3.2.9) and some rearrangement gives

$$\frac{dT_1}{dt} = -\frac{Ua}{\rho_2 V_2 \hat{C}_{p2}}[(1 + \Theta) T_1 - (T_{1i} + \Theta T_{2i})]. \tag{3.2.12}$$

This equation is separable, and the resultant integration gives

$$\ln\left[\frac{(1 + \Theta) T_1 - (T_{1i} + \Theta T_{2i})}{\Theta (T_{1i} - T_{2i})}\right] = -\frac{Ua (1 + \Theta)}{\rho_2 V_2 \hat{C}_{p2}}t. \tag{3.2.13}$$

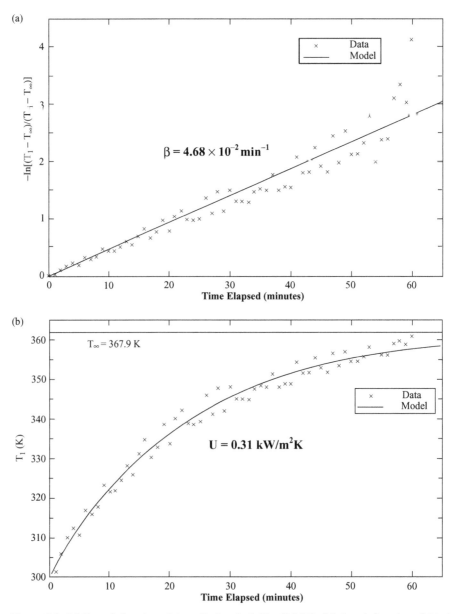

Figure 3.2. (a) Data ($\times$) and model prediction (—), Eq. (3.2.15); (b) data ($\times$) and model behavior (—), Eq. (3.2.14).

We can rearrange this to obtain T_1 as a function of time. The equation relating T_1 and t is

$$T_1(t) = \left(\frac{T_{1i} + \Theta T_{2i}}{1 + \Theta}\right) + \left(\frac{\Theta}{1 + \Theta}\right)(T_{1i} - T_{2i})\,e^{-\beta t}, \qquad (3.2.14)$$

where

$$\beta = \left(\frac{Ua}{\rho_1 V_1 \hat{C}_{p1}} + \frac{Ua}{\rho_2 V_2 \hat{C}_{p2}}\right).$$

$T_2(t)$ can be found from Eq. (3.2.11), given $T_1(t)$.

Equation (3.2.14) has a constant term and a term that is a function of time. At large times the second term approaches zero. The equilibrium temperature $T_1 = T_2$ is given by the first term and reproduces Eq. (3.1.5), the Level II analysis.

The experiment is carried out to check the validity of the constitutive relation [Eq. (3.2.8)]. To do so we need to check the functional form and obtain a value for the heat transfer coefficient U. This is similar to what we did in Chapter 2, Subsection 2.2.2. The most convenient way to obtain U is to rearrange Eq. (3.2.14) into a form that yields a straight-line plot. After some considerable algebra, we can obtain

$$\ln\left(\frac{T_1(t) - T_\infty}{T_{1i} - T_\infty}\right) = -\beta t. \tag{3.2.15}$$

A semilog plot of the left-hand side of (3.2.15) versus time is shown in Figure 3.2(a). The experimental data are shown with ×'s. The heat transfer coefficient U contained in the β term is calculated as 0.31 kW/m² K. With this value of U, T_1 is plotted against time in Figure 3.2(b). Note that using the model equation to determine a simple linear form enables a more direct comparison with experiment and a more critical test of the model.

It is instructive to compare this model behavior with that in Subsection 2.2.2, which illustrated how a specific reaction-rate constant k was calculated. The model equation manipulations and data comparison are the same. The key issue is the use of U or k in situations other than a batch laboratory experiment. Experience with the use of k determined as described for single-liquid-phase reactions shows that it can be used with considerable confidence in designing a large commercial-scale reactor. The U determined in a batch heat exchanger experiment depends on the geometry of the vessel and the mixing equipment that influences the microscale fluid mechanics. This did not affect the determination of k. To use the U obtained in a batch experiment for other exchanger analysis may require additional experiments.

3.2.2 Approach to Equilibrium

Equation (3.2.14) can be used to estimate the approximate time for T_1 to approach the equilibrium value T_∞. To simplify algebraic manipulations, we assume that the control volumes are equal and that both fluids have similar properties. This produces the following simplified form of Eq. (3.2.14):

$$T_1(t) = 0.5\,(T_{1i} + T_{2i}) + 0.5\,(T_{1i} - T_{2i})\,e^{-\beta t}, \tag{3.2.16}$$

with β now defined as $\beta = 2(Ua/\rho V\hat{C}_p)$.

Thermal equilibrium is approximately achieved at e^{-3} ($e^{-3} = 0.05$), when the temperature difference to equilibrium is reduced by 95% of the initial value:

$$\left(2\frac{Ua}{\rho V\hat{C}_p}\right)t_{eq} \approx 3.$$

Solving for t_{eq}, we have

$$t_{eq} \approx \frac{3}{2}\left(\frac{\rho V\hat{C}_p}{Ua}\right). \tag{3.2.17}$$

As the thermal capacity $\rho V \hat{C}_p$ increases, it lengthens the time required for achieving equilibrium. Conversely, a large heat transfer coefficient U or a large heat transfer area will decrease this time. The time for $T_1 \Rightarrow T_{eq}$ is proportional to the volume. If the exchanger volume is doubled, the time will increase by a factor of 2.

EXAMPLE 3.1 CONTINUED. Equation (3.2.15) can be used to estimate the heat transfer coefficient U if an experiment is carried out on tank A. Such an experiment yields a T_1 value of 65 °C after 5.5 min. Θ in Vessel A is equal to one:

$$\ln\left(\frac{65 - 42.5}{80 - 42.5}\right) = -\beta t,$$

$$\beta = 1.55 \times 10^{-3}\,\text{s}^{-1}.$$

The area for heat transfer in a vessel like that illustrated in Figure 3.1 is 4.55 m². The liquid height in Vessel A corresponding to the liquid volume of 1.4 m³ is 1.45 m. (The total liquid volume is made up of 0.26 m³ in the bottom dished head and 1.14 m in the vessel above the tangent line).

β and U are related by the following definitions:

$$\beta = \frac{2Ua}{\rho_1 V_1 \hat{C}_{p1}},$$

$$1.55 \times 10^{-3}\,\text{s}^{-1} = \frac{2(4.55\,\text{m}^2)U}{(1000\,\text{kg/m}^3)(1.4\,\text{m}^3)(4.184\,\text{kJ/kg\,K})},$$

$$U = 1\frac{\text{kJ}}{\text{s\,m}^2\,\text{K}} = 1\frac{\text{kW}}{\text{m}^2\,\text{K}}.$$

Using this U value we can plot T_1 and T_2 as functions of time [Figure E3.1(a)] and $Ua(T_1 - T_2)$ as a function of time [Figure E3.1(b)]. With U determined experimentally, we can find the time to cool the process fluid from 80 to 50 °C:

$$\ln\left(\frac{50 - 42.5}{80 - 42.5}\right) = -1.55 \times 10^{-3}\,t,$$

$$t = 1038\,\text{s} = 17.3\,\text{min}.$$

Figure E3.1(a) is prepared with Eqs. (3.2.14) and (3.2.11) used to calculate T_1 and T_2 as functions of time t for the tank-type exchanger described in Example 3.1. Figure E3.1(b) illustrates how $Ua[T_1 - T_2]$ varies with time in this batch exchanger; $Ua(T_1 - T_2)$ approaches zero at sufficiently long times. This is evident in both figures. Note that $(T_1 - T_2)$ is the *driving force for heat transfer*, and, as equilibrium is reached, this driving force approaches zero [Figure E3.1(a)].

The U value obtained for the pilot-scale batch exchanger (tank A) is 1 kW/m² K whereas that obtained in the laboratory-scale exchanger experiment described in Table 3.3 is 0.31 kW/m² h. This factor of ~3 difference is due partially to the different turbulence created in the exchanger by the mechanical mixing, the different materials of construction and the physical properties of the two fluids. In Chapters 5 and 6 we

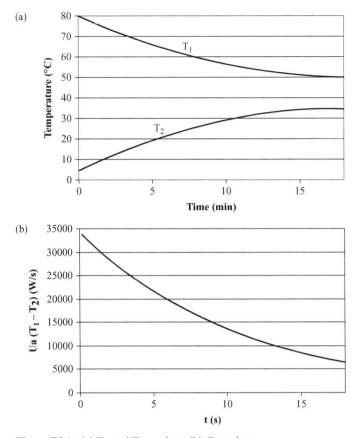

Figure E3.1. (a) T_1 and T_2 vs. time; (b) Q vs. time.

will show how the overall heat transfer coefficient depends on these factors, which will enable understanding and predicting such differences.

3.3 Tank-Type Heat Exchangers

When stirred tanks are used in a heat transfer process, there are two basic modes of operating for each tank — batch or continuous-flow operation. When both tanks are batch vessels, the heat transfer system is a batch heat exchanger. If only one tank is operated in batch mode, whereas the other is a continuous-flow system, the system is classified as a semibatch heat exchanger. The term semibatch is commonly used to describe a reactor into which one raw material is introduced over a period of time while the reaction is taking place. The volume of the reacting mixture thus changes with time. A semibatch exchanger is one in which the heating or cooling of some process fluid is batch with no change of volume with time, but in which the heat exchange medium flows continuously. The heat exchange can be by direct contact as with steam sparging into a liquid, but it is much more common to have two control volumes. If both tanks have continuous flows, we have a continuous-flow heat exchanger.

Batch and semibatch heat exchangers are used when the heating or cooling of a fixed volume of fluid is required, whereas continuous systems are used when the temperature of a continuous stream must be changed. Batch systems are rarely employed even at the laboratory scale, but they can be used to estimate the heat transfer coefficient.

Semibatch systems are more common but this terminology is not in common use. In a semibatch operation, the batch process fluid is charged to the vessel at $t = 0$, and then the utility heating–cooling stream is pumped through the jacket. The flow rate of heating–cooling fluid can be adjusted in order to maintain control over the temperature in the batch vessel. We often do this by constructing a jacket around the process vessel, through which a heating or cooling fluid is passed. We designate this as a semibatch exchanger. This jacketed vessel can be operated in a batch mode as a mixed–mixed system. The process material temperature will vary as a function of time whereas the jacket temperature can be assumed constant if it is well mixed. If the jacket is not well mixed, the analysis will produce a useful limit on what can be achieved.

Temperature control in a batch process can also be achieved with a pipe coil inside the process vessel. This is a semibatch system, which has mixed–plug fluid motions. For a process that operates continuously, the same vessels as those just described can be used with a flow into and out of the process side. We have mixed–mixed and mixed–plug fluid motions. Temperature control of either a batch or a continuously operated process vessel can also be achieved by circulation of the process fluid to a tubular exchanger operating as a separate entity. Each of these types of exchangers is discussed next.

3.3.1 Batch Heat Exchanger

The batch heat exchanger system was discussed extensively in Subsection 3.2.1 with the development of methods to compare experimental data with a model.

3.3.2 Semibatch Heat Exchanger

Semibatch heat exchangers take one of two forms. The usual operation has the process fluid as the (1) well-mixed batch system, which has the same model equations as the well-mixed tank discussed in the previous section. The utility fluid is usually (2) the continuously flowing fluid, either on the "shell" side, which is considered to be well mixed, or contained in a pipe or coil, which is considered to be in plug flow. These are idealized fluid motions, but can be approximately realized in practice and greatly simplify the model equations describing the heat exchanger performance. Mixing is achieved in the jacket by the proper design of the inlet jacket nozzle, often called an agitation nozzle. The utility fluid issuing from a properly designed agitation nozzle causes a swirling motion in the jacket. The nozzle power is $q\Delta P$, where ΔP is the nozzle pressure drop (in pascals). Nozzle power in the range of 50–100 W/m^2 or jacket/vessel area should ensure adequate mixing for waterlike materials. As both fluids involved in the former type of heat exchanger are well mixed, we classify this system as mixed–mixed fluid motions. For the latter case the flow fluid is pumped

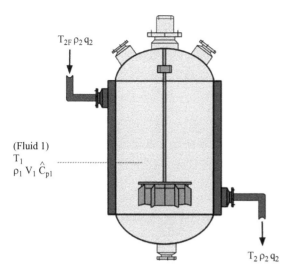

Figure 3.3. Semibatch jacketed heat exchanger.

through a coil that is in contact with the batch fluid; the fluid motions are classified as mixed–plug. The model equations for these two semibatch systems are derived in the following two subsections.

3.3.2.1 Mixed–Mixed Fluid Motions

A jacketed vessel that can be used as a heat exchanger is shown in Figure 3.3. A Level I analysis applied to the entire vessel relates the change in enthalpy of the batch process fluid and the utility fluid in the shell to the flows in and out of the exchanger as

$$(H_1 + H_2)_{t+\Delta t} = (H_1 + H_2)_t - (\rho_2 q_2 \hat{H}_2)_{out} \Delta t + (\rho_2 q_2 \hat{H}_2)_F \Delta t,$$

$$\frac{d(H_1 + H_2)}{dt} = \rho_2 q_2 (\hat{H}_{2F} - \hat{H}_2), \tag{3.3.1}$$

where the exit (out) and feed or entrance (F) flow rates of the utility stream are equal by conservation of mass. Furthermore, because of the well-mixed assumption for stream (2), the exit properties are the same as the properties of the fluid in the device. The same assumptions used in the batch exchanger model development, Section 3.1, are made to express the enthalpy in terms of the fluid temperatures and properties, yielding

$$\rho_1 V_1 \hat{C}_{p1} \frac{dT_1}{dt} + \rho_2 V_2 \hat{C}_{p2} \frac{dT_2}{dt} = \rho_2 q_2 \hat{C}_{p2} (T_{2F} - T_2). \tag{3.3.2}$$

If the time derivative of the utility stream flowing on the shell side, dT_2/dt, is small enough relative to the other terms in Eq. (3.3.2) it can be neglected. This is valid when the flow rate through the exchanger jacket gives a holding time in the jacket that is small compared with the time required for carrying out the batch heat exchange. Under this assumption, the Level I balance relates the two temperatures through the following differential equation:

$$\rho_1 V_1 \hat{C}_{p1} \frac{dT_1}{dt} = \rho_2 q_2 \hat{C}_{p2} (T_{2F} - T_2). \tag{3.3.3}$$

This equation cannot be solved to yield a simple algebraic relationship between T_1 and T_2 because both are functions of time.

We can obtain the heat load Q_{Load} for this vessel by integrating the first term on the left-hand side of Eq. (3.3.2):

$$Q_{Load} = \rho_1 V_1 \hat{C}_{p1} (T_1 - T_{1i}). \tag{3.3.4}$$

We achieved a Level II analysis in the batch case by realizing that, given enough time, T_1 would be equal to T_2. In this semibatch case, equilibrium will occur given enough time when T_1 is equal to T_{2F}. This, as it was in the batch case, gives a limit on what can be achieved for cooling the process fluid, but it is a value of little practical use.

EXAMPLE 3.1 CONTINUED. Vessel A is now equipped with inlet and outlet nozzles on the jacket so the cooling water available at $T_{2F} = 5\,°C$ can be pumped continuously through the jacket at the flow rate q_2 (in cubic meters per second). The 1.4-m^3 process batch has to be cooled from 80 to 50 °C, as indicated before. Both the process and the utility fluid can be assumed to have waterlike properties.

For this technically feasible analysis problem, specifying a q_2 allows T_2 to be obtained. This needs to be carried out with a Level III analysis because Eq. (3.3.3) cannot be integrated as both T_1 and T_2 are functions of time.

A Level III analysis with Eq. (3.2.8) used to describe the heat exchange between the process vessel and the jacket is required. This is subsequently developed.

The energy balance for the process fluid in the tank is derived in the same manner as for the batch exchanger in Section 3.2, yielding

$$\rho_1 V_1 \hat{C}_{p1} \frac{dT_1}{dt} = -Ua(T_1 - T_2). \tag{3.3.5}$$

We derive the energy balance for the continuously flowing utility stream by using the word statement of the conservation of energy and accounting for the flows in and out of the control volume represented by the jacket in Figure 3.3:

$$\rho_2 V_2 \hat{C}_{p2} \frac{dT_2}{dt} = \rho_2 q_2 \hat{C}_{p2} (T_{2F} - T_2) + Ua(T_1 - T_2). \tag{3.3.6}$$

Equations (3.3.5) and (3.3.6) are coupled because T_1 and T_2, which are both functions of time, appear in each equation. These equations can be readily solved numerically, and there are many software packages that will do so.

Figure 3.4(a) is a plot of T_1 and T_2 as functions of time for the parameters given in Example 3.1. The initial conditions on T_1 and T_2 that were used to prepare the plot were $T_{1i} = 80\,°C$ and $T_{2i} = 5\,°C$. These initial conditions assume that the process fluid is charged to the vessel after the chilled water in the jacket has been flowing long enough to fill the jacket and establish a steady flow.

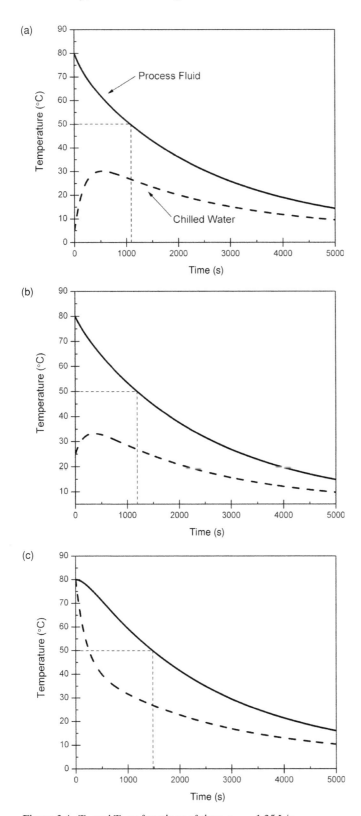

Figure 3.4. T_1 and T_2 as functions of time. $q_2 = 1.35$ L/s.

The calculations show that about 1100 s are required for cooling the process fluid to 50 °C. Other initial conditions for the utility fluid are possible, and these affect the process time, as shown next.

If the process fluid is charged to the tank at 80 °C and the chilled water in the jacket is at 25 °C before the chilled water flow is started, the temperature profiles shown in Figure 3.4(b) are obtained; q_2 is 1.35 L/s, as before, and it is supplied at 5 °C. The time required increases to nearly 1200 s.

If the process fluid and the chilled water in the jacket are both at 80 °C before the chilled water flow is begun, the temperature profiles shown in Figure 3.4(c) are obtained. Again, q_2 is 1.35 L/s and $T_{2F} = 5$ °C. The time required increases to nearly 1500 s.

3.3.2.2 Mixed–Plug Fluid Motions

A semibatch design of this type is shown in Figure 3.5. The Level I analysis is identical to the mixed–mixed semibatch exchanger. The process fluid is in the tank and is assumed to be well mixed. However, unlike the mixed–mixed case, the utility stream in the tube, which flows in a coil or multiple tubes, is assumed to be in plug flow. The temperature of the utility stream will vary along the length of the tube, and hence there is no unique temperature to use in the constitutive equation in the Level III analysis of the process fluid in the tank. As the tank is well mixed, the average driving force for heat transfer along the tube $T_1 - \langle T_2 \rangle$ is used in defining Q. Again, by use of the word statement of the conservation of energy, the governing system equation for the batch vessel becomes

$$\rho_1 V_1 \hat{C}_{p1} \frac{dT_1}{dt} = -Ua(T_1 - \langle T_2 \rangle),\tag{3.3.7}$$

where $\langle T_2 \rangle$ is the average temperature of the plug-flow fluid to be defined shortly. We can derive the corresponding model equation for the utility stream in plug flow by taking a control volume element, shown in Figure 3.6, of thickness Δz and the

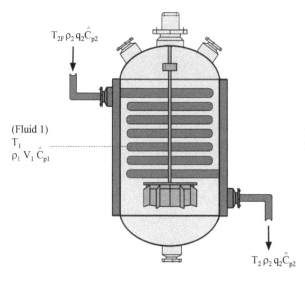

Figure 3.5. Semibatch process vessel coil heat exchanger.

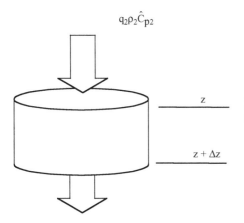

$q_2 \rho_2 \hat{C}_{p2}$

z

$z + \Delta z$

Figure 3.6. Control volume for analysis of utility stream in semibatch exchanger.

coil's cross-sectional area A_c. The model equation in the control value at distance L along the tube is

$$\frac{\partial}{\partial t}\left(\rho_2 A_c \hat{C}_{p2} T_2\right) = -\frac{\partial}{\partial z}\left(\rho_2 q_2 \hat{C}_{p2} T_2\right) + \frac{Ua}{L}(T_1 - T_2). \qquad (3.3.8)$$

Following the same procedure we performed with the mixed–mixed case, we assume the derivative of T_2 with time is small to derive the following model equations. This implies that the residence time in the coil is small compared with the time required for the process fluids to be either heated or cooled:

$$\rho_1 V_1 \hat{C}_{p1} \frac{dT_1}{dt} = -Ua(T_1 - \langle T_2 \rangle), \qquad (3.3.9)$$

$$\rho_2 q_2 \hat{C}_{p2} \frac{dT_2}{dz} = \frac{Ua}{L}(T_1 - T_2). \qquad (3.3.10)$$

Within this level of approximation, Eq. (3.3.10) is already separable and can be integrated, because, with regard to the flow of fluid 2 through the coil, the tank temperature is approximately constant. Because the batch vessel is well mixed, T_1 is only a function of time and has no spatial dependence. The utility steam enters the jacket, $z = 0$, at T_{2F}. Separating and integrating, we obtain

$$T_2 = T_1 - (T_1 - T_{2F})\,e^{-\beta_1 z}, \qquad (3.3.11)$$

with

$$\beta_1 = \frac{Ua}{\rho_2 q_2 \hat{C}_{p2}} \frac{1}{L}.$$

The average coil temperature, $\langle T_2 \rangle$, in batch equation (3.3.9) needs to be replaced with a function involving T_1 so that it too can be integrated. The average value of a function with respect to a variable x over the interval [a, b] is given by

$$\langle f \rangle = \left(\frac{1}{b-a}\right) \int_a^b f(x)\,dx. \qquad (3.3.12)$$

In this case, the independent variable of interest is z, and we are interested in the average value of the temperature over the length of the pipe in contact with the batch fluid, so we use the interval [0, L]. Using (3.3.11) we find that the average temperature inside the pipe is

$$\langle T_2 \rangle = T_1 - \left(\frac{\rho_2 q_2 \hat{C}_{p2}}{U_{?a}} \right) (T_1 - T_{2F}) (1 - e^{-\beta_1 L}). \tag{3.3.13}$$

$\langle T_2 \rangle$ can be substituted in (3.3.9). Separation of the variables and integration yields an expression for the temperature in the tank. The coil temperature can also be determined as a function of time and position. The model equations are

$$T_1(t) = T_{2F} + (T_{1i} - T_{2F}) e^{-\beta_2 t}, \tag{3.3.14}$$

$$T_2(t, z) = T_{2F} + (T_{1i} - T_{2F}) (1 - e^{-\beta_1 z}) e^{-\beta_2 t}, \tag{3.3.15}$$

with

$$\beta_2 = (1 - e^{-\beta_1 L}) \left(\frac{\rho_2 q_2 \hat{C}_{p2}}{\rho_1 V_1 \hat{C}_{p1}} \right). \tag{3.3.16}$$

Although this model is capable of describing the fluid temperature inside the coil, we are generally interested only in the temperature at the outlet. We obtain this temperature simply by substituting $z = L$ into (3.3.15). Once again, the temperature of the process fluid is observed to approach that of the inlet utility stream with an exponential dependence on time. The solution is nearly the same as that for the mixed–mixed semibatch case, with an additional factor in the exponential that accounts for the average temperature in the utility stream. Model equation (3.3.7) or (3.3.8) can be solved numerically to yield temperature versus time and position if Eq. (3.3.13) is employed to obtain $\langle T_2 \rangle$.

3.3.3 Continuous-Flow Tank-Type Heat Exchangers

In the previous subsection, we discussed two exchanger configurations in which a batch system could exchange heat with a continuously flowing fluid. If we have a process in which a continuously flowing fluid must be heated or cooled, we can repeat the modeling of the previous sections by using the mixed–mixed and plug–mixed motions.

3.3.3.1 Mixed–Mixed Fluid Motions

The Level I analysis for the mixed–mixed system shown in Figure 3.7 at steady state is given by the following balance:

Amount of energy in control volume at time $t + \Delta t$	=	Amount of energy in control volume at time t	+	Amount of energy added to control volume in Δt	−	Amount of energy removed from control volume in Δt

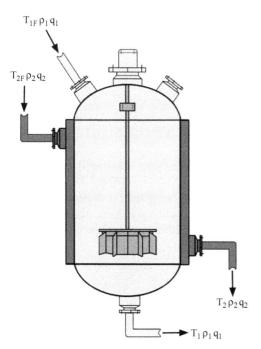

Figure 3.7. Continuous-flow jacketed heat exchanger.

The amount of energy added and removed depends on the volumetric flow rate q and time Δt, so we need to know the enthalpy per volume. We can express the energy in terms of the enthalpy per mass ($\hat{H}$) and density (ρ) to yield the mathematical form of the word statement:

$$\left(\rho_1 V_1 \hat{H}_1 + \rho_2 V_2 \hat{H}_2\right)_{t+\Delta t} = \left(\rho_1 V_1 \hat{H}_1 + \rho_2 V_2 \hat{H}_2\right)_t + q_1 \rho_1 \hat{H}_{1F} \Delta t$$
$$+ q_2 \rho_2 \hat{H}_{2F} \Delta t - q_1 \rho_1 \hat{H}_1 \Delta t - q_2 \rho_2 \hat{H}_2 \Delta t$$

$$\frac{d\left(\rho_1 V_1 \hat{H}_1 + \rho_2 V_2 \hat{H}_2\right)}{dt} = q_1 \rho_1 \hat{H}_{1F} + q_2 \rho_2 \hat{H}_{2F} - q_1 \rho_1 \hat{H}_1 - q_2 \rho_2 \hat{H}_2.$$

At steady state

$$\frac{d\left(\rho_1 V_1 \hat{H}_1 + \rho_2 V_2 \hat{H}_2\right)}{dt} = 0.$$

Therefore,

$$q_1 \rho_1 \left(\hat{H}_{1F} - \hat{H}_1\right) = -q_2 \rho_2 \left(\hat{H}_{2F} - \hat{H}_2\right),$$
$$\rho_1 q_1 \hat{C}_{p1} \left(T_{1F} - T_1\right) = -q_2 \rho_2 \hat{C}_{p2}(T_{2F} - T_2). \tag{3.3.17}$$

In the preceding equations, the average heat capacities are used and the mass balances are used to equate the entrance and exit mass flow rates. The heat load in the exchanger is typically set by the process flow stream designated by the subscript 1. For a desired change in process flow, the heat load is calculated as

$$Q_{\text{Load}} = \rho_1 q_1 \hat{C}_{p1}(T_1 - T_{1F}). \tag{3.3.18}$$

The two streams are related by an algebraic equation. The Level II analysis of equilibrium operation follows directly, in which the two exit stream temperatures are equal. The solution yields

$$T_{1\infty} = T_{2\infty} = \frac{\rho_1 q_1 \hat{C}_{p1} T_{1F} + \rho_2 q_2 \hat{C}_{p2} T_{2F}}{\rho_1 q_1 \hat{C}_{p1} + \rho_2 q_2 \hat{C}_{p2}}. \qquad (3.3.19)$$

Equations (3.3.17)–(3.3.19) can be manipulated to carry out a technically feasible analysis on existing equipment.

EXAMPLE 3.2. Vessel A described in Example 3.1 has been supplied with the necessary piping and pumps so it can be operated to continuously heat or cool a process fluid flow of 100 L/min (1.67×10^{-3} m³/s). We can assume waterlike properties for both the process fluid and the utility fluid. The process fluid needs to be cooled from 80 to 50 °C. The chilled water utility stream is available at 5 °C.

If there are no constraints on the chilled water exit temperature, what is the theoretical minimum flow rate of a chilled water stream?

$$T_{1F} = 80\,°C,$$
$$T_1 = 50\,°C,$$
$$T_{2F} = 5\,°C$$
$$q_1 = 1.67 \times 10^{-3}\ m^3/s,$$
$$\hat{C}_{p1} = \hat{C}_{p2} = 4.184\ kJ/kg\ K.$$

SOLUTION. The minimum utility flow rate possible occurs when $T_1 = T_2$. With the Level I analysis, Eq. (3.3.17) yields

$$q_2 = q_1 \frac{(T_{1F} - T_1)}{(T_2 - T_{2F})}$$

$$= q_1 \left(\frac{80 - 50}{50 - 5}\right)$$

$$= \frac{2}{3}(100)$$

$$= 66.7\ L/min.$$

This operation cannot be attained in practice as it will take an infinitely long residence time in each compartment to reach thermal equilibrium between the two streams. This implies an infinite exchanger area. A higher flow rate and lower value of T_2 must be specified if Vessel A and its attendant pumps and piping are to be used. At a flow rate of 100 L/min

$$T_2 = T_{2F} + \frac{q_1}{q_2}(T_{1F} - T_1)$$

$$= 35\,°C.$$

A technically feasible analysis is a utility flow at 100 L/min with a chilled water exit temperature T_2 of 35 °C.

We derive the Level III analysis by deriving an energy balance on the two mixed control volumes of the tank-type exchanger illustrated in Figure 3.7. Process stream (1) is assumed to flow through a vessel surrounded by the jacket, which contains continuously flowing utility fluid (2):

$$\frac{d}{dt}\left(\rho_1 V_1 \hat{C}_{p1} T_1\right) = \rho_1 q_1 \hat{C}_{p1}\left(T_{1F} - T_1\right) - Ua\left(T_1 - T_2\right), \tag{3.3.20}$$

$$\frac{d}{dt}\left(\rho_2 V_2 \hat{C}_{p2} T_2\right) = \rho_2 q_2 \hat{C}_{p2}\left(T_{2F} - T_2\right) + Ua\left(T_1 - T_2\right). \tag{3.3.21}$$

Although we have used a jacketed vessel to illustrate the physical situation, Eqs. (3.3.20) and (3.3.21) will apply to any two well-mixed control volumes in thermal contact with each other.

Unless we are interested in a start-up or shutdown transient, we can assume steady-state operation, which is how the unit will operate unless disturbed. The model reduces to two algebraic equations:

$$0 = \rho_1 q_1 \hat{C}_{p1}\left(T_{1F} - T_1\right) - Ua\left(T_1 - T_2\right), \tag{3.3.22}$$
$$0 = \rho_2 q_2 \hat{C}_{p2}\left(T_{2F} - T_2\right) + Ua\left(T_1 - T_2\right). \tag{3.3.23}$$

This set of algebraic equations is similar to that developed in Chapter 2 for the CFSTR. Because this is a continuous system, we are concerned with flows instead of batch volumes. The thermal capacity ratio Θ is defined with volumetric flow rates in place of a volume:

$$\Theta = \frac{\rho_2 q_2 \hat{C}_{p2}}{\rho_1 q_1 \hat{C}_{p1}}.$$

The outlet temperatures are then given by

$$T_1 = \frac{\left(1 + \dfrac{\rho_2 q_2 \hat{C}_{p2}}{Ua}\right) T_{1F} + \Theta T_{2F}}{1 + \dfrac{\rho_2 q_2 \hat{C}_{p2}}{Ua} + \Theta}, \tag{3.3.24}$$

$$T_2 = \frac{T_{1F} + \left(\dfrac{\rho_2 q_2 \hat{C}_{p2}}{Ua} + \Theta\right) T_{2F}}{1 + \dfrac{\rho_2 q_2 \hat{C}_{p2}}{Ua} + \Theta}. \tag{3.3.25}$$

As the rate of heat exchange characterized by the term Ua becomes large relative to the thermal capacities, this system will approach equilibrium, as defined by Level II analysis equation (3.3.19).

EXAMPLE 3.2 CONTINUED. If we run an experiment on Vessel A we can determine the heat transfer coefficient by using the Level III analysis.

The experiment is carried out with $q_1 = q_2 = 100$ L/min. Inlet and outlet temperatures of both the process stream and the chilled water stream need to be measured:

$$T_{1F} = 80\,°C, \qquad T_1 = 65\,°C,$$
$$T_{2F} = 5\,°C, \qquad T_2 = 20\,°C.$$

When the vessel is operated continuously, and is full, the area for heat transfer $a = \pi Dh$, where for Vessel A, $D = 1$ m and $h = 2$ m. Therefore, $a = 6.28$ m^2.

U can be calculated with Equation (3.3.22):

$$U = \frac{\rho_1 q_1 \hat{C}_{p1}(T_{1F} - T_1)}{(T_1 - T_2)a}$$

$$= \frac{(1000\,\text{kg/m}^3)(1.67 \times 10^{-3}\,\text{m}^3/\text{s})(4.186\,\text{kJ/kg K})\left[\frac{(80-65\,^\circ\text{C})}{(65-20\,^\circ\text{C})}\right]}{6.283\,\text{m}^2}$$

$$= 0.74\,\frac{\text{kJ}}{\text{m}^2\,\text{K s}} = 0.74\,\frac{\text{kW}}{\text{m}^2\,\text{K}}.$$

3.3.3.2 Mixed–Plug Fluid Motions

The other design scheme in this category uses a pipe rather than another tank as the means of exchanging heat between two fluids. The Level I and II analyses of the mixed–plug exchanger are identical to those for the mixed–mixed case just described.

The Level III analysis of this system is similar to the plug–mixed semibatch case considered in Section 3.3 as the utility stream has a spatially varying temperature profile, and the process stream is well mixed. As before, the energy balance for the process fluid in the tank is derived with the word statement of the conservation of energy, and the driving force uses the average temperature of the utility stream:

$$\frac{d}{dt}(\rho_1 V_1 \hat{C}_{p1} T_1) = \rho_1 q_1 \hat{C}_{p1}(T_{1F} - T_1) - Ua(T_1 - \langle T_2 \rangle). \tag{3.3.26}$$

The only difference between this equation and the one developed for the mixed–mixed continuous-flow case is the use of the term $\langle T_2 \rangle$, the average temperature of the fluid inside the pipe at a given time. Assuming a turbulent flow pattern within the pipe, the fluid can be assumed to be well mixed within a cross section, with only axial temperature gradients forming. Under these assumptions, the energy balance for the utility stream in the pipe is the same as that derived for the plug flow pipe stream in the semibatch case derived previously:

$$\frac{\partial}{\partial t}\left(\rho_2 A_c \hat{C}_{p2} T_2\right) = -\frac{\partial}{\partial z}\left(\rho_2 q_2 \hat{C}_{p2} T_2\right) + \frac{Ua}{L}(T_1 - T_2). \tag{3.3.27}$$

A relatively simple example of this type of heat exchanger is shown in Figure 3.8.

Steady-state operation is assumed for design purposes, and the time derivative terms are therefore zero. The simplified system of equations is then:

$$0 = \rho_1 q_1 \hat{C}_{p1}(T_{1F} - T_1) - Ua(T_1 - \langle T_2 \rangle), \tag{3.3.28}$$

$$\rho_2 q_2 \hat{C}_{p2}\frac{dT_2}{dz} = \frac{Ua}{L}(T_1 - T_2). \tag{3.3.29}$$

This system can be solved in an almost identical manner as that of the equations describing the semibatch heat exchanger. The only difference is that we can find T_1 by solving an algebraic equation instead of an ordinary differential equation. Once again, it is convenient to use the ratio of thermal capacities:

$$T_1 = \frac{T_{1F} + \Theta\left(1 - e^{-\beta L}\right)T_{2F}}{1 + \Theta\left(1 - e^{-\beta L}\right)}, \tag{3.3.30}$$

$$T_2(z) = \frac{T_{1F} + \Theta\left(1 - e^{-\beta L}\right)T_{2F}}{1 + \Theta\left(1 - e^{-\beta L}\right)} - \left[\frac{T_{1F} - T_{2F}}{1 + \Theta\left(1 - e^{-\beta L}\right)}\right]e^{-\beta z}, \tag{3.3.31}$$

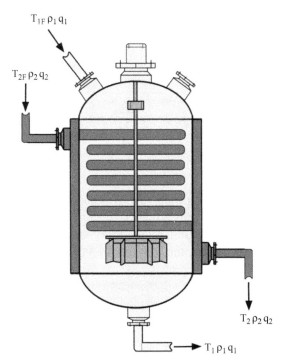

$T_{1F}\,\rho_1\,q_1$

$T_{2F}\,\rho_2\,q_2$

Figure 3.8. Continuous-flow coil heat exchanger.

$T_2\,\rho_2\,q_2$

$T_1\,\rho_1\,q_1$

with

$$\beta = \frac{Ua}{\rho_2 q_2 \hat{C}_{p2}} \frac{1}{L}.$$

The performance limit of this system is given by the Level II analysis, Equation (3.3.19), or, equivalently, by taking the limit of the preceding equations when $\beta \to \infty$, which is the same limit observed in continuous-flow jacketed design.

3.4 Tubular Heat Exchangers

A very effective way to heat or cool fluids is to flow one fluid through a pipe or pipes that are inside a pipe or cylinder of larger diameter through which another fluid can be pumped. The simplest of such exchangers is a double-pipe exchanger in which one pipe is placed inside a pipe of slightly larger diameter. A sketch of this type of device is shown in Figure 3.9.

When more than one pipe or tube is used for the inner part of an exchanger, we designate this as a shell-and-tube exchanger. A photograph of a typical shell-and-tube exchanger showing the tube bundle and the shell is shown in Figure 3.10. Shell-and-tube exchangers are widely used, and standard tube sizes of 1/4, 3/8, 1/2, 5/8, 3/4, 1, 1 1/4, and 1 1/2 in. outside diameter are employed. The wall thickness of heat exchanger tubes can be found in tables and are measured in Birmingham wire gauge (BWG) (*Chemical Engineers' Handbook*, 1997).

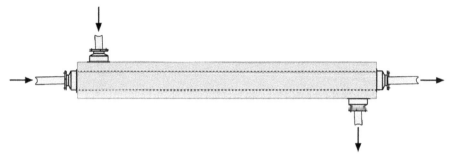

Figure 3.9. Double–pipe heat exchanger (cocurrent flow).

The shells can be standard pipe of the appropriate size to accommodate the required number of tubes. If a shell diameter of more than 12 in. is required, the shell is usually constructed of rolled steel plate. Standard lengths of shell-and-tube exchangers are 8, 10, 12, 16, and 20 ft., with 20 ft. the most common length if the space available can accommodate.

Figure 3.10. Shell-and-tube exchanger showing the tube bundle.

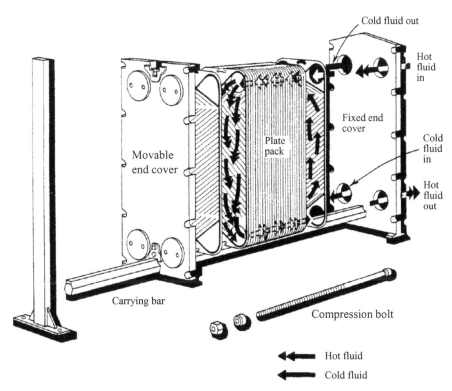

Figure 3.11. Sketch of plate-and-frame heat exchanger (Thermal Division, Alfa Laval Inc.) (Figure 11.49 in the *Chemical Engineers' Handbook*, 1997. Reproduced with permission.)

There are various other geometries to effect heat transfer between two fluids. One such is a plate-and-frame heat exchanger, an exploded view of which is sketched in Figure 3.11. Two plates are shown to illustrate the countercurrent flow. The plates of a plate-and-frame exchanger are held within a frame separated by gaskets that direct the process flow through every second channel and the utility flow is directed into the channels between.

We use the term plug flow to designate the fluid motion within the two control volumes of the exchanger (Figure 3.9). Double-pipe exchangers can clearly be modeled as tubular–tubular plug flow. Shell-and-tube exchangers can also be modeled this way, but sometimes, depending on exchanger geometry and the flow in the shell, it is more realistic to consider the shell side fluid as well mixed. Heat exchangers such as the plate and frame can be successfully modeled by assuming plug-flow fluid motions in both control volumes.

All heat exchangers with plug fluid motions can be operated as cocurrent or countercurrent devices. If the fluids flow in the same direction we have a cocurrent heat exchanger. If the two fluids flow in opposite directions we have a countercurrent heat exchanger. We will develop the model equations for both cases.

3.4.1 Cocurrent Flow

A cocurrent double-pipe heat exchanger is shown in Figure 3.12. This is an exchanger configuration in which both fluids enter the exchanger at the same position.

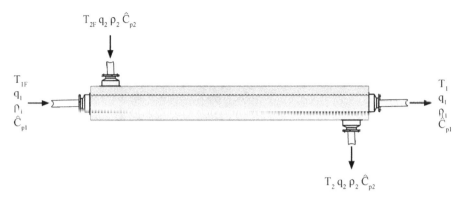

Figure 3.12. Cocurrent double-pipe exchanger.

A Level I analysis assuming constant physical properties yields

$$\rho_1 q_1 \hat{C}_{p1} (T_{1F} - T_1) + \rho_2 q_2 \hat{C}_{p2} (T_{2F} - T_2) = 0,$$

$$T_2 = T_{2F} + \frac{\rho_1 q_1 \hat{C}_{p1}}{\rho_2 q_2 \hat{C}_{p2}} (T_{1F} - T_1). \tag{3.4.1}$$

A Level II analysis yields the maximum amount of heat transferred for the given flow rates if the exchanger is infinitely long:

$$T_{1\infty} = T_{2\infty} = \frac{\rho_1 q_1 \hat{C}_{p1} T_{1F} + \rho_2 q_2 \hat{C}_{p2} T_{2F}}{\rho_1 q_1 \hat{C}_{p1} + \rho_2 q_2 \hat{C}_{p2}}. \tag{3.4.2}$$

Equations (3.4.1) and (3.4.2) are identical to those developed for the continuous mixed–mixed exchanger, Eqs. (3.3.17) and (3.3.19). We will soon see that the Level I balance will also hold true for the countercurrent exchanger.

Typical temperature profiles for a cocurrent exchanger are sketched in Figure 3.13 for the case in which a process fluid at T_{1F} is to be heated to T_1 by a utility stream available at T_{2F}.

As noted before, $\Delta T(z) = T_2(z) - T_1(z)$ is the driving force for heat transfer. Most exchangers operate with $T_2 - T_1$ greater than ~10 °C. As seen in Figure 3.13, increasing the length of the exchanger beyond a ΔT of this value will result in little additional heat exchange and capital costs will increase.

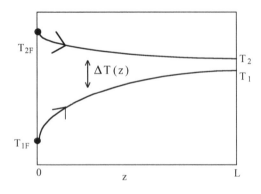

Figure 3.13. Temperature vs. position for a cocurrent exchanger.

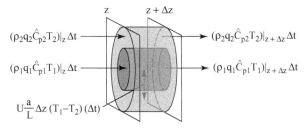

Figure 3.14. Cross-sectional slice of a cocurrent heat exchanger.

The heat load expressed in terms of the process stream is the required rate of heat transfer for the continuous exchanger operating at steady state:

$$Q_{Load} = q_1 \rho_1 \hat{C}_{p1}(T_1 - T_{1F}). \tag{3.4.3}$$

In a typical heat exchange problem, the flow rate q_1, the temperatures T_1 and T_{1F}, and the physical properties are known so Q_{Load} can be readily determined. If the stream is being cooled, $T_1 < T_{1F}$ and Q_{Load} is negative. For heating, Q_{Load} is positive. Equation (3.4.1) then shows that there is a range of values of q_2 and T_2 that will meet the Q_{Load} for a heat exchange fluid whose physical properties and T_{2F} are known.

For a given utility fluid, the minimum possible flow rate q_2 can be obtained from Eq. (3.4.2) ($T_1 = T_{1\infty} = T_{2\infty}$). This gives a lower limit on q_2. To obtain a technically feasible exchanger design, q_2 must be above this minimum as this corresponds to infinite exchanger length.

The Level I and Level II balances cannot tell us what area is required to satisfy Q_{Load}. To determine the exchanger area, i.e., the length and diameter of the pipes, a Level III analysis is required.

A Level III analysis applies the law of conservation of energy to control volumes within the tubular system. Such a control volume is depicted in Figure 3.14.

The model equation development proceeds as it did for the tubular reactor, except that we now have to deal with two control volumes. As shown in Figure 3.14 we write our energy balance in terms of temperatures. Expressing the word statement for the fluid in the inner pipe (Figure 1.6) in symbols yields

$$\left(\rho_1 \hat{C}_{p1} \frac{\pi}{4} D_{IN}^2 \Delta z T_1\right)\Big|_{t+\Delta t} = \left(\rho_1 \hat{C}_{p1} \frac{\pi}{4} D_{IN}^2 \Delta z T_1\right)\Big|_t + q_1 \rho_1 \hat{C}_{p1} T_1 \Delta t|_z$$
$$- q_1 \rho_1 \hat{C}_{p1} T_1 \Delta t|_{z+\Delta z} - Ua \frac{\Delta z}{L}(T_1 - T_2)\,\Delta t.$$

If we divide by Δt, Δz, we obtain

$$\frac{\partial}{\partial t}\left(\rho_1 A_{c1} \hat{C}_{p1} T_1\right) = -\frac{\partial}{\partial z}\left(\rho_1 q_1 \hat{C}_{p1} T_1\right) - \frac{Ua}{L}(T_1 - T_2). \tag{3.4.4}$$

In a similar fashion we obtain for the fluid in the outer pipe

$$\frac{\partial}{\partial t}\left(\rho_2 A_{c2} \hat{C}_{p2} T_2\right) = -\frac{\partial}{\partial z}\left(\rho_2 q_2 \hat{C}_{p2} T_2\right) + \frac{Ua}{L}(T_1 - T_2), \tag{3.4.5}$$

where

$$A_{c1} = \frac{\pi}{4} D_{IN}^2,$$

$$A_{c2} = \frac{\pi}{4} \left(D_{OUT}^2 - D_{IN}^2 \right),$$

$$a = \pi D_{IN} L.$$

Like the other continuous systems discussed in Section 3.3, tubular systems are designed to be operated at steady state. Neglecting the time derivatives and assuming the physical properties do not vary with z, we find that these equations reduce to

$$\rho_1 q_1 \hat{C}_{p1} \frac{dT_1}{dz} = -\frac{Ua}{L}(T_1 - T_2), \tag{3.4.6}$$

$$\rho_2 q_2 \hat{C}_{p2} \frac{dT_2}{dz} = \frac{Ua}{L}(T_1 - T_2). \tag{3.4.7}$$

Equation (3.4.1) provides the Level I relationship between temperatures T_1 and T_2. Equation (3.4.6) or (3.4.7) can then be integrated to solve for either T_1 or T_2 as a function of z. At $z = 0$, $T_1 = T_{1F}$, $T_2 = T_{2F}$.

Integration then leads to the model equations describing a cocurrent-flow heat exchanger.

$$T_1(z) = \frac{1}{\Theta + 1}(T_{1F} + \Theta T_{2F}) + \frac{\Theta}{\Theta + 1}(T_{1F} - T_{2F})e^{-\beta z}, \tag{3.4.8}$$

$$T_2 = T_{2F} + \frac{1}{\Theta}(T_{1F} - T_1). \tag{3.4.9}$$

Equation (3.4.8), which expresses T_1 as a function of tubular exchanger length z, is directly comparable with Eq. (3.2.14) for the batch exchanger in which T_1 is expressed as a function of t.

The parameter Θ is the ratio of the thermal capacities for continuous systems:

$$\Theta = \frac{\rho_2 q_2 \hat{C}_{p2}}{\rho_1 q_1 \hat{C}_{p1}},$$

$$\beta = \left(\frac{Ua}{\rho_1 q_1 \hat{C}_{p1}} + \frac{Ua}{\rho_2 q_2 \hat{C}_{p2}} \right) \frac{1}{L}.$$

The limiting behavior of a cocurrent heat exchanger can be found if a limit is taken as the term Ua approaches ∞ relative to the thermal capacities of both streams. The resultant limiting temperatures for infinite area are

$$T_{1\infty} = T_{2\infty} = \frac{\rho_1 q_1 \hat{C}_{p1} T_{1F} + \rho_2 q_2 \hat{C}_{p2} T_{2F}}{\rho_1 q_1 \hat{C}_{p1} + \rho_2 q_2 \hat{C}_{p2}},$$

which is Eq. (3.4.2) obtained from the Level II analysis.

Equation (3.4.8) can be used for both technically feasible analysis and technically feasible design problems. If an exchanger exists, U, a, L, and the physical properties of the fluids will be known. If both flow rates are assumed, then Eq. (3.4.8) can be used to calculate T_1 given the utility fluid inlet temperature T_{2F}.

The solutions to the energy balances for this exchanger, Eqs. (3.4.6) and (3.4.7), are a bit awkward to use even with modern software packages. An alternative solution to Eqs. (3.4.6) and (3.4.7) can be developed that has proven useful for both technically feasible analysis and design problems.

We begin by subtracting Eqs. (3.4.6) and (3.4.7):

$$\frac{d(T_1 - T_2)}{dz} = \left(\frac{-Ua}{\rho_1 q_1 \hat{C}_{p1}} - \frac{Ua}{\rho_2 q_2 \hat{C}_{p2}} \right) \frac{1}{L} (T_1 - T_2).$$

This relation can be integrated as follows:

$$\int_{(T_1-T_2)_{z=0}}^{(T_1-T_2)_{z=L}} \frac{d(T_1 - T_2)}{(T_1 - T_2)} = \int_{z=0}^{z=L} \left(\frac{-Ua}{\rho_1 q_1 \hat{C}_{p1}} - \frac{Ua}{\rho_2 q_2 \hat{C}_{p2}} \right) \frac{1}{L} dz.$$

$(T_1 - T_2)$ at $z = 0$, the inlet of the exchanger, is $(T_{1F} - T_{2F})$ for cocurrent flow, and $(T_1 - T_2)$ at $z = L$, the outlet of the exchanger, is $(T_1 - T_2)$.

The right-hand side can be integrated to L if we assume that U, a, ρ, q, and $\hat{C}_p$ are constant with distance along the exchanger. This is usually a pretty good assumption. But if not true, the equations can be solved numerically if the properties are known. Of course, a relationship with these parameters expressed as a function of z would be required. For constant parameters we find:

$$\ln \frac{(T_1 - T_2)}{(T_{1F} - T_{2F})} = \left(-\frac{Ua}{\rho_1 q_1 \hat{C}_{p1}} - \frac{Ua}{\rho_2 q_2 \hat{C}_{p2}} \right). \tag{3.4.10}$$

The right-hand side of this equation can be expressed in terms of Q_{Load}, the heat load for process stream 1. We can obtain this by integrating the left-hand side of Eq. (3.4.6):

$$Q_{Load} = \rho_1 q_1 \hat{C}_{p1} (T_1 - T_{1F}).$$

The heat load for utility stream 2 is Q_{Load}:

$$\rho_1 q_1 \hat{C}_{p1} = \frac{Q_{Load}}{T_1 - T_{1F}},$$

$$\rho_2 q_2 \hat{C}_{p2} = \frac{-Q_{Load}}{T_2 - T_{2F}}.$$

These substitutions yield an expression relating Q_{Load} to the inlet and outlet temperatures of both the process and utility streams:

$$\ln \left(\frac{T_1 - T_2}{T_{1F} - T_{2F}} \right) = \frac{Ua}{Q_{Load}} (-T_1 + T_{1F} + T_2 - T_{2F}),$$

$$Q_{Load} = Ua \left[\frac{(T_2 - T_1) - (T_{2F} - T_{1F})}{\ln \left(\frac{T_2 - T_1}{T_{2F} - T_{1F}} \right)} \right] = Ua \, \Delta T_{LM}. \tag{3.4.11}$$

The term containing the exchanger temperatures on the right-hand side of Eq. (3.4.10) is commonly designated as ΔT_{LM}, the log-mean temperature difference, a quantity that has proven useful for solving a variety of exchanger problems.

Recall that Q_{load} is defined by the process stream properties and temperature as $\rho_1 q_1 \hat{C}_{p1} (T_1 - T_{1F})$. Equation (3.4.11) is particularly convenient if U needs to be determined for an exchanger that has been operating for some period. Such

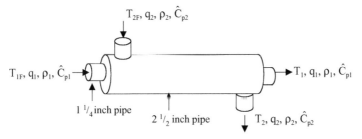

Figure E3.2. Double-pipe heat exchanger.

exchangers may have deposits that reduce the heat transfer rate. An experiment that measures T_1, T_{1F}, T_2, T_{2F}, and q_1 allows U to be estimated if the densities and heat capacities are known.

EXAMPLE 3.3. A cocurrent double-pipe exchanger (see Figure E3.2) exists to cool the process stream described previously ($q_1 = 100$ L/min $= 1.67 \times 10^{-3}$ m^3/s) from 80 to 50 °C. The process fluid flows in the inner pipe, and the chilled water flows in the outer pipe (Figure 3.10), $T_{2F} = 5$ °C. Determine the flow rate and exchanger exit temperature of the chilled water that will be needed.

The following table shows the pipe characteristics for the exchanger.

Schedule 40 S steel pipe as inner pipe in double-pipe heat exchanger

$D_{Nominal}$	1.25 in.	
D_{ID}	1.38 in.	3.5×10^{-2} m
D_{OD}	1.66 in.	4.2×10^{-2} m
v (linear velocity)	5.5 ft./s (at q_1)	1.67 m/s

Schedule 40 S steel pipe used as outer pipe in double-pipe heat exchanger

$D_{Nominal}$	2.5 in.	
D_{ID}	2.47 in.	6.27×10^{-2} m
D_{OD}	2.875 in.	7.3×10^{-2} m
v (linear velocity)	6.28 ft./s through annulus (at 267 L/min flow)	1.91 m/s

There are 67 m of 1.25-in. pipe (nominal diameter) configured as a double-pipe exchanger stretched below. The exchanger is not one long pipe, but a series of 20-ft. sections (6.07 m). The total exchanger area is $(67 \text{ m})(\pi \times 3.5 \times 10^{-2} \text{ m}) = 7.37$ m^2.

SOLUTION. Because both fluids have waterlike properties the Level I analysis is simplified to

$$T_2 = T_{2F} - \frac{q_1}{q_2}(T_1 - T_{1F}),$$

$$T_2 = 5\,°C - \frac{100}{q_2}(50 - 80\,°C).$$

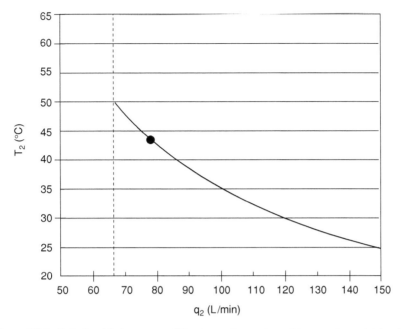

Figure E3.3. Relationship between utility stream flow rate and temperature obtained with a Level I analysis. The • represents the solution to this problem, and the dashed line is q_{2Min}.

The Level I balance relates the utility stream exit temperature to the utility stream flow rate, but does not provide a unique solution.

The theoretical minimum flow rate can be found if the exchanger is operated such that the utility stream exists at 50 °C, which is the maximum allowable value. This is given by a Level II analysis as

$$q_{2Min} = \frac{q_1 (T_{1F} - T_{1\infty})}{T_{2\infty} - T_{2F}},$$

where $T_{1\infty} = T_{2\infty} = 50\,°C$.

Thus,

$$q_{2Min} = \frac{100(80 - 50)}{50 - 5} = 67\ L/min.$$

The relationship between T_2 and q_2 given by $T_2 = (5\,°C - (100/q_2)(-30\,°C))$ can be seen in Figure E3.3.

However, only one combination will provide the desired cooling with the existing exchanger. To proceed, a Level III analysis is required, which necessitates an estimate of the overall heat transfer coefficient U.

A previous experiment allowed U to be estimated as

$$U = 1\frac{kW}{m^2\,K}.$$

The Level III analysis, Equation (3.4.8), which is one solution to the coupled energy balance equations describing this exchanger, can be used or we can use Eq. (3.4.10), which is another solution to the coupled energy balance equation.

Equation (3.4.8) can be solved for q_2 because we know β and Θ and the process fluid inlet and exit temperatures:

$$\Theta = \frac{q_2}{1.67 \times 10^{-3}},$$

$$\beta L = \frac{(1\,kJ/m^2\,s\,K)(7.37m^2)}{(1000\,kg/m^3)(4.184\,kJ/kg\,K)(1.67 \times 10^{-3}\,m^3/s)}\left(1+\frac{1}{\Theta}\right)$$

$$= 1.05\left(1+\frac{1}{\Theta}\right),$$

and, because $L = 67\,m$,

$$\beta = 0.016\left(1+\frac{1}{\Theta}\right)\,m^{-1}.$$

Equation (3.4.8) is an implicit equation and must be solved by any method that selects Θ and then computes the right-hand side and compares it with $T_1 = 50\,°C$.
Such a procedure yields $q_2 = 78\,L/min$. T_2 is calculated from the Level I balance:

$$T_2 = 5 - \frac{100}{78}(50-80)$$

$$= 43\,°C$$

Alternatively, Equation (3.4.10) can be used to obtain T_2 as follows:

$$Q_{Load} = (1.67/1000\,m^3/s)(1000\,kg/m^3)(4.185\,kJ/kg\,K)(50-80\,K)$$

$$= -210\,kJ/s.$$

Equation (3.4.10) becomes

$$-210\,kJ/s = (1.0\,kJ/s\,m^2\,K)(7.27\,m^2)\,\Delta T_{LM},$$

$$\Delta T_{LM} = \frac{(T_2-50)-(5-80)}{\ln\left(\frac{T_2-50}{5-80}\right)}.$$

An iterative solution yields $T_2 = 43\,°C$.
q_2 is found from the Level I balance:

$$q_2 = q_1\frac{(T_{1F}-T_1)}{(T_2-T_{2F})}$$

$$= 78\,L/min.$$

This solution is shown as the • in Figure E3.3.

3.4.2 Countercurrent Flow—Double-Pipe Heat Exchanger

A countercurrent double-pipe exchanger is shown in Figure 3.15 with the inlet and outlet steam identified. Notice that it is similar to the cocurrent exchanger except that the flow direction of stream 2 is reversed. Utility fluid (2) enters at the end of the exchanger where process fluid (1) exits.

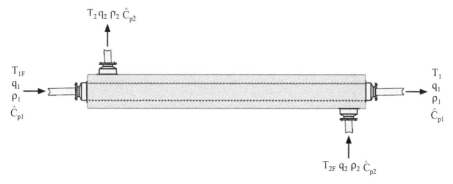

Figure 3.15. Countercurrent double-pipe exchanger.

A Level I analysis gives the same expression we obtained for the cocurrent exchanger and the mixed–mixed exchanger:

$$\rho_1 q_1 \hat{C}_{p1} (T_{1F} - T_1) + \rho_2 q_2 \hat{C}_{p2} (T_{2F} - T_2) = 0,$$

$$T_2 = T_{2F} + \frac{1}{\Theta}(T_{1F} - T_1). \qquad (3.4.1)$$

A Level II analysis is more complex for the countercurrent exchanger because the flows are in the opposite direction to each other. There are two limiting cases for the infinite-length exchanger shown in Figure 3.16 for the case in which stream 1 is heated from T_{1F} to T_1:

a. $T_{1\infty} \Rightarrow T_{2F}$. Here excess utility fluid (2) enables heating the process fluid to T_{2F}.
b. $T_{2\infty} \Rightarrow T_{1F}$. Here utility side (2) is limiting the maximum allowable heating of the process fluid ($T_1 < T_{2F}$).

Note that in both cases it is possible to heat the process fluid hotter than the exit of the utility stream ($T_1 > T_2$). This is impossible in a cocurrent configuration.

The curvature of the temperature profile in the preceding figure depends on the flow rates q_1 or q_2 and the physical properties of each stream.

As with the cocurrent double-pipe exchanger, Q_{Load} is determined as before:

$$Q_{Load} = q_1 \rho_1 \hat{C}_{p1} (T_1 - T_{1F}).$$

With Q_{Load} specified Level I balance equation (3.4.1) shows a range of T_2, q_2 values for any exchanger with known physical properties for the fluids and a specified T_{2F}.

In most cases, the process fluid exit temperature is specified (T_1); hence to find the minimum q_2 if the countercurrent exchanger is infinite in length, we obtain a limiting value by considering the case in which $T_2 \Rightarrow T_{1F}$.

The Level I balance can then be solved for q_2 minimum:

$$q_{2Min} = \frac{-Q_{Load}}{\rho_2 \hat{C}_{p2} (T_{1F} - T_{2F})}. \qquad (3.4.12)$$

A Level III analysis of the control volumes given in Figure 3.17 proceeds by application of the conservation of energy.

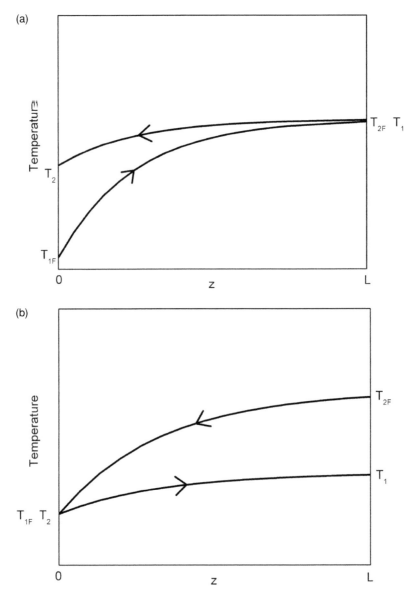

Figure 3.16. Countercurrent temperature profiles.

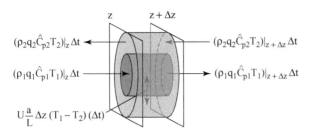

Figure 3.17. Cross-sectional slice of a countercurrent heat exchanger.

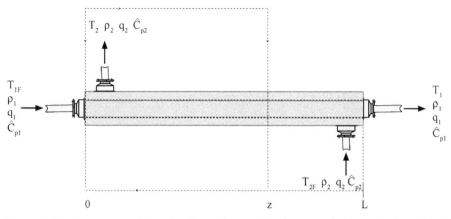

Figure 3.18. Countercurrent double-pipe exchanger showing control volume for Level I balance.

The energy balance equations for the countercurrent system are

$$\frac{\partial}{\partial t}\left(\rho_1 A_{c1} \hat{C}_{p1} T_1\right) = -\frac{\partial}{\partial z}\left(\rho_1 q_1 \hat{C}_{p1} T_1\right) - \frac{Ua}{L}(T_1 - T_2), \qquad (3.4.13)$$

$$\frac{\partial}{\partial t}\left(\rho_2 A_{c2} \hat{C}_{p2} T_2\right) = \frac{\partial}{\partial z}\left(\rho_2 q_2 \hat{C}_{p2} T_2\right) + \frac{Ua}{L}(T_1 - T_2). \qquad (3.4.14)$$

Notice that there is a sign difference in the spatial derivative of Eq. (3.4.14) in comparison with the cocurrent equation, (3.4.5). This is a consequence of the different directions of flow.

For steady state behavior, these equations can be reduced to the following system of differential equations (and their respective boundary conditions):

$$\rho_1 q_1 \hat{C}_{p1} \frac{dT_1}{dz} = -\frac{Ua}{L}(T_1 - T_2) \quad T_1 (z = 0) = T_{1F}, \qquad (3.4.15)$$

$$\rho_2 q_2 \hat{C}_{p2} \frac{dT_2}{dz} = -\frac{Ua}{L}(T_1 - T_2) \quad T_2 (z = L) = T_{2F}. \qquad (3.4.16)$$

It is important to note that the boundary conditions for this system are located at opposite ends of the heat exchanger. This introduces an uncoupling problem that we have not encountered thus far. Nevertheless, an analytic solution is possible.

To integrate with Eq. (3.4.15) or (3.4.16), we need a relationship between T_1 and T_2 in the control volume. We can not use our Level I balance on the total exchanger to obtain T_2 in terms of T_1 because it applied only at the two ends of the exchanger, not as is the case for the cocurrent exchanger for any control volume inside.

However, with the coordinates defined in Figure 3.18, the Level I balance yields the temperatures of both streams at position z as

$$\rho_1 q_1 \hat{C}_{p1} \left[T_1(z) - T_{1F}\right] = \rho_2 q_2 \hat{C}_{p2} \left[T_2(z) - T_2(0)\right]. \qquad (3.4.17)$$

Substitution of $T_2(z)$ into Eq. (3.4.14) allows it to be solved:

$$T_1(z) = \left\{\frac{\Theta\left[T_2(0) - T_{1F}\right]}{\Theta - 1}\right\} + \left(\frac{\Theta}{\Theta - 1}\right)\left[T_{1F} - T_2(0)\right] e^{-\beta z},$$

where

$$\beta = \left(\frac{Ua}{\rho_1 q_1 \hat{C}_{p1}} - \frac{Ua}{\rho_2 q_2 \hat{C}_{p2}} \right) \frac{1}{L}.$$

Because $T_2(0)$ is the temperature of stream 2 leaving the exchanger, it can be expressed in terms of T_1 by use of Level I balance equation (3.4.1):

$$T_2(0) = T_{2F} + \frac{1}{\Theta} [T_{1F} - T_1(L)].$$

The final forms for T_1 and T_2 as functions of exchanger length z are as follows. For $\Theta \neq 1$,

$$T_1(z) = \left(\frac{\Theta T_{2F} - e^{-\beta L} T_{1F}}{\Theta - e^{-\beta L}} \right) + \frac{\Theta}{\Theta - e^{-\beta L}} (T_{1F} - T_{2F}) e^{-\beta z}. \qquad (3.4.18)$$

where

$$\Theta = \frac{\rho_2 q_2 \hat{C}_{p2}}{\rho_1 q_1 \hat{C}_{p1}} \quad \text{and} \quad \beta = \left(\frac{Ua}{\rho_1 q_1 \hat{C}_{p1}} - \frac{Ua}{\rho_2 q_2 \hat{C}_{p2}} \right) \frac{1}{L}.$$

In the case in which $\Theta = 1$ (exactly), the solution is instead

$$T_1(z) = T_{1F} - \frac{Ua}{\rho_2 q_2 \hat{C}_{p2} + Ua} (T_{1F} - T_{2F}) (z/L). \qquad (3.4.19)$$

$T_2(z)$ is obtained from $T_1(z)$ by use of Eq. (3.4.17).

As with the cocurrent exchanger, Eq. (3.1.18) can be used to solve technically feasible operating and design problems. In almost all cases the countercurrent exchanger will require less area to meet a given Q_{Load}. Equation (3.4.19) is presented for completeness in a situation in which $\Theta = 1$, an unlikely situation.

As we did with the cocurrent exchanger, we can obtain ΔT_{LM} for the countercurrent exchanger. The derivation is similar, but we must be careful with the boundary conditions. We begin by subtracting Eqs. (3.4.15) and (3.4.16):

$$\frac{d(T_1 - T_2)}{dz} = \left(\frac{-Ua}{\rho_1 q_1 \hat{C}_{p1}} + \frac{Ua}{\rho_2 q_2 \hat{C}_{p2}} \right) (T_1 - T_2) \frac{1}{L}.$$

The limits on the right-hand side of this equation are

$$z = 0, \quad T_1 = T_{1F}, \quad T_2 = T_2,$$
$$z = L, \quad T_1 = T_1, \quad T_2 = T_{2F}.$$

Using the same substitutions as in the cocurrent case leads to

$$Q_{Load} = Ua \left[\frac{(T_{2F} - T_1) - (T_2 - T_{1F})}{\ln \left(\frac{T_{2F} - T_1}{T_2 - T_{1F}} \right)} \right] = Ua \, \Delta T_{LM}. \qquad (3.4.20)$$

Although similar to Eq. (3.4.11), the calculation of ΔT_{LM} requires keeping careful bookkeeping of the entrance and exit temperatures at each end of the tubular exchanger.

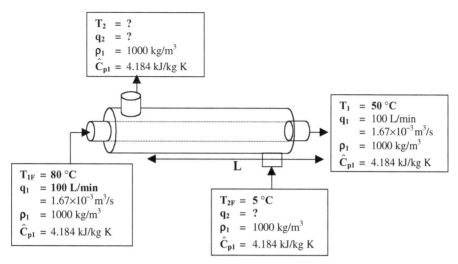

Figure E3.4. Countercurrent heat exchanger.

EXAMPLE 3.3 CONTINUED. If the double-pipe exchanger described before is operated as a countercurrent device (see Figure E3.4), what flow rate q_2 will be needed? The exchanger is as follows:

$U = 1\,kW/m^2\,K$,
$L = 67\,m$ ($\sim$200 ft. $\rightarrow$ 10 sections of 20 ft. pipe),
$a = 7.37\,m^2$.

Level I Anaylsis

This is the same set of equations as that used for the cocurrent operation. For the water–water case, this is

$$q_1(T_{1F} - T_1) = q_2(T_2 - T_{2F}).$$

Level II Analysis

The limiting case of interest is when $T_2 = T_{1F}$:

$$q_{2\,Min} = \frac{q_1(T_{1F} - T_1)}{T_{1F} - T_{2F}}$$

$$= 100 \times \frac{30}{75}$$

$$= 40\frac{L}{min}.$$

This is considerably less flow than required for the cocurrent case in which $q_{2min} = 67$ L/min.

The Level III analysis is summarized by Eqs. (3.4.17) and (3.4.18). Equation (3.4.18) can be solved for Θ by trial-and-error iteration for q_2 until T_1 is 50 °C. The result of the iteration yields a q_2 of 51 L/min. T_2 is obtained with Eq. (3.4.17):

$$T_2 = T_{2F} - \frac{1}{\Theta}(T_1 - T_{1F}) = 5 - \frac{1}{0.51}(50 - 80) = 64\,°C.$$

Alternatively, Eq. (3.4.20) can be used. The heat load is obtained as before:

$$Q_{Load} = (1.67 \times 10^{-3} \text{ m}^3/\text{s})(1000 \text{ kg/m}^3)(4.184 \text{ kJ/kg C})$$
$$(50 - 80\,^\circ\text{C}) = -2.1 \times 10^5 \text{ W}.$$

Because Q_{Load}, U, and a are known, T_2 can be obtained with

$$a = \frac{Q_{Load}}{U \Delta T_{LM}}.$$

An iterative solution yields $T_2 = 64\,^\circ\text{C}$.

Knowing the exit temperature allows the flow rate to be calculated from Equation (3.4.17):

$$q_2 = q_1 \frac{T_{1F} - T_1}{T_2 - T_{2F}} = 100 \frac{80 - 50}{65 - 5} = 50 \frac{\text{L}}{\text{min}}.$$

To cool the process fluid with a countercurrent heat exchanger, q_2 must be 50 L/min, and T_2 must be 64 °C. Note that q_2 is considerably lower than the cocurrent case in which q_2 is 78 L/min.

So, for the identical exchange considered in the cocurrent case, the countercurrent operation requires a much lower utility flow rate. If we compare the temperature profiles for the cocurrent and countercurrent cases along the length of the heat exchanger, shown in Figure E3.5, a larger ΔT is observed, on average, for the countercurrent equipment. Cocurrent operation requires 78 L/min of cooling water. Simply changing the direction of flow of the utility fluid to countercurrent reduces the required utility flow rate to 51 L/min, showing the better performance of the countercurrent exchanger.

Having analyzed the operation of the major classes of heat transfer equipment, we now turn our attention from analysis to design.

3.5   Technically Feasible Heat Exchanger Design

There are many excellent books that have been written on heat exchanger design, and much of what exists has been summarized in the seventh edition of the *Chemical Engineers' Handbook* (1997). The Heat Transfer Research Institute (HTRI) has the latest information on heat transfer but access is limited to members. Our aim here in discussing the technically feasible design is to illustrate to chemical professionals how the design of a heat exchanger is initiated.

Thus far, we have developed models that describe the behavior of a number of heat exchangers. For each heat exchanger, we used a Level III analysis to determine the temperature of both streams. The model equations were rearranged to solve for the area required for heat transfer for a given desired temperature difference and stream flow rates and properties. This solution of the model equations is the basis for developing a technically feasible design. In this section we develop a step-by-step procedure for constructing a technically feasible design of an exchanger for heating

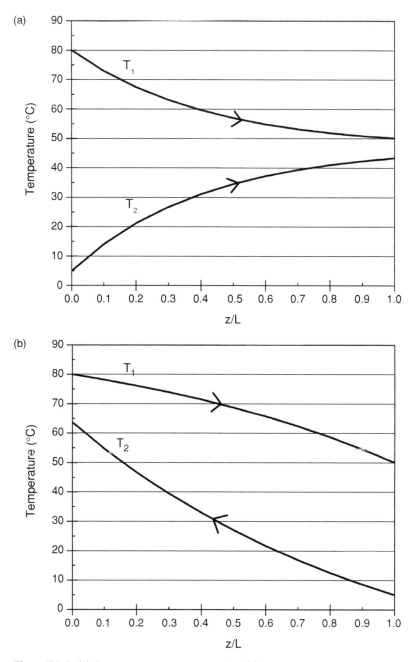

Figure E3.5. (a) Cocurrent temperature profile; (b) countercurrent temperature profile.

or cooling a process stream. A technically feasible design is a necessary first step in the iterative process of the design of any exchanger. It provides the designer with a quantitative understanding of the critical issues that affect the optimal design and allows the effect of the exchanger design and operation on the rest of the process to be examined.

3.5.1 Design Procedure

Step 1. Determine Objective. The typical heat exchanger design problem requires that an exchanger be specified to change the temperature of some process stream given the flow rate and inlet and outlet temperatures. A technically feasible design is one which will meet the required objective.

Step 2. Calculation of Heat Load (Level I Analysis). For a given objective, we can directly calculate the heat load of the heat exchanger. This is given by the Level I analysis:

$$Q_{Load} = \rho_1 q_1 (\hat{H}_1 - \hat{H}_{1F}) \approx \rho_1 q_1 \hat{C}_{p1}(T_1 - T_{1F}). \qquad (3.5.1)$$

For exchangers with any phase change (generation or steam) the enthalpies must be used. If there is no phase change, the average material properties such as density and specific heat capacity can be used for this calculation. Usually, they can be easily found in reference sources such as the *Chemical Engineers' Handbook* (1997). The density and volumetric flow rate of a fluid may vary with temperature, but the product of the two is the mass flow rate, which does not vary with temperature. Therefore, it is acceptable to use values corresponding to the feed conditions in (3.5.1). The heat capacity should be chosen as the average value over the applicable temperature range. Note that, for equipment to be put out for bid on performance, this completes the specifications required if the inlet temperature of the heating or cooling medium is known.

Step 3. Choose a Heating–Cooling Utility Fluid and Utility Exit Temperature. All heat exchangers are part of some process, and the choice of a heating or cooling medium is often constrained to those available in the processing plant. Cooling water, steam, chilled water, or some process stream are examples of utility fluids that are frequently used to satisfy the heat load. The density, specific heat capacity, and feed temperature (utilities generally are available at a given temperature) are known once the heat exchange fluid is selected.

A Level II analysis can be used to estimate the minimum flow rate of heating–cooling fluid. For most continuous systems (except countercurrent exchangers), the Level II analysis results in

$$q_{2\,Min} = q_1 \frac{\rho_1 \hat{C}_{p1}}{\rho_2 \hat{C}_{p2}} \left(\frac{T_{1F} - T_1}{T_1 - T_{2F}} \right).$$

For the countercurrent case there were three different cases that could occur, but here we are interested in only the one that gives the minimum utility flow rate (or $\Theta < 1$). The limiting expression is then

$$q_{2\,Min} = q_1 \frac{\rho_1 \hat{C}_{p1}}{\rho_2 \hat{C}_{p2}} \left(\frac{T_{1F} - T_1}{T_{1F} - T_{2F}} \right).$$

Note that, although these equations are similar in form, they give different results. A countercurrent operation will have a lower minimum flow rate than other designs. These relations provide minimum values for the utility stream flow rate. The utility

flow rate is bounded by this minimum. The temperatures are bounded by the temperatures of the process stream (T_1, except for countercurrent operation, which is
T_{1F}).

We know T_2 only as a function of q_2 from our Level I analysis. T_{2F}, the utility
stream feed temperature is specified once the utility fluid is chosen. Hence,

$$\rho_1 q_1 \hat{C}_{p1}(T_{1F} - T_1) + \rho_2 q_2 \hat{C}_{p2}(T_{2F} - T_2) = 0,$$

$$T_2 = T_{2F} + \frac{1}{\Theta}(T_{1F} - T_1).$$

The technically feasible design requires that either T_2 or q_2 be specified to begin
an iterative design process. Although the minimum utility flow rate is known from
the Level II analysis, the exchanger area corresponding to this minimum flow rate is
infinite. This provides little guidance for estimating a technically feasible flow rate.
On the other hand, the exit utility temperature is a parameter that is more strictly
bound. In fact, using the "rule of thumb" that temperature differences of at least 10
°C are required for effective heat transfer, the exit utility stream temperature T_2 can
be set to be 10 °C above (heating) or below (cooling) of that of the exit process stream
temperature T_1 in a cocurrent or tank-type exchanger. In countercurrent exchangers
T_2 can be set 10 °C above or below the process stream inlet temperature, T_{1F}. Such
initial selections of T_2 can always be modified as the iterative process proceeds to
obtain a technically feasible design.

Step 4. Heat Transfer Coefficient Estimation. Once an appropriate heating–cooling
fluid has been chosen and its temperatures and flow rate specified, an estimate of
the area required for heat transfer must be made. To do so, we need to select a heat
transfer coefficient to begin a trial-and-error procedure that will adjust our initial
selection until a final value is obtained. Tables 3.2 and 3.4 provide some guidance but
if nothing has been decided about the type of exchanger a value of 1 kW/m^2 K can
be used to get an initial estimate of area by use of Equation (3.5.2).

After an initial exchanger design has been achieved with an estimated U, we
can calculate the heat transfer coefficient from correlations that relate the rate of
heat transfer to the fluid properties, materials of construction, and the fluid flow
rates in the device. We can then compare this heat transfer coefficient with our
initial selection and iterate to a final value. This will be illustrated in Part II of this
book.

Table 3.4. Tubular exchangers, selected U values (adapted
from *Chemical Engineers' Handbook*, 1997, Table 11-3)

Shell side	Tube side	U (kW/m^{2}K)
Water	Water	1.135–1.520
Fuel oil	Water	0.085–0.142
Gasoline	Water	0.340–0.568
Monoethanolamine or Diethanolamine	Water	0.795–1.135

Table 3.5. *Temperature difference expressions for exchanger area determination*

Continuous-flow model	Average temperature difference
Two tank	$T_2 - T_1$
Heating–cooling coil	$\dfrac{(T_2 - T_1) - (T_{2F} - T_1)}{\ln\left(\dfrac{T_2 - T_1}{T_{2F} - T_1}\right)}$
Cocurrent	$\dfrac{(T_2 - T_1) - (T_{2F} - T_{1F})}{\ln\left(\dfrac{T_2 - T_1}{T_{2F} - T_{1F}}\right)}$
Countercurrent ($\Theta \neq 1$)	$\dfrac{(T_{2F} - T_1) - (T_2 - T_{1F})}{\ln\left(\dfrac{T_{2F} - T_1}{T_2 - T_{1F}}\right)}$
Countercurrent ($\Theta = 1$)	$\dfrac{(T_{2F} - T_1) - (T_2 - T_{1F})}{\ln\left(\dfrac{T_{2F} - T_1}{T_2 - T_{1F}}\right)}$

Step 5. Exchanger Area Estimation (Level III Analysis). At this point in the design, the outlet temperatures of both sides of the heat exchanger are specified. For a given exchanger type, the area can be determined from the model equations derived in this chapter. These are rearranged to yield an expression relating the heat load of the exchanger to the average temperature difference between the fluids in the exchanger:

$$Q_{Load} = \rho_1 q_1 \hat{C}_{p1}(T_1 - T_{1F}) = Ua \langle \Delta T_{21} \rangle. \qquad (3.5.2)$$

In this equation, $\langle \Delta T_{21} \rangle$ is the average temperature difference between the two streams in the device. The inlet and outlet temperatures of both the process stream and the utility stream must be specified as previously described.

Expressions for $\langle \Delta T_{21} \rangle$ for continuous-flow tubular exchangers were derived in this chapter. They are summarized in Table 3.5.

This temperature difference is often denoted the "log-mean" temperature difference when referred to cocurrent and countercurrent heat exchangers, and its logarithmic form is another way of expressing the solution of the coupled model equations derived in this chapter.

Equation (3.5.2) is solved for the exchanger area a. Further design is required to determine the actual dimensions of the exchanger.

Step 6. Preliminary Heat Exchanger Selection. A tank-type vessel would rarely be designed based on the required heat load alone. Heat transfer to a process fluid from a jacket or a coil is useful to control temperatures for reacting or other processes that must be carried out in a tank-type vessel. The process requirements other than its heat load may set vessel size and geometry, and so a jacketed vessel may not be feasible. A coil can be used to independently control the area for heat transfer. If a coil is to be considered, the design proceeds as if for a tubular exchanger with a constraint on coil length because the vessel size is fixed. The coil diameter should be between 0.25 and 1 in., and a diameter within this range should be selected to begin the iterative process. The velocity can be determined with q_2. The area is determined with Eq. (3.5.2) and the entry in Table 3.5 for the heating–cooling coil. If the length required cannot be fit into the vessel, an exchanger, external to the vessel, may need to be considered.

To begin the iterative process to design a tubular exchanger, the length and diameter of the pipes or channels through which the process and utility fluid will flow must be specified. The iterative process is most easily demonstrated with a double-pipe exchanger design, which is just a shell-and-tube exchanger with a single tube. A pipe velocity or a pipe diameter must be selected. Because q_1 and q_2 are both known, the internal diameter of a pipe can be estimated if a velocity is assumed or a velocity can be calculated if a diameter is selected. Velocities in exchangers are usually of the order of magnitude of 1 m/s. Tube diameters range from 0.25 to 1.50 in. This defines the diameter and length of both the inner and outer pipes of a double pipe exchanger for both countercurrent and cocurrent exchangers, as the area has been estimated with Eq. (3.5.2) and Table 3.5.

Step 7. Iteration to a Technically Feasible Design. To reach this point, three parameters were specified:

1. T_2,
2. U,
3. Pipe diameters or velocities.

With the exchanger geometry known, the overall heat transfer coefficient U can be obtained from the following equation, which will be derived in Chapter 5.

$$\therefore U = \frac{1}{\dfrac{1}{h_{out}} + \dfrac{r_{out}\ln(r_{out}/r_{in})}{k_{wall}} + \dfrac{r_{out}}{h_{in}r_{in}}}. \tag{5.4.12}$$

The heat transfer coefficients h can be estimated from correlations of the following form

$$Nu = a\,Re^b\,Pr^c \tag{6.3.3}$$

where a, b, and c are determined by experiment and are known for pipes, for the shell side of shell-and-tube exchangers and for other geometries such as a plate-and-frame exchanger. Tabulations are presented in Chapter 6. The dimensionless groups are defined in the nomenclature section of Chapter 5.

U can be calculated and compared with the estimated value used to initiate the iterative process. The calculated U is now used to obtain a new area and exchanger length. It may be necessary to iterate on pipe diameters and/or velocities to achieve a feasible design, as will be discussed in more detail in Chapter 8.

EXAMPLE 3.4. In the examples we have examined so far in this chapter we were able to perform a technically feasible analysis for an existing exchanger. The following calculations illustrate a technically feasible design following the steps outlined in this section.

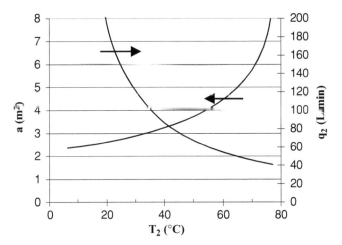

Figure E3.6. Necessary utility flow rate q_2 and area required a, at different exit temperature T_2.

SOLUTION

Step 1. Determine Objective. The objective is to cool 100 L/min (1.67×10^{-3} m^3 /s) of process stream from 80 to 50 °C. The process fluid has waterlike properties.

Step 2. Calculation of Heat Load. The heat load can be found by use of a Level I analysis:

$$Q_{Load} = \rho_1 q_1 \hat{C}_{p1} (T_1 - T_{1F}) = (1 \text{ kg/L}) (100 \text{ L/min}) (4184 \text{ J/kg K}),$$
$$(50 - 80 °C) = -2.1 \times 10^5 \text{ W}.$$

Step 3. Choose a Heating–Cooling Fluid. In most heat exchanger design problems the heating or cooling mediums available are known. This may be cooling water, steam, another process stream, chilled water, or some special heat transfer fluid such as Dowtherm. For this problem there is chilled water available at 5 °C.

The minimum flow rate of the cooling fluid for a countercurrent heat exchanger is found by use of a Level II analysis:

$$q_{2\,Min} = q_1 \left(\frac{T_{1F} - T_1}{T_{1F} - T_{2F}} \right) = (100 \text{ L/min}) \left(\frac{80 - 50}{80 - 5} \right) = 40 \text{ L/min}.$$

At this flow rate the contact area must be infinitely large. There are also an infinite number of combinations of temperatures and flow rates that can achieve the heat duty, as can be seen in Figure E3.6, which is prepared from a Level I energy balance:

$$\rho_1 q_1 \hat{C}_{P1} (T_1 - T_{1F}) = -\rho_2 q_2 \hat{C}_{p2}(T_2 - T_{2F}).$$

Selecting a T_2 will allow q_2 to be determined and also allow an estimate of the exchanger area to be made. A reasonable first choice is to pick a chilled water exit temperature that is 10 °C below the process stream inlet temperature, T_{1F}, of 80 °C. Using a Level I balance, we can obtain a flow rate:

$$q_2 = \frac{Q_{Load}}{\hat{C}_{p}\rho (T_{2F} - T_2)} = \frac{-2.09 \times 10^5 \text{ W}}{4184 \text{ J/kg °C})(1 \text{ kg/L})(5 - 70 °C)} = 46 \text{ L/min}.$$

Step 4. Heat Transfer Coefficient Estimation. In Table 3.4 one can see that, in general, the overall heat transfer coefficient for a water–water double-pipe heat exchanger should be between 1.0 and 1.5 kW/m² K. To obtain an initial estimate of the area, the heat transfer coefficient will be assumed to be 1.5 kW/m² K. We will check this assumption by using the analysis outlined in Part II of the book.

Step 5. Equipment Type and Area (Level III Analysis). To find the contact area a Level III analysis should be employed:

$$Q_{Load} = Ua \langle \Delta T_{21} \rangle.$$

The log-mean temperature difference for a countercurrent heat exchanger is

$$\langle \Delta T_{21} \rangle = \frac{(T_{2F} - T_1) - (T_2 - T_{1F})}{\ln \left(\dfrac{T_{2F} - T_1}{T_2 - T_{1F}} \right)}.$$

The relationship between the T_2 and area can be seen in Figure E3.6. After $T_2 = 70\,°C$ the area required begins to increase sharply. With an exit temperature of the utility stream, $T_2 = 70\,°C$, the heat exchanger areas must be 9 m²(U = 1.0 kW/m²K) and 6 m²(U = 1.5 kW/m²K). Further considerations of economics (capital and operation) are required for going beyond a technically feasible design.

Step 6. Preliminary Heat Exchanger Selection. For most double-pipe heat exchangers the inner pipe diameter D_1 should be between 0.5 and 2 in. Because there is a limited number of pipe sizes in common use, evaluating the design for nominal pipe diameters between 0.5 and 4.0 in. should produce a technically feasible design. For example, a 2-in. nominal diameter schedule 40 pipe is used (see Table E3.1):

$$v_1 = \frac{q_1}{Ac_1} = \frac{4q_1}{\pi D_{1,in}^2} = \frac{4(0.00167 \text{ m}^3/\text{s})}{\pi (0.0525 \text{ m})^2} = 0.77 \text{ m/s}.$$

Using this velocity for the outer pipe yields D_{2in}, the internal diameter of the outer pipe:

$$v_2 = \frac{q_2}{A_{C2}} = \frac{4q_2}{\pi D_{2,in}^2 - \pi D_{1,out}^2}.$$

Table E3.1. Pipe schedule for 40 S pipe

Nominal diameter (in.)	D_{in} [inch (cm)]	D_{out} [inch (cm)]	Wall thickness x_w [inch (cm)]
12	12.00 (30.48)	12.75 (32.4)	0.375 (0.953)
10	10.02 (25.45)	10.75 (27.3)	0.365 (0.928)
8	7.98 (20.27)	8.6 (21.9)	0.322 (0.818)
6	6.07 (15.42)	6.6 (16.8)	0.28 (0.712)
4	4.03 (10.24)	4.5 (11.4)	0.237 (0.602)
3	3.07 (7.80)	3.5 (8.89)	0.216 (0.549)
2.5	2.47 (6.27)	2.9 (7.3)	0.203 (0.516)
2	2.07 (5.26)	2.38 (6.03)	0.154 (0.392)
1.5	1.61 (4.09)	1.9 (4.83)	0.145 (0.369)
1	1.05 (2.67)	1.31 (3.34)	0.133 (0.338)
0.5	0.662 (1.58)	0.84 (2.13)	0.109 (0.277)

Table E3.2. *Velocity and Reynolds numbers*

$D_{1, nom}$ (in)	v_1 (m/s)	$D_{2, nom}$ (in)	v_2 (m/s)	Re_1	Re_2
4	0.20	5	0.29	41 511	5 751
3	0.35	4	0.38	54 474	7 295
2.5	0.54	3.5	0.35	67 689	8 548
2	0.77	3	0.40	80 854	10 087
1.5	1.27	2.5	0.61	103 804	12 567
1	2.99	1.5	1.75	159 319	18 771
0.5	8.50	1	3.83	268 690	29 065

Solving for D_{2in} yields

$$D_{2, in} = \sqrt{\frac{4q_2}{\pi v_2} + D_{1, out}^2} = \sqrt{\frac{4\,(0.000766)}{\pi 0.77} + 0.0525^2}$$

$$= 0.063 \, m = 2.48 \, in.$$

Either a 3- or 2.5-in. nominal diameter could be used. To find which is more appropriate, the velocity in each tube was calculated.

For a 3-in.-diameter pipe,

$$v_2 = \frac{q_2}{A_{C2}} = \frac{4q_2}{\pi D_{2, in}^2 - \pi D_{1, out}^2} = \frac{4(0.000766)}{\pi(0.0779^2 - 0.0603^2)} = 0.40 \, m/s.$$

The Reynolds number calculation for the outer pipe (Re_2) can be found on page 330. For a 2.5-in. nominal-diameter pipe the velocity is 0.61 m/s.

The result of such calculation for a schedule 40 pipe with an inner pipe nominal diameter between 0.5 and 4 in. can be seen in Table E3.2.

When the outer diameter of the 2-in. pipe is known, the length L can be found for both 6 m^2 and 9 m^2:

$$L = \frac{a}{\pi D_i} = \frac{6.0 \, m^2}{\pi 0.0525 \, m} = 36.4 \, m = 119.4 \, ft.$$

$$L = \frac{a}{\pi D_{i, in}} = \frac{9.0 \, m^2}{\pi 0.0525 \, m} = 54.6 \, m = 179 \, ft.$$

Piping comes in 20-ft. (6.1-m) segments. For the two cases under consideration, we will need 6 and 9 20-ft. pipe segments.

Using these values for the diameters and flow rate, we can calculate the heat transfer coefficient by using the methods outlined in Part II, which will be illustrated in Chapter 8.

REFERENCES

Chemical Engineers' Handbook (7th ed.). New York: McGraw-Hill Professional, 1997.

Sandler SI. *Chemical and Biochemical Engineering Thermodynamics* (4th ed.). New York: Wiley, 2006.

The Heat Transfer Research Institute (HTRI) also provides resources and links to the manufacturers of heat exchanger equipment, www.htri-net.com.

PROBLEMS

3.1 The jacketed batch vessel as described in Example 3.1 is filled with 1.4 m^3 of water, and must be heated from 20 to 60 °C by use of a 1.4-m^3 jacket loaded with DOWTHERM^{TM} ($\rho = 1000$ kg/m^3 and $\hat{C}_p = 3.6$ kJ/kg K), initially at 120 °C. Determine the following:

   a. process heat load (Q_{Load}),
   b. equilibrium temperature (T_∞). Can this operation be carried out?
   c. Final temperature of the DOWTHERM (T_2).

3.2 To avoid the difficulties of pumping and removing a utility fluid, it has been suggested putting a 30-kW heating coil in the vessel (see Problem 3.1). How long will it take to heat the process fluid with this coil?

3.3 If we continue to use DOWTHERM instead of the heating coil, how long will it take to heat the process fluid from 20 to 60 °C? The heat transfer area is 4.55 m^2. Carry out this calculation for U values of 100 W/(m^2 K), 500 W/(m^2 K), 1000 W/(m^2 K).

3.4 To reduce the heating time, we decided to continuously flow the DOWTHERM through the heating jacket. If the utility fluid is available at 120 °C, find the value for the utility fluid flow rate (q_2) and exit temperature (T_2) required for heating the water from 20 to 60 °C in 15 min. You can assume that U $= 1000$ W/(m^2 K).

3.5 The vessel described in Figure 3.7 is supplied with the necessary piping and pumps so it can be operated to continuously heat a water-like process fluid with a flow of 100 L/min. DOWTHERM, which is available at 120 °C, will flow continuously through the heating jacket of this continuous mixed–mixed exchanger (Figure 3.7) in order to heat the process fluid from 20 to 60 °C. DOWTHERM properties are included in Problem 3.1.

   a. Calculate Q_{Load}.
   b. Find the theoretical minimum utility fluid flow rate (q_2).
   c. Determine a technically feasible value of q_2 that will allow operation with the parameters specified, and calculate T_2 at the exit.
   d. Estimate the U needed to achieve this value of T_2.

3.6 The data shown in the following table were taken with the constant temperature bath apparatus shown in Figure 3.1(b). Both the fluid in the bath and in the bottle are water. Find the value of U for this laboratory-scale heat exchanger if the bath temperature is constant at 35 °C and the area for transfer between the bath and the bottle is 3.8×10^{-2} m^2. The volume of the glass bottle in the bath is 0.8 L.

Time (s)	T_1 (°C)
30	14.3
90	18.8
150	22.5
210	25.5
240	26.7
300	28.9
360	30.8
420	32.3

3.7 At the lab scale, batch heat exchange is commonly achieved by submerging the system (i.e., well-mixed beaker of liquid) in a large, constant-temperature bath. Consider a well-mixed batch system initially at a temperature $T_1 = T_{1i}$ with fluid properties $\hat{C}_{p1}$ [kJ/(kg K)], T_1 (K), ρ_1 (kg/L), and V_1 (L) submerged in a constant-temperature bath having material properties $\hat{C}_{p2}$, T_2, ρ_2.

a. Derive the model equation describing how the temperature of the batch system changes with time.

b. What is the average rate of heat transfer between the bath and the batch system from $t = 0$ to t?

3.8 A well-built home is well insulated, and loses heat only through its 12 single-pane glass windows, which on average measure 1.1×1.3 m. Each window has a thickness of 0.7 cm. During a Delaware winter, the average temperature is roughly 4 °C, whereas the preferable indoor temperature is 20 °C. You may assume that the internal air is well mixed, the external heat transfer coefficient is 10 W/(m^2 K), the heat transfer resistance of the window is negligible, and the heating system is approximately 42% efficient. If the price of heating is $0.036/(kW h), what is the cost of heating this home in the winter (90 days)?

3.9 Liquid nitrogen is stored in a thin-walled metallic sphere of radius $r_1 = 0.25$ m. The sphere is covered with an insulating material of thickness 25 mm and $U = 0.068$ W/(m^2 K). The exposed surface of the insulation is surrounded by air at 300 K. The convective heat transfer coefficient to the air is $U = 20$ W/(m^2 K). The latent heat of vaporization and density of liquid nitrogen are 200 kJ/kg and 804 kg/m^3, respectively. Liquid nitrogen boils at 77 K at 1 atm. What is the rate of liquid boil-off in kilograms per day?

3.10 Water flows through a long tube of circular cross section at a mass flow rate of 0.01 kg/s. The heat flux through the wall is maintained uniformly at 3000 W/m^2. The water enters the tube at 20 °C. The inside diameter is 7.5 cm.

a. Find the required length L so that $T_{out} = 80$ °C.

b. Plot the temperature of the tube as a function of length.

3.11 Styrene, initially at room temperature (25 °C), is heated in a double-pipe cocurrent heat exchanger prior to entering a polymerization reactor. To carry out this operation, a hot stream from another operation in the process with a density of 1000 kg/m^3 and a heat capacity of 3.6 kJ/kg K is fed into the outer pipe at 120 °C. Thermal initiation of the polymerization requires the styrene stream to enter the reactor at a temperature of 70 °C. For a 500-kg/h styrene stream flowing through the inner pipe (1 in. inside diameter) the overall heat transfer coefficient has been measured to be 230 W/m^2 K. If the heating stream flow rate is 800 L/h and the heat capacity and density of styrene are 1.75 kJ/kg K and 906 kg/m^3, respectively, answer the following questions:

a. What length of heat exchanger is required for performing the necessary heating of the styrene stream?

b. Repeat part a if the styrene flow rate is doubled and the utility stream flow rate is held constant. (*Hint*: The heat transfer coefficient varies with the fluid velocity as $U \propto v^{0.8}$.)

c. Can the specified operation be performed at double the flow rate in the same heat exchanger as that in part (a) by altering the utility fluid flow rate?

3.12 A well-stirred batch polymerization reaction is to be run isothermally in a well-mixed vessel at a temperature of 60 °C. The rate of disappearance of the styrene monomer, C_M, from any control volume is

$$2k_pC_{Ii}[1 - \exp(-k_dt)]C_M.$$

In this case, 1-M styrene monomer is polymerized in benzene, with the initial initiator concentration C_{Ii} being 0.01 M. The polymerization kinetic rate constants k_p and k_d are 1.87×10^2 (M s)$^{-1}$ and 8.45×10^{-6} s^{-1}, respectively. The heat of reaction is –73 kJ/mol, and the cooling water is provided at a temperature of 25 °C. If the reactor volume is 1 L, and the heat transfer coefficient has been reported to be 120 W/(m^2 K), determine

a. the heat load as a function of time,
b. a feasible design of a cooling jacket. The design should specify the jacket volume, the effective cooling area, and the flow rate of cooling water.

3.13 A copper sphere of diameter 0.3 m is initially at a uniform temperature of 80 °C. Cool 15 °C air flows around the sphere, cooling it everywhere on the surface. The external heat transfer coefficient U is 6.3 W/(m^2 K). Copper has a density of 8930 kg/m^3, and its specific heat capacity is 385 J/(kg K).

a. Write the model equations for this process.
b. Determine the time required to cool the ball to 18 °C.
c. If the ball was instead made out of rubber and the heat transfer coefficient was the same, would it require longer, shorter, or the same amount of time to reach an average temperature of 18 °C? Explain.

3.14 Two volumes of liquid water at different temperatures are in well-stirred adjacent chambers as shown in Figure P3.1 and the subsequent table. The initial water temperatures in Chambers 1 and 2 are 20 and 70 °C, respectively. There is no mass flow into, out of, or

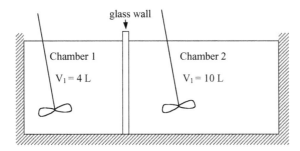

Figure P3.1.

between the tanks, and the two-tank system is completely insulated from the surroundings. The tanks can exchange heat through the glass wall separating the two chambers, which has a thickness of 2.4 cm and a total surface area of 0.25 m^2. The following data were collected:

t (s)	T_1 (°C)
0	20
500	30

a. Determine the system equilibrium temperature.
b. Write the basic energy balance equations for this system.
c. Estimate the heat transfer coefficient of the system.

3.15 Besides being applied in engineering, countercurrent heat exchangers find use in nature. The cardiovascular system, which serves an additional purpose as a biological heat exchanger, is a good example of this phenomenon. The arterial blood, which starts out at the body's core temperature, runs countercurrent to the cooled venous blood, causing heat exchange to occur.

In recent years, this principle has been applied in sports medicine. During hot weather, professional athletes are exposed to the risk of hyperthermia, which could potentially lead to organ failure or death. The traditional method for curing hyperthermia had been to spray cold water over the athlete. This method, however, was problematic in light of the fact that it caused blood vessels to contract. This, in turn, reduced the blood flow between the core of the body and the extremities, resulting in even greater trauma to the athlete. To fix this problem, a clever engineer decided to limit the external cooling to the hand by means of a glove and allow the blood in the body core to cool by means of blood circulation. This device works in such a way that the arterial blood enters the arm at the body core temperature and, after being cooled by the glove, returns to the vein at a temperature of 10 °C.

You are faced with the following situation: After trying out with the San Francisco 49ers, Joe Blue Hen leaves the field with a dangerous body temperature of 41.3 °C. He immediately puts on the glove, and now needs to know how long he should keep it on his hand. Being an expert on heat transfer, and having access to Joe's medical records, you decide to help him. In doing this, you may assume that the rate of blood perfusion is negligible, and the heat transfer from the rest of the body is much slower than through the hand. Compute the following:

a. The heat loss through the glove as a function of time. You may assume that the rate of blood circulation is much greater than that of heat transfer.
b. The length of time it would take to cool the body core temperature to 37 °C.
c. How the time calculated in part b would change if the blood were returned to the vein at temperatures of 5, 15, 20, 25, and 30 °C?
d. Comment on the accuracy of the plug-flow assumption. The viscosity of blood can be taken as $\sim 3 \times 10^{-3}$ Pa s.

3.16 Beer is made with the yeast *Saccharomyces cerevisiae*, which produces ethanol from the sugar glucose. Dissolved ammonia is used as the nitrogen source in an industrial production of ethanol by use of an isothermal semi-batch fermenter of 1400 L. You must design a cooling jacket for the fermenter such that the temperature inside the vessel is maintained at 25 °C.

In the reactor, 0.451 C mol of ethanol and 0.235 C mol of biomass with elemental composition $CH_{1.8}O_{0.56}N_{0.17}$ were produced per C mol of glucose by use of 0.0399 mol of ammonia.

The heats of combustion of the biomass are calculated from the "regularization method" to be 519 kJ/C mol, and the heats of combustion of ethanol, glucose, and ammonia are 684, 468, and 348 kJ/C mol, respectively (see Sandler, 4th ed. Chapter 15).

a. Sketch the operation of a continuous fermenter into which ammonia, water, and glucose are fed in continuously at 25 °C and air is fed at 25 °C to supply oxygen.
b. Compute the heat of reaction per C mol of glucose consumed.
c. Compute the *minimum* rate of cooling water required for using Vessel B (in Example 3.1, which has been outfitted with the necessary piping to operate continuously) with

cooling water available at 5 °C if the residence time in the fermenter is 4 h and the glucose concentration in the feed is 50 g glucose/L.

d. Calculate the required cooling water flow and exit temperature from the jacket if the overall heat transfer coefficient is 309 W/m² K.

3.17 Reconsider the design problem requiring the cooling of a process stream (water) flowing at q_1 = 100 L/min from an inlet temperature of 80 to 50 °C by use of a chilled water utility stream available at 5 °C. The utility company will accept returned utility water only if it is less than 60 °C.

a. Given that the heat transfer coefficient is U = 1000 W/m² K, complete a technically feasible design of a continuous-flow coil heat exchanger (mixed–plug fluid motions). To complete the design, draw and label a diagram of the apparatus and create a stream table specifying all stream conditions. Be sure to *specify (and justify)* all critical design details (e.g., the coil and tank dimensions, etc.), and discuss the implications of the design.

b. For the utility flow rate q_2, calculated in part a, calculate (i) the outlet utility temperature (T_2) and (ii) the area required for heat transfer for each of the four continuous-flow heat exchanger designs: tank type (mixed–mixed, mixed–plug) and tubular (cocurrent, countercurrent). Tabulate these values and provide a brief written comparison of these designs. Namely, discuss the differences in area required among the designs in light of the driving force for heat transfer.

3.18 Consider the design of a heat exchanger for a continuous-flow holding tank for a process stream (a process flow sheet is given in Figure P3.2). The tank is used both as a buffer against fluctuations in the flow rate and as a method to reduce the temperature for the next step in the processing. The stream enters the tank at 80 °C and must be cooled to 50 °C. Cooling water is available at 5 °C and must be returned no warmer than 60 °C.

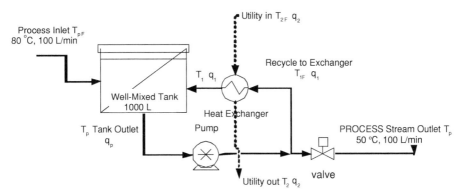

Figure P3.2.

The proposed design is to use an external cooling loop with a recycle. If the tank volume is 1000 L and can be considered to be well mixed, and if the process stream is 100 L/min, propose a technically feasible design for the heat exchanger indicated in the diagram in the figure (i.e., specify the area, type, and give the area and geometry of the device). Assume that the overall heat transfer coefficient is 500 W/m² K and that the process and utility streams have the properties of water. Label all stream flow rates and temperatures by making a stream table. Comment on the advantages or disadvantages of your choice of design.

3.19 Oil is to be cooled in a schedule 10 S pipe by a countercurrent heat exchanger from 120 to 20 °C at a flow rate of 100 L/min. Previous experience suggests the overall heat transfer coefficient for this exchanger will be 1000 W/m² K. The oil has a density of 800 kg/m³ and a heat capacity of 2 kJ/kg K. Water is available for cooling at 5 °C, and is useful in the plant in heat integration if returned as saturated water at 100 °C.

a. Explain why a countercurrent heat exchanger is chosen for this problem, and the possible limits on the maximum temperature of the utility outlet.

b. What is the flow rate of water required and the area and size of the exchanger? Provide a flow diagram and label all streams.

c. Another design group suggests that, to reduce the required area and pressure drop on the utility side, two exchangers of equal size are to be employed as configured in Figure P3.3. Determine the required area and size the exchangers. Use the same pipe sizes as in Problem 3.18. Specify the outlet temperature of the utility stream for Figure P3.3. Assume the overall heat transfer coefficient, fluid properties, and limits on temperature are the same as previously specified.

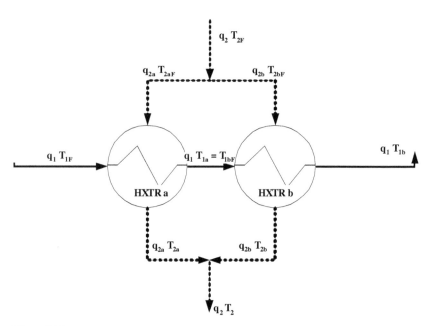

Figure P3.3.

d. Plot out the temperature profile along the exchangers for parts b and c, and explain why the total area is less for the configuration provided in the diagram.

3.20 A typical procedure for sizing a shell-and-tube heat exchanger with multiple tubes (single pass) is as follows:

- Compute the heat load using Eq. (3.5.1) assuming countercurrent flow.
- Use a chart provided for that type of exchanger to determine the correction factor **F** for the heat transfer, and multiply the log-mean temperature difference by this correction factor to estimate the actual exchanger performance [$Q_{Load} = ha(\Delta T_{12})F$].

An example of such a chart is subsequently reproduced from the *Chemical Engineers' Handbook* (1997, Section 10) in Figure P3.4. To use these charts, the following dimensionless temperature differences must be calculated:

$$R = \frac{T_1 - T_2}{t_2 - t_1}, \qquad S = \frac{t_2 - t_1}{T_1 - t_1}.$$

[*Note*: The nomenclature for the flows is indicated on the diagram and is different from those used in the text. The process stream (stream 1 in text) is on the tube side (t) and the utility stream (stream 2 in text) is on the shell side (T).]

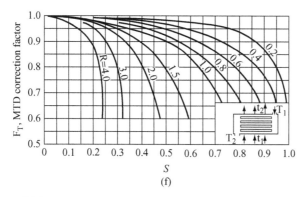

Figure P3.4.

A process stream having the properties of water on the tube side (stream 1) must be cooled from 80 to 50 °C at a flow rate of 100 L/min. Utility water is available at 5 °C and must be returned no hotter than 40 °C. You may assume the heat transfer coefficient is 1000 W/m^2 K.

a. Determine the minimum flow rate and area of a plug–plug countercurrent heat exchanger required for accomplishing this task.
b. Using the chart provided, determine the minimal flow rate and area for a shell-and-tube heat exchanger required for performing the same task.
c. If the shell side flow is such that the shell side is well mixed and the tube side (process stream) is plug flow, reanalyze the shell-and-tube heat exchanger as if it were a plug–mixed continuous-flow heat exchanger. Compare with the preceding part b and comment.

APPENDIX: ENERGY BALANCE

Consider the mixture of A and B in a well-stirred tank at constant pressure (i.e., open to the atmosphere). In the following discussion, we consider the energy balance for the system composed of the material in the tank [see Figure 3.1(a)].

The first law of thermodynamics, i.e., conservation of energy, takes the following form for a system with negligible changes in kinetic and potential energy and no nuclear processes. Here we follow the form nomenclature found in Chapter 15 of Sandler (2006):

$$\frac{dU}{dt} = \sum_{k=i}^{K} \dot{N}_k \underline{H}_k + \dot{Q} + \dot{W}_s - p\frac{dV}{dt}. \qquad (3.A1)$$

For the system consisting of the well-mixed tank with components A and B, at constant pressure (p), and no inlet or outlet streams (K = 0), this equation simplifies to

$$\frac{dU}{dt} + p\frac{dV}{dt} = \frac{d(U + pV)}{dt} = \frac{dH}{dt} = \dot{Q} + \dot{W}_s.$$

Further, the shaft work that is due to stirring is generally negligible. Application of this balance equation to the system defined by the tank in Figure 3.1(a), ignoring any shaft work and using Newton's constitutive equation for cooling yields

$$\frac{dH_i}{dt} = -Ua(T_1 - T_2). \tag{3.A2}$$

Note that, for liquids, $C_v \approx C_p$, which is equivalent to setting $(dV/dt) \approx 0$. This results in the left-hand side of the energy balance consisting of just (dU/dt). Such an equation is valid for *constant-volume* systems, which would be more typical of a gas-phase reactor. Consequently, the derivation here for liquids can also be applied for gases in constant-volume reactors, with the noted change. Furthermore, it is often possible to use an approximation for small temperature differences,

$$\frac{dH_1}{dt} \approx \rho_1 V_1 \hat{C}_{p1} \frac{dT_1}{dt},$$

$$\frac{dU_1}{dt} \approx \rho_1 V_1 \hat{C}_{v1} \frac{dT_1}{dt}, \tag{3.A3}$$

where the values of the heat capacity and density are evaluated at the mean temperature of the process. Thus, for liquids with this approximation, we can write

$$\frac{dH_1}{dt} = \rho_1 V_1 \hat{C}_{p1} \frac{dT_1}{dt} = -Ua(T_1 - T_2). \tag{3.A4}$$

Given modern equations of state for mixtures and computational programs with built-in mixture thermodynamic property calculations, it is possible to avoid the approximation of constant physical properties and calculate the enthalpy change directly. Further simplifications for mixtures used in the analysis are provided in the following section.

Useful Approximations for Mixtures

Assuming that the mixture remains single phase (liquid) and is well mixed, the system enthalpy in Eq. (3.A4) can be expressed explicitly in terms of partial molar properties as

$$H(T, p, x_A) = \sum_{i=1}^{C} N_i \overline{H}_i(T, p, x_A)$$

$$= \sum_{i=1}^{C} N_i \underline{H}_i(T, p) + \Delta H_{mix}(T, p, x_A), \tag{3.A5}$$

where the distinction is made between pure component properties (underline) and the partial molar properties (overbar), which are themselves mixture properties.

Here, the subscript i refers to a chemical species. For many mixtures of interest here, we can assume *ideal mixture properties*, that is, $\Delta H_{mix} \approx 0$. Further, we recognize that

$$\underline{H}_i(T, p) - \underline{H}_i(T_{ref}, p) = \int_{T_{ref}}^{T} C_p dT.$$

Again, for small temperature changes, one can use an approximate calculation such as

$$\underline{H}_i(T, p) - \underline{H}_i(T_{ref}, p) = \int_{T_{ref}}^{T} C_p(T')(dT') \approx C_p(T_{avg})(T - T_{ref}),$$

where the average temperature will be an estimate of the average temperature of the process of interest. Note that the reference temperature will not matter in this problem as the derivative of the enthalpy is all that is actually required and not its absolute value; hence, T_{ref} can be set to 0.

Using the approximation of ideal mixing and approximating the heat capacity as constant allows us to rewrite the exact energy balance for the tank (3.A2) as

$$\frac{dH_1}{dt} \approx \left(\sum_{i=1}^{C} N_i C_{p_i} \right)_1 \frac{dT_1}{dt} = -Ua(T_1 - T_2). \tag{3.A6}$$

Alternatively, we can write this in more compact form, recognizing that

$$\sum_{i=1}^{C} N_i C_{p_i} = \rho \hat{C}_p V,$$

$$\rho \hat{C}_p V \frac{dT_1}{dt} = -Ua(T_1 - T_2),$$

where the mass density and the mixture heat capacity ($\hat{C}_p$) are defined by the summation in the first line.

In summary, the approximate energy balance equations employed in this chapter assume physical property data evaluated at the average temperature of the process. Furthermore, the evaluation of the mixture properties can often be simplified by the assumption of ideal mixing (i.e., negligible volume and enthalpy change on mixing). A more rigorous calculation can always be performed if sufficient thermodynamic data (partial molar properties) are available by either estimation or measurement.

Batch System with Reaction

For reacting systems, the heat of reaction appears in the energy balance model equation. The evolution of the species concentration evolves with time because of chemical reaction, which will have an associated heat of reaction. As shown here, this is already built into the rigorous energy balance defined in Eq. (3.A1).

Consider the reaction $A + B \rightarrow D$. We define the stoichiometric coefficients (v_i) by writing the reaction as products minus reactants: $D - A - B = 0$. Then, in terms of the *molar extent of reaction* (X), the mole numbers are

$$N_i(t) = N_i^0 + v_i X,$$

$$\frac{dN_i}{dt} = v_i \frac{dX}{dt},$$ (3.A7)

or

$$r_A V = \frac{dX}{dt}.$$

Then the energy balance equation for a homogeneous system of these components with reaction becomes

$$\frac{dH_1}{dt} = -Ua(T_1 - T_2)$$

$$= \frac{d\left(\sum_{i=1}^{C} N_i \overline{H}_i\right)_1}{dt} = \left(\sum_{i=1}^{C} N_i \frac{d\overline{H}_i}{dt} + \sum_{i=1}^{C} \overline{H}_i \frac{dN_i}{dt}\right)_1$$

$$\left(\sum_{i=1}^{C} N_i \frac{d\overline{H}_i}{dt} + \sum_{i=1}^{C} v_i \overline{H}_i \frac{dX}{dt}\right)_1 = -Ua(T_1 - T_2).$$

As seen in the preceding equation, the variation in mole number with time that is due to the reaction leads to two terms on the left-hand side of the equation. The first term is what we had previously for constant mole number (i.e., no reaction), whereas the second term is the variation in enthalpy to the chemical reaction. This can be written in terms of the extent of reaction if the enthalpy of mixing is assumed to be zero:

$$\sum_{i=1}^{C} v_i \underline{H}_i \approx \sum_{i=1}^{C} v_i \underline{H}_i = \Delta \underline{H}_{rxn}.$$ (3.A8)

This leads to the following equations for reacting systems, as shown in the next section.

Useful Approximations for the Reacting Batch System

Again, assuming *ideal mixing*, the energy balance becomes

$$\frac{dH_1}{dt} \approx \left(\sum_{i=1}^{C} N_i \frac{d\underline{H}_i}{dt} + \sum_{i=1}^{C} v_i \underline{H}_i \frac{dX}{dt}\right)_1 = -Ua(T_1 - T_2).$$ (3.A9)

Again, the average heat capacity can be used, and, further, the heat of reaction can be used as given in Eq. (A3.8). With these simplifications, the energy balance becomes

$$(\rho \hat{C}_p V)_1 \frac{dT_1}{dt} + N_1 \Delta \underline{H}_{1,rxn} \frac{dX}{dt} = -Ua(T_1 - T_2),$$ (3.A10)

where $(\rho \hat{C}_p V)_1$ is the total heat capacity of the fluid in system 1. Further, recognizing that $(dX/dt) = r_{A-} V$ yields the working equation

$$(\rho \hat{C}_p V)_1 \frac{dT_1}{dt} = -Ua(T_1 - T_2) - \Delta \underline{H}_{rxn} r_{A-} V_1.$$ (3.A11)

Note that one can check the sign: An *exothermic reaction* is one that evolves heat. If there is no heat transfer, the temperature change will have the sign of the rate of reaction of A (which is written as a positive rate) times the molar enthalpy of reaction (which is negative for an exothermic reaction), leading to a positive increase in temperature, as expected.

Plug-Flow Tubular Reactor

As a fluid element moving through a plug-flow reactor can be analyzed as a batch reactor, the results are the same as those for the batch reactor, where t is the residence time in the tubular reactor. Note that the area for heat transfer is the wetted perimeter of the pipe times a differential length along the pipe, Δz, and the volume is the cross-sectional area times this same differential length. Consequently, the differential length drops out of the energy equation for negligible change in volume with reaction.

CSTR with Reaction

The same analysis holds, except now the flow terms in Eq. (3.A1) are kept. As an example, consider the steady-state operation of a well-mixed tank reactor operated in continuous mode (CSTR) with the reaction given previously. The balance equation for the system in the CSTR becomes

$$0 = \left(\sum_{k=1}^{K} \dot{N}_k \underline{H}_k \right)_1 + \dot{Q}. \tag{3.A12}$$

Now, Newton's constitutive equation for cooling is used to express the rate of heat transfer to and from a cooling bath or environment (system 2), as before. The outlet molar flow rates are expressed in terms of the extent of reaction (3.A7), and the mixed stream enthalpies are taken to be ideal mixtures (i.e., negligible heats of mixing). Then the balance can be written as

$$\sum_{k=1}^{K} \dot{N}_k \underline{H}_k = (\dot{N}\underline{H})_{out} - (\dot{N}\underline{H})_{in},$$

$$(\dot{N}\underline{H})_{out} \xrightarrow{\text{ideal mixing}} \left(\sum_{i=1}^{C} \dot{N}_i \underline{H}_i \right)_{out}$$

$$= \left[\sum_{i=1}^{C} \left(\dot{N}_{i,in} + v_i \frac{dX}{dt} \right) \underline{H}_i \right]_{out}.$$

Therefore

$$-Ua(T_1 - T_2) = \left\{ \sum_{i=1}^{C} \dot{N}_{i,in} [\underline{H}_i(T_{out}) - \underline{H}_i(T_{in})] \right\}_1 + V_1 r_A - \Delta \underline{H}_{i,rxn}. \tag{3.A13}$$

A flow calorimeter works on this principle, whereby the heat of reaction can be measured by the heat flow required for maintaining isothermal, steady-state reaction.

4 Mass Contactor Analysis

Portions of this chapter are taken from *Introduction to Chemical Engineering Analysis* by Russell and Denn (1972) and are used with permission.

In Chapter 2, a constitutive equation for reaction rate was introduced, and the experimental means of verifying it was discussed for some simple systems. The use of the verified reaction-rate expression in some introductory design problems was illustrated in Chapter 2. Chapter 3 expanded on the analysis of reactors presented in Chapter 2 by dealing with heat exchangers and showing how the analysis is carried out for systems with two control volumes. A constitutive rate expression for heat transfer was presented, and experiments to verify it were discussed.

This chapter considers the analysis of mass contactors, devices in which there are at least two phases and in which some species are transferred between the phases. The analysis will produce a set of equations for two control volumes just as it did for heat exchangers. The rate expression for mass transfer is similar to that for heat transfer; both have a term to account for the area between the two control volumes. In heat exchangers this area is determined by the geometry of the exchanger and is readily obtained. In a mass contactor this area is determined by multiphase fluid mechanics, and its estimation requires more effort. In mass contactors in which transfer occurs across a membrane the nominal area determination is readily done just as for heat exchangers, but the actual area for transfer may be less well defined.

We direct our attention to the analysis of simple two-phase systems in which a single species is transferred from one phase to the other. Isothermal two-phase systems represent the next stage in complexity over isothermal single-phase systems, providing an opportunity to treat complicated modeling situations while still basing our analysis solely on the law of conservation of mass and the appropriate constitutive equations. Two-phase systems are encountered in a wide variety of situations. We examine two basic issues, the rate of transfer between phases and the equilibrium relationship relating the concentrations of a species in each phase.

The structure of this chapter follows the modeling and analysis paradigm of Chapter 3 and is shown in Table 4.1. We begin with an analysis of mixed–mixed batch systems to gain an appreciation of the experimental techniques and then extend the study to tank-type mixed–mixed continuous-flow systems, tank-type plug–mixed continuous-flow systems, and tubular plug–plug continuous-flow systems. Before any

Table 4.1. Equipment fluid motion classification

Reactors: Single phase	Reactors: Two phase	Heat exchangers	**Mass contactors**
Single control volume	Two control volumes	Two control volumes	**Two control volumes**
Tank type	Tank type	Tank type	**Tank type**
Mixed	*Mixed–mixed*	*Mixed–mixed*	*Mixed–mixed*
• Batch	• Batch	• Batch	• **Batch**
• Semibatch	• Semibatch	• Semibatch	• **Semibatch**
• Continuous	• Continuous	• Continuous	• **Continuous**
	Mixed–plug	*Mixed–plug*	*Mixed–plug*
	• Semibatch	• Semibatch	• **Semibatch**
	• Continuous	• Continuous	• **Continuous**
Tubular	Tubular	Tubular	**Tubular**
Plug flow	*Plug flow*	*Plug–plug flow*	*Plug–plug flow*
	• Cocurrent	• Cocurrent	• **Cocurrent**
		• Countercurrent	• **Countercurrent**

of the mathematical descriptions are developed, however, it is helpful to describe the two-phase systems that may be encountered.

Table 4.1 shows that we can classify the fluid motions within the contactor as we did for heat exchangers. Mass contactor analysis requires two control volumes, just as for heat exchanger analysis, but there is no equipment boundary between the control volumes in a mass contactor (except for membrane contactors). There are two control volumes because there are two distinct phases present in the mass contactors we examine in this chapter. The two-phase systems that we need to study are

- solid–fluid (gas or liquid),
- liquid–liquid,
- liquid–gas.

Our analysis is restricted to isothermal systems in which there is no phase change during the operation. This constitutes a large class of operations as the isothermal analysis is pragmatically useful. Exceptions include distillation and crystallization operations. Parts of the isothermal analysis presented in this chapter can be applied, but distillation requires additional application of the conservation of energy, and crystallization requires population balance equations and a quantitative description of nucleation. A complete analysis of these operations will be left for further study.

Solid–Fluid Systems

Solid–fluid contactors effect the transfer of a soluble component from the fluid phase to the solid phase or from the solid to the fluid. The unit operations terminologies for such operations are *adsorption*, if material is transferred from the fluid to the solid, and *desorption*, *washing*, or *leaching*, if the transfer is from the solid to the fluid phase. Note that the solute may not be necessarily soluble in the solid phase, but may be adsorbed on the surface.

Experiments to determine feasibility and to obtain equilibrium and/or rate of transfer data can be carried out at the laboratory scale in a tank-type batch mixed–mixed contactor. Well-mixed tank-type solid–liquid contactors require properly designed mixers. Mixed–mixed semibatch solid–fluid contactors effecting material transfer to or from the solid phase with a continuously flowing fluid are used for *washing and or leaching.* Examples are the extraction of sugar from solid sugar beets, the recovery of copper from low-grade copper ores with sulfuric acid, or the removal of caffeine from coffee beans with supercritical carbon dioxide. In all these examples the solid is removed after the operation is complete. Solid–gas batch mixed–mixed contactors are not so easy to design and operate, and such operations are less common. Continuously operated mixed–mixed tank-type contactors are rare.

Tank-type semibatch mixed–plug mass contactors are often configured with a solid phase as the batch phase, with a fluid phase flowing through the solids, fluidizing the bed of solids. If sufficiently fluidized, the solid phase can be considered well mixed. Tank-type fluidized-bed mixed–plug contactors cannot be continuously operated.

Cocurrent tubular solid–fluid systems can be either horizontal or vertical and are employed for transporting solids but are not common as mass contactors. Countercurrent solid–fluid tubular contactors can be designed and operated if the tube or pipe is vertical and the solids flow downward through an upward rising gas or liquid. There are constraints imposed on the geometry of all tubular contactors by the need to satisfy two-phase fluid mechanics.

Many absorbers contain a fixed bed of solids and the fluid flows through the bed. The solids are regenerated in place or removed, treated, and replaced. Such operations are batch with respect to the solid phase and are quite common in water treatment to remove impurities. Household water filters that attach to a faucet or are part of a drinking water pitcher are common. Larger-scale operations for absorption often have two beds of solids with one treating a fluid while the other is regenerated.

Liquid–Liquid Systems

Liquid–liquid mass contactors are employed when a component, the *solute*, from one liquid, the *solvent*, is transferred to another liquid. One commonly employs two immiscible liquids for such *solvent extraction* unit operations. Tank-type systems are most frequently employed with enough agitation to control droplet size and ensure adequate mixing. Batch mixed–mixed vessels are used for laboratory-scale experiments to obtain equilibrium and rate data. Semibatch operations are not feasible, but continuous mixed–mixed operations are common. The mass contactor must be followed by a separator to complete any operation. With solid–fluid and liquid–gas systems the mass contactor acts as the separator and an additional vessel (a separator) is not required. The two immiscible liquids are most frequently separated by density difference in a gravity separator. This mass contactor separator is often referred to as a *mixer–settler*, and the stream exiting from the settler from which some solute has been extracted is termed the *raffinate*. Some examples of *solvent extractions* are the removal of adipic acid from water by diethyl ether, the removal of naphthenes from lubricating oil by furfural, and the removal of mercaptans from naphtha by a NaOH solution. Tank-type mixed–plug extraction units are possible if there is sufficient density difference between the two liquids, but there are no practical operations of this sort, either semibatch or continuous.

Tubular cocurrent extractors are possible in both a horizontal and a vertical configuration, but there are few, if any, practical applications at present. Countercurrent tubular mass contactors can be achieved only in vertical pipes or towers in which the heavier fluid flows down as drops into an upward-flowing lighter fluid, or the lighter fluid flows up as drops into the downward flowing heavier fluid. Such towers often have proprietary internals. This configuration is more common in liquid–gas systems, discussed next.

Liquid–Gas Systems

Liquid–gas mass contactors are some of the most commonly encountered pieces of equipment in the process industries and in both municipal and industrial waste treatment facilities. When a component is transferred from a gas to a liquid, it is referred to as a *scrubbing* or *absorption* unit operation; when a component is transferred from the liquid to the gas it is a *stripping* or *desorption* operation. Natural gas is *scrubbed* with monoethanolamine or diethanolamine solutions to remove hydrogen sulfide. Oxygen is *stripped* from water by contacting it with a nitrogen stream to deoxygenate it so the water can be used in electronic material processing. Oxygen is transferred from airstreams to water solutions to supply the oxygen needed to support biomass growth in fermentors or large-industrial-scale waste treatment reactors.

Tank-type mixed–mixed mass contactors can be used in a batch mode to study systems at the laboratory scale, but it can be difficult to get a well-mixed liquid–gas system. Mixed–mixed semibatch and continuous tank-type systems are not feasible because the phases separate easily and poor contacting is achieved. Tank-type mixed–plug mass contactors operating in a semibatch or continuous mode are very common. Tubular cocurrent mass contactors are possible in a horizontal or vertical configuration. The most common tubular configuration is countercurrent vertical. Such towers or columns are always packed or have tray internals.

At the molecular level, heat and mass transfer are linked by similar constitutive relations and conservation equations. The constitutive relations for equipment control volume analyses are also similar for both heat and mass transfer. The exact nature of these similarities will be explored in Chapters 5 and 6, but the consequence is that the model equations governing a mass contactor are very similar to those governing a heat exchanger. The behavior of mass contactors can be more difficult to rationalize conceptually than that of heat exchangers. Heat transfer is intuitively easier to understand from everyday experience, whereas the central role of phase equilibrium and the issues surrounding the area available for mass transfer are distinctions that make mass transfer analyses more challenging. As an aid to the student, we frequently refer back to the heat exchanger analysis presented in Chapter 3 to build analogies between heat and mass transfer unit operation analysis.

4.1   Batch Mass Contactors

A batch process is one in which there is no flow to or from the system during the operation. As shown in Chapter 2, it is a convenient and useful system for obtaining experimental information. A batch two-phase system is one in which both phases are charged at time $t = 0$ and allowed to interact for a specified period of time so that data may be collected. The physical situation is sketched in Figure 4.1. We will assume that the vessel contents are well mixed and that the two phases are in intimate contact, with one phase, the dispersed phase (II), uniformly distributed throughout the other, which is denoted as the continuous phase (I).

Homogeneous mixing is possible to achieve in practice, and it is essential that the batch two-phase system have this characteristic if the analysis is to reflect the physical behavior. Proper mixing or agitation ensures that the concentrations of the two phases are spatially homogeneous, i.e., the dispersed phase amount and droplet size are uniform throughout the vessel. The consequences of this homogeneous mixing assumption are that samples chosen from one phase in the tank at any given location will be identical to those drawn from the same phase at any other location.

Although there are both isothermal and nonisothermal two-phase systems, we restrict our analysis in this chapter to the isothermal case. There are many situations of interest that are isothermal, and in those that are not, the isothermal analysis is always part of the more complex problem. With this restriction, mass becomes the fundamental variable in the analysis, and the principle of conservation of mass and experimentally verifiable constitutive relationships are the primary tools of analysis. As before, we use as characterizing variables V for volume, ρ for density, and C for concentration. Concentration will be measured in mass units (e.g., kg/m^3). The choice of control volume is important, as selection of the entire vessel is not useful

Phase II (Dispersed)
V^{II}
ρ^{II}
C_A^{II}

Phase I (Continuous)
V^{I}
ρ^{I}
C_A^{I}

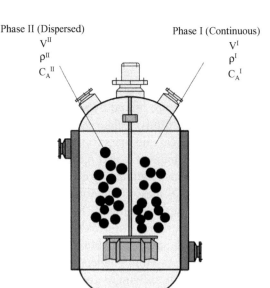

Figure 4.1. Batch mass contactor.

in a two-phase system. Because we are interested in transfer between the phases, we must work with two control volumes, one for each phase.

We arbitrarily designate one phase as the continuous phase (Phase I) and one as the dispersed (Phase II). The concentration of any species in the continuous phase is designated with a superscript as C_i^I. Other properties of the continuous phase, volume and density, are denoted by V^I and ρ^I, respectively. The dispersed-phase control volume, V^{II}, is made up of all the elements of the other phase, and although we consider it a single control volume for modeling purposes, it may consist physically of a number of distinct volumes. Concentrations in the dispersed phase are designated with a superscript II, as C_i^{II}, and density as ρ^{II}. As with most situations, learning the nomenclature requires some effort; it helps if one can visualize how each quantity is measured.

Mathematical descriptions are developed for the general two-phase case, recognizing that with appropriate identification of continuous and dispersed phases the basic model equations will apply for batch solid–liquid, solid–gas, liquid–liquid, or liquid–gas systems. To simplify the algebraic manipulation, we develop the model equations for the physical situation in which a single component, species A, is transferred between the phases. Species A is present with concentration C_A^I in the continuous phase and C_A^{II} in the dispersed phase and does not react with any component in either phase. It is not conceptually difficult to extend the analysis to any number of species or to include reaction in one or both phases. This would only increase the number of equations and the amount of manipulation required to determine behavior.

4.1.1 Level I Analysis

As with reactors and heat exchangers, we begin with a Level I analysis on the control volume consisting of both the continuous and dispersed phases. As just discussed, we assume that we are primarily interested in component A and how it is transported between Phase I and Phase II. Applying the word statement (Figure 1.6) yields

$$(C_A^I V^I + C_A^{II} V^{II})_{t+\Delta t} = (C_A^I V^I + C_A^{II} V^{II})_t, \tag{4.1.1}$$

or, in the limit $\Delta t \to 0$,

$$\frac{dC_A^I V^I}{dt} + \frac{dC_A^{II} V^{II}}{dt} = 0.$$

The superscripts designate the particular phase.

Application of the law of conservation of mass to all species gives

$$\frac{d}{dt}(\rho^I V^I) + \frac{d}{dt}(\rho^{II} V^{II}) = 0. \tag{4.1.2}$$

Equations (4.1.1) and (4.1.2) can be integrated to complete the Level I analysis for a batch mass contactor. Denoting the initial concentrations, volumes, and densities with the subscript "i," the integrated equations express conservation of species A and total mass as

$$\left(C_A^I V^I - C_{Ai}^I V_i^I\right) = -\left(C_A^{II} V^{II} - C_{Ai}^{II} V_i^{II}\right),$$
$$\left(\rho^I V^I - \rho_i^I V_i^I\right) = -\left(\rho^{II} V^{II} - \rho_i^{II} V_i^{II}\right). \qquad (4.1.3)$$

Notice that the Level I analysis yields two equations, whereas in Chapter 3 the analysis consisted of only one equation, (3.13). This is the first major difference between heat and mass transfer. In many cases overall mass balance equations (4.1.3) can be neglected if the quantity of mass being transferred between phases is small and does not significantly change the density or volume of either phase. This assumption greatly simplifies the equations, reducing this system to

$$C_A^I = C_{Ai}^I - \frac{V^{II}}{V^I}\left(C_A^{II} - C_{Ai}^{II}\right),$$
$$C_A^{II} = C_{Ai}^{II} - \frac{V^I}{V^{II}}\left(C_A^I - C_{Ai}^I\right). \qquad (4.1.4)$$

However, these equations relate only the final concentrations in both phases – we must specify one of these concentrations in order to calculate the other. The appropriate Level II and III analyses are required for completing any analysis.

4.1.2 Level II Analysis, Phase Equilibrium

We developed the Level II analysis in Chapter 3 by recognizing that the temperature in the two control volumes of any heat exchanger eventually reached the same equilibrium temperature, T_∞. The equivalent situation for mass contactors is achieved when the concentration of the species being transferred has reached a constant value in each of the control volumes. Unlike temperature equilibrium, the concentrations are not equal but are related to each other such that the chemical potentials of species A are equal in each phase:

$$C_{A,eq}^I = f\left(C_{A,eq}^{II}\right). \qquad (4.1.5)$$

For solid–liquid mixtures and some liquid–gas mixtures the equilibrium concentration of A in the continuous phase, C_A^I, is equal to the saturated concentration, a value readily obtained experimentally for such systems. For liquid–liquid systems it is frequently found that over the concentration range of interest the concentrations in each of the phases are related by a constant.

Phase equilibrium is easily achieved for many systems, and we make use of our familiarity with the concept in developing the form of the rate expression for mass transfer, r_A. An enormous amount of data on equilibrium concentrations in two-phase systems of all types has been obtained and recorded in the technical literature, with a variety of techniques used to measure phase concentrations. In most chemical engineering curricula there is a thermodynamics course devoted entirely to phase equilibrium and another course concerned with Level II analysis of distillation, absorption, stripping, and liquid–liquid extraction operations. Such courses are a critical part of chemical engineering, but they do not consider the rate of mass transfer and thus do not provide a means to complete the design of a mass contactor.

Table 4.2. Solubility and solid density of some
inorganic compounds (20 °C)

Substance	$C_{A,eq}^{I}(kg/m^3)$	ρ^{II} (kg/m^3)
$BaCl_2$	338	3900
$FeCl_3$	730	2900
$KMnO_4$	61.9	2700
K_2CO_3	821	1988
K_2NO_3	282	2110
NH_4Cl	293	1519
$NaCl$	360	2170
$NaOH$	686	2130

Solid–Liquid Systems

The experimental procedure for determining equilibrium concentrations in solid–liquid systems is relatively simple. The solid particles are placed into the agitated liquid until addition reaches the point at which more solid cannot be dissolved. At this point at which there is solid phase present in the liquid, the solution is saturated and the concentration of solid in the liquid phase can be determined. Because the phases are easily separated, we can measure equilibrium concentrations of the solid in the liquid by weighing the solid before the phases are contacted and again after equilibrium is established.

A liquid phase with its equilibrium concentration of dissolved solid is often referred to as saturated, and this equilibrium concentration of solute is called the solubility limit or saturation concentration. We denote it by $C_{A,eq}^{I}$ when dealing with solid–liquid systems:

$$C_{A,eq}^{I} = C_{A,sat}^{I}. \tag{4.1.6}$$

Saturation concentrations for several inorganic salts in water are shown in Table 4.2.

Liquid–Gas Systems

Liquids that are continually in intimate contact with some gases can absorb the gas and reach equilibrium. The experiments require that the liquid and gas be in contact for a sufficient period of time that phase equilibrium is reached. A common practical example is water and air. The equilibrium concentrations of oxygen in water, $C_{A,eq}^{I}$, as functions of temperature for air are presented in Table 4.3. $C_{A,eq}^{I}$ at 25 °C is 8.2×10^{-3} kg/m^3.

Table 4.3. Oxygen-saturated concentrations in water in contact with air

Temperature (°C)	0	5	10	15	20	25	30	35	40
$H \times 10^{-4}$ (atm)	2.55	2.91	3.27	3.64	4.01	4.38	4.7	5.07	5.35
$C_{A,eq} \times 10^3$ (kg/m^3)	14.1	12.4	11.0	9.9	8.9	8.2	7.6	7.1	6.7

We obtained the values of $C_{A,eq}^I$ in Table 4.3 assuming Henry's "Law" holds:

$$H = p_A/x_A \tag{4.1.7}$$

= partial pressure of A in atm/mol fraction of A in solution.

We can illustrate how the values in Table 4.3 were obtained using Henry's "Law" with the following example.

If we assume that oxygen makes up approximately 20.5% of air, the partial pressure of the oxygen at atmospheric pressure (1 atm) is 0.205 atm. The liquid mole fraction of oxygen at 0 °C is readily calculated through use of Henry's "Law" constant at that temperature from Table 4.3:

$$x_A = p_A/H = \frac{0.205 \text{ atm}}{2.55 \times 10^4 \text{ atm}},$$

$$x_A = 8.04 \times 10^{-6}.$$

The molar concentration of oxygen in the liquid phase can be calculated through conversion of the obtained mole fraction.

$$C_A^I = \frac{x_A \rho^I}{MW^I} = \frac{8.04 \times 10^{-6}(1000 \text{ kg/m}^3)}{18.02 \text{ kg/kmol}},$$

$$C_A^I = 4.5 \times 10^{-4} \text{ kmol/m}^3.$$

Finally, the concentration of oxygen in the water at 0 °C may be calculated through multiplying by the molecular weight of oxygen:

$$C_A^I = (4.5 \times 10^{-4} \text{ kmol/m}^3)(32 \text{ kg/kmol}),$$

$$C_A^I = 14 \times 10^{-3} \text{ kg/m}^3.$$

This is the equilibrium concentration of oxygen in water at 0 °C listed in Table 4.3.

This relationship is valid if the partial pressure of the gas does not greatly exceed 1 atm. Solubility tables for gases in water and other liquids as functions of temperature can be found in the *Chemical Engineers' Handbook* (1997). Selected values of H for some common gases in water at 20 °C are presented in Table 4.4.

Liquid–Liquid Systems

Most liquid–liquid systems of pragmatic interest are more complicated than solid–liquid or liquid–gas systems, for which the saturation concentration alone is sufficient

Table 4.4. Henry's "Law" constants at 20 °C
for some common gases in water

Gas	H (atm)
Carbon monoxide	5.36×10^4
Carbon dioxide	1.42×10^3
Hydrogen	6.83×10^4
Bromine	0.59×10^2
Nitrogen	8.04×10^4

to perform a Level II analysis. In the liquid–liquid case, the component of interest is distributed between two chemically distinct liquids, which themselves may be partially miscible over a range of compositions. Equilibrium data on many liquid–liquid systems (often *ternary systems*) are available in the literature (see the reference list at the end of this chapter) and in technical bulletins issued by solvent manufacturers, though it is frequently necessary to collect one's own data on the system of interest. The simplest liquid–liquid system, which is fairly common in practice in the chemical industry, is one in which the two liquids are immiscible and the component being transferred is often at a low concentration.

Generally, there exists a thermodynamic relationship between the equilibrium concentrations in each phase. For relatively simple systems, this may be a relationship of the form shown in Eq. (4.1.5). The function describing the equilibrium relationship is usually not linear, though it may be nearly linear at low concentrations. A common form is

$$C^{I}_{A,eq} = M C^{II}_{A,eq},\qquad(4.1.8)$$

where M is a constant (dependent on temperature and pressure, however). This constant for solute partitioning between phases is known as a *distribution coefficient*. Rigorously, the distribution coefficient can be shown to be equal to the reciprocal ratio of the activity coefficients of the solute in each phase (Sandler, 2006). If these activity coefficients are relatively constant over the solute concentrations of interest, the distribution coefficient will be a constant. This behavior is termed *ideal* (in the sense of Henry's "Law," not that the mixture has ideal mixture properties). Few ternary systems are ideal over a wide range of concentrations, although M may be nearly constant over a particular concentration range.

The experiments to determine the distribution coefficient are easy to carry out when the liquid phases are immiscible. A known amount of solute A is added to the two solvents and the system is thoroughly mixed and allowed to reach equilibrium. If the solvents are of different densities, then the phases are readily separated and the amount of A in each is measured by some appropriate analytical method. M is determined by dividing $C^{I}_{A,eq}$ by $C^{II}_{A,eq}$. The experiment is then repeated with a different amount of A. The data shown in Figure 4.2 for the system acetone–water–(1,1,2)-trichloroethane (TCE) at 25 °C were obtained in this way.

It is evident from this figure that M is not a constant over the full range of concentrations. Still, we can see from this figure that this system is nearly ideal for aqueous concentrations of less than 100 g/L. The Level II analysis proceeds by setting the concentration of A in the two phases to be in equilibrium, which can be expressed by relationship (4.1.8). Notice that, unlike heat exchangers in which the final equilibrium temperatures of the process and utility fluid are equal, here the concentrations of A in both phases will generally not be equal. Substitution into the second of simplified Level I balance equations (4.1.4), we obtain

$$C^{II}_{A,eq} = \frac{V^{I} C^{I}_{Ai} + V^{II} C^{II}_{Ai}}{V^{I} M + V^{II}}.\qquad(4.1.9)$$

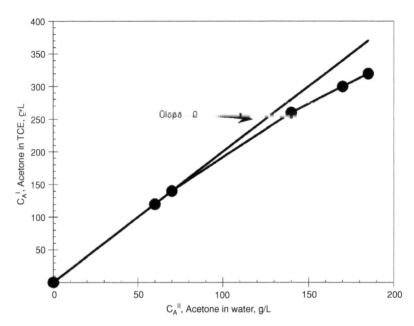

Figure 4.2. Distribution of acetone in (1,1,2)-trichloroethane (TCE) and water (25 °C) (adapted from Treybal RE, Weber LD, Daley IF. The system acetone-water-1,1,2-trichloroethane. Ternary liquid and binary vapor equilibria. Ind. Eng. Chem. 1946; 38: 817).

If $V^I = V^{II}$, (4.1.9) simplifies to

$$C^{II}_{A,eq} = \frac{C^I_{Ai} + C^{II}_{Ai}}{M+1}. \tag{4.1.10}$$

This result is similar to the batch heat exchanger result. In that case we found that the equilibrium temperature was an average of the temperatures, weighted by the thermal capacities of each fluid. It is logical to extend the idea of a phase "capacity" here as well, defining the "mass capacity" of each phase. For the first phase, this is $V^I M$. For the second phase, it is simply the volume of Phase II, or V^{II}. The mass capacity of Phase I contains the partitioning coefficient because, at equilibrium, Phase I will hold M times as much mass as Phase II per unit volume.

The true utility of this Level II analysis is that it provides an upper limit on the amount of mass of a species that can be transferred between phases. Although this analysis provided helpful information in the design of heat exchangers, it is especially valuable for understanding the feasibility limits of mass contactors.

EXAMPLE 4.1. A 700-L batch of process water, V^{II}, is contaminated with 100 kg/m^3 (100 g/L) of acetone, C^{II}_{Ai}. Laboratory experiments have shown that acetone can be removed from water by contacting the water–acetone solution with 1,1,2-TCE, an organic solvent, which must be carefully handled because of its toxic properties.

700 L, V^I of TCE is available with $C^I_{Ai} = 0$, i.e., there is no acetone initially in the TCE. How much acetone can be removed from the water? A pilot plant vessel (Figure 4.1) is available for this operation and we can use either Vessel A or B described in Example 3.1.

SOLUTION. As we did with our batch heat exchanger problem, we first need to see what value of $C_{A,eq}^I$ can be obtained if equilibrium is reached (a Level II problem). Equation (4.1.9) can be used to compute the equilibrium concentration of acetone in the TCE, $C_{A,eq}^{II}$.

Any operation needs to be carried out carefully with the proper safety considerations. The vessel must be closed to the atmosphere when the contactor is operating. Material Safety Data Sheets (MSDSs), which give a great deal of critical information for anyone dealing with chemicals, are readily available and can be downloaded from the Web. TCE has a "time-weighted" average (TWA) concentration for a normal 8-h workday to which nearly all workers may be repeatedly exposed (threshold limit value [TLV]-TWA) of 10 ppm, 55 mg/m^3. Acetone has a TLV-TWA of 750 ppm, 1780 mg/m^3.

As a first approximation we assume TCE and water are insoluble and that the amount of species A transferred is small enough to neglect changes in phase volumes. From Figure 4.2, M = 2. Hence Equation (4.1.9) yields

$$C_{A,eq}^{II} = \frac{700\,(100) + 700\,(0)}{700\,(2) + 700}$$

$$= 33 \text{ g/L acetone in water.}$$

Because we have assumed equilibrium,

$$C_{A,eq}^I = MC_{A,eq}^{II}$$

$$= 66 \text{ g/L acetone in TCE.}$$

If we desire to remove more of the acetone from the water, we need to contact the water–acetone from this batch with additional TCE. A critical factor is the time required to come close to the equilibrium value calculated. A Level IV analysis is required for determining the batch time, as discussed next. Please note that we will have to use this vessel as a separator to separate the water from the TCE in this batch operation. We are concerned with two times, the time for mass contacting and the time for separation.

4.2 Rate of Mass Transfer and Determination by Experiment

As demonstrated in the analysis of both reactors and heat exchangers, a constitutive equation is required for the rate expression to complete the analysis. Here, a constitutive equation relating the rate of mass transfer to measurable quantities in the system, such as concentrations, is required. Then it will be possible to use the model equations to extract the rate parameters from an appropriately designed experiment. Mass transfer represents a step up in complexity over conventional heat transfer analysis in several ways.

Application of the law of conservation of mass to species A in each control volume (applying the word statement of the conservation law in Figure 1.6) produces the following development taken directly from *Russell and Denn* (1972):

$$\left(C_A^I V^I\right)\big|_{t+\Delta t} = \left(C_A^I V^I\right)\big|_t + \left(\begin{array}{l}\text{amount of A transferred from Phase II}\\ \text{to Phase I from time t to t} + \Delta t\end{array}\right)$$

$$- \left(\begin{array}{l}\text{amount of A transferred from Phase I}\\ \text{to Phase II from time t to t} + \Delta t\end{array}\right)$$

$$\left(C_A^{II} V^{II}\right)\big|_{t+\Delta t} = \left(C_A^{II} V^{II}\right)\big|_t + \left(\begin{array}{l}\text{amount of A transferred from Phase I}\\ \text{to Phase II from time t to t} + \Delta t\end{array}\right)$$

$$- \left(\begin{array}{l}\text{amount of A transferred from Phase II}\\ \text{to Phase I from time t to t} + \Delta t\end{array}\right). \tag{4.2.1}$$

As with chemical reaction rates discussed in Chapter 2, it is necessary to define the rate of mass transfer. To distinguish this rate from the rate of reaction r with units of mole/(time × volume) we denote it by the boldface symbol **r**. Thus the rate per unit surface area at which A accumulates in Phase I is designated $\mathbf{r}_{A+}^I$, while the rate per unit surface area at which A is depleted from Phase I by mass transfer is $\mathbf{r}_{A-}^I$. Similar conventions hold for Phase II, for which the rates per unit area are written $\mathbf{r}_{A+}^{II}$ and $\mathbf{r}_{A-}^{II}$. The units of a rate of mass transfer per unit area are that of a *flux*, i.e., mass per area per time. The transfer of mass between phases is written on a per-area basis because, all other things being equal, an increase in the area between phases will lead to a proportional increase in the total rate of mass transferred. Notice that this is the same logic that leads to the writing of reaction rate on a per-volume basis.

Let "a" denote the total interfacial area between the two phases. Introduction of the rate of mass transfer into (4.2.1) leads to

$$\left(C_A^I V^I\right)\big|_{t+\Delta t} = \left(C_A^I V^I\right)\big|_t + a\left(\mathbf{r}_{A+}^I - \mathbf{r}_{A-}^I\right)\Delta t,$$

$$\left(C_A^{II} V^{II}\right)\big|_{t+\Delta t} = \left(C_A^{II} V^{II}\right)\big|_t + a\left(\mathbf{r}_{A+}^{II} - \mathbf{r}_{A-}^{II}\right)\Delta t. \tag{4.2.2}$$

Dividing by Δt and taking the limit as $\Delta t \to 0$ yields

$$\frac{d}{dt}\left(C_A^I V^I\right) = +a\left(\mathbf{r}_{A+}^I - \mathbf{r}_{A-}^I\right),$$

$$\frac{d}{dt}\left(C_A^{II} V^{II}\right) = +a\left(\mathbf{r}_{A+}^{II} - \mathbf{r}_{A-}^{II}\right). \tag{4.2.3}$$

It is evident for the batch system that the mass leaving Phase I must enter Phase II, and vice versa. Thus $\mathbf{r}_{A+}^I = \mathbf{r}_{A-}^{II}$ and $\mathbf{r}_{A-}^I = \mathbf{r}_{A+}^{II}$. For convenience we define an overall mass transfer rate per area, $\mathbf{r}_A$, with the same units:

$$\mathbf{r}_A = \mathbf{r}_{A+}^{II} - \mathbf{r}_{A-}^{II}. \tag{4.2.4}$$

Then

$$\frac{d}{dt}\left(C_A^I V^I\right) = -a\mathbf{r}_A \qquad \frac{d}{dt}\left(C_A^{II} V^{II}\right) = a\mathbf{r}_A. \tag{4.2.5}$$

A similar analysis applied to the total mass in each phase leads to the following relations:

$$\frac{d}{dt}(\rho^{I} V^{I}) = -a\mathbf{r}_A,$$

$$\frac{d}{dt}(\rho^{II} V^{II}) = a\mathbf{r}_A. \qquad (4.2.6)$$

Equations (4.2.5) and (4.2.6) constitute the Level IV analysis of the batch mixed–mixed mass contactor. These equations are similar to the Level III analysis of the batch mixed–mixed heat exchanger, Eqs. (3.2.9) and (3.2.10). However, the mass contactor is a multiphase mixture and the volumes of each phase will also change as mass is transferred, necessitating the additional calculation expressed by Eqs. (4.2.6). Further, the area for contact is generally not specified by the device geometry, as for heat exchangers, and must be calculated or measured. Consideration of the area calculation is deferred to Chapter 7. To proceed with the Level IV analysis, a rate law must be specified, as shown next.

4.2.1 Rate Expression

As we did with our analysis of reactors and heat exchangers, we must formulate a constitutive equation relating the rate $\mathbf{r}_A$ to the measured variables of the system, particularly the concentrations in each phase, C_A^{I} and C_A^{II}. As with all constitutive relations, the one we propose for mass transfer between phases must be verified by experiment, and its use in situations other than the experiments to determine it must be done with care.

We know from our experience with batch systems that after some time there will be no changes in the concentration of species in either phase and that phase equilibrium is reached. The time derivatives will be equal to zero in Eqs. (4.2.5) and (4.2.6) and thus $\mathbf{r}_A$ must equal zero:

$$\left[C_{A,\,eq}^{I} = f\left(C_{A,\,eq}^{II}\right) \right] \Leftrightarrow \mathbf{r}_A = 0. \qquad (4.2.7)$$

Our proposed rate law $\mathbf{r}_A$ must have a form such that equilibrium implies that the rate law vanishes and vice versa.

A further observation about the form of the rate follows from the way in which $\mathbf{r}_A$ is defined [(4.1.4)] in terms of $\mathbf{r}_{A+}^{II}$ and $\mathbf{r}_{A-}^{II}$. If C_A^{I} is greater than its equilibrium value for a given C_A^{II}, then there must be a net transfer of species A from Phase I to Phase II, so $\mathbf{r}_{A+}^{II} > \mathbf{r}_{A-}^{II}$ and $\mathbf{r}_A > 0$. Thus

$$\left[C_A^{I} - f\left(C_A^{II}\right) > 0 \right] \Rightarrow \mathbf{r}_A > 0. \qquad (4.2.8)$$

Similarly, if A is in excess in Phase II,

$$\left[C_A^{I} - f\left(C_A^{II}\right) < 0 \right] \Rightarrow \mathbf{r}_A < 0. \qquad (4.2.9)$$

Therefore mass transfer occurs from the phase with a higher concentration (relative to that which would be in equilibrium between the phases). Note that this discussion parallels the development of the rate law for heat transfer and the same conventions of sign apply here. A positive rate of species mass transfer leaves Phase I and enters Phase II. We choose the simplest form compatible with the properties highlighted in (4.2.7)–(4.2.9), namely, the rate is linearly proportional to the concentration difference driving mass transfer:

$$\mathbf{r}_A a = K_m a \left[C_A^I - f\left(C_A^{II}\right) \right],$$

where the overall mass transfer coefficient K_m has dimensions of a velocity, or length per time. As stated previously, the equilibrium relationship can often be simplified to that shown in Eq. (4.1.8), and the rate expression is written as

$$\mathbf{r}_A a = K_m a \left(C_A^I - M C_A^{II} \right). \tag{4.2.10}$$

If K_m and M are both allowed to be functions of concentration, then this form is quite general. For simplicity of illustration we generally assume that the coefficients are constants. This expression reduces further for solid–liquid and many liquid–gas systems, in which $M C_A^{II} = C_{A,eq}^I$. Substituting into the rate expression, the rate of mass transfer becomes dependent on the concentration only in the liquid phase and the saturation concentration, and an experimentally determined quantity that is constant for isothermal systems,

$$\mathbf{r}_A a = K_m a \left(C_A^I - C_{A,eq}^I \right). \tag{4.2.11}$$

The rate expression for mass transfer, Eq. (4.2.10), has a driving force characterized by a term representing the displacement from equilibrium, $C_A^I - M C_A^{II}$, divided by a resistance, K_m^{-1}, much like Ohm's "Law" in electrical circuits. It will be shown in Chapter 5 that the overall resistance K_m^{-1} can be expressed in terms of the individual phase mass transfer coefficients, denoted by k_m^I and k_m^{II}. Further, just as for an electrical circuit with resistances in series, the overall resistance is related to the sum of the individual resistances:

$$\frac{1}{K_m} = \frac{1}{k_m^I} + \frac{M}{k_m^{II}}. \tag{4.2.12}$$

It is important that the driving force used with the mass transfer coefficient in the constitutive equation be in accord with the way the distribution coefficient M is defined. In Eq. (4.2.10), M was defined by $C_{A,eq}^I = M C_{A,eq}^{II}$ [Eq. (4.1.8)]. It is possible to define the driving force for mass transfer in terms of the concentration of species A in Phase II. Keeping careful track of the signs, we obtain the equivalent rate expression written in terms of this driving force:

$$-\mathbf{r}_A a = \left(\mathbf{r}_{A-}^{II} - \mathbf{r}_{A+}^{II} \right) a = K_m' a \left(C_A^{II} - \frac{1}{M} C_A^I \right).$$

It is straightforward to show that $K_m' = M K_m$. In the rest of the text we define the rate of mass transfer consistently with the driving force expressed in terms of the concentration difference in Phase I, Eq. (4.2.10) or Eq. (4.2.11).

We can now write the Level IV equations for the batch mass contactor by adding our constitutive equation to Eqs. (4.2.5) and (4.2.6):

$$\frac{d}{dt}\left(C_A^I V^I\right) = -K_m a\left(C_A^I - MC_A^{II}\right), \tag{4.2.13}$$

$$\frac{d}{dt}\left(C_A^{II} V^{II}\right) = K_m a\left(C_A^I - MC_A^{II}\right), \tag{4.2.14}$$

$$\frac{d}{dt}\left(\rho^I V^I\right) = -K_m a\left(C_A^I - MC_A^{II}\right), \tag{4.2.15}$$

$$\frac{d}{dt}\left(\rho^{II} V^{II}\right) = K_m a\left(C_A^I - MC_A^{II}\right). \tag{4.2.16}$$

Batch mass transfer experiments to determine K_m are usually difficult to carry out because equilibrium is reached too rapidly to obtain accurate data. However, as the total rate is proportional to the area available for mass transfer, an experiment can be designed to measure the rate of mass transfer and extract the mass transfer coefficient K_m if a sufficiently small area is employed. Such an experiment is subsequently described. It is designed to measure K_m by use of tablets of table salt (sodium chloride, NaCl), in which the tablet surface area is such that it allows mass transfer to take place over several minutes. The aqueous phase is initially distilled water, and the dissolved salt concentration is measured with a conductivity probe.

A Level I analysis for this batch salt-tablet water experiment is

$$\rho^{II}\left(V^{II} - V_i^{II}\right) = V_i^I C_{Ai}^I - V^I C_A^I. \tag{4.2.17}$$

For solid–liquid systems at equilibrium,

$$C_A^I = C_{A,eq}^I.$$

The Level II analysis can then be written as follows if our saltwater experiment is at phase equilibrium:

$$\rho^{II}\left(V^{II} - V_i^{II}\right) = V_i^I C_{A_i}^I - V^I C_{A,eq}^I. \tag{4.2.18}$$

The system is described by Eqs. (4.2.13)–(4.2.16). Because $C_A^{II} = \rho^{II}$ for pure salt, the species balance equation for Phase II is redundant. Substituting in the rate expression, Eq. (4.2.11), the Level IV balances for our saltwater experiment become

$$\frac{d}{dt}\left(\rho^I V^I\right) = -K_m a\left(C_A^I - C_{A,eq}^I\right) \quad \text{water,} \tag{4.2.19}$$

$$\frac{d}{dt}\left(\rho^{II} V^{II}\right) = K_m a\left(C_A^I - C_{A,eq}^I\right) \quad \text{salt,} \tag{4.2.20}$$

$$\frac{d}{dt}\left(C_A^I V^I\right) = -K_m a\left(C_A^I - C_{A,eq}^I\right) \quad \text{salt in water.} \tag{4.2.21}$$

The thermodynamic data for the experiment are shown in Table 4.2. The experimental data shown in Table 4.5 lead to some further simplifications. For example, the total mass of salt is 0.33% of the mass of water. Thus there will not be a significant change in the density or volume of the aqueous phase, so we can disregard the overall mass balance for Phase I, Eq. (4.2.19), and take V^I as a constant, allowing this term to be removed from the time derivative. Furthermore, C_A^I is at all times more than

Table 4.5. Experimental results: Concentration of salt dissolved vs. time

		Phase properties	
Volume	V_i^I		6.000 L
Surface-to-volume factor	α		5.32
Saturation concentration	$C_{A,eq}^I$		360 g/L
Number of tablets	N		30
Salt density	ρ^{II}		2160 g/L
Total mass	m^{II}		19.2 g

Experimental Data

t (s)	C_A^I (g/L)	t (s)	C_A^I (g/L)
0	0.00	120	1.40
15	0.30	135	1.49
30	0.35	150	1.68
45	0.64	165	1.76
60	0.89	195	2.06
75	1.08	200	2.14
90	1.10	240	2.31
105	1.24	270	2.43

two orders of magnitude less than $C_{A,eq}^I$ so the driving force inside the parentheses in Eqs. (4.2.20) and (4.2.21) can be reduced as follows:

$$\left(C_A^I - C_{A,eq}^I\right) \cong -C_{A,eq}^I.$$

Also, ρ^{II} is a constant, so it can be removed from the differential in (4.2.20). These assumptions lead to the following two model equations:

$$\rho^{II}\frac{dV^{II}}{dt} = -K_m a\left(C_{A,eq}^I\right), \tag{4.2.22}$$

$$V^I\frac{dC_A^I}{dt} = K_m a\left(C_{A,eq}^I\right). \tag{4.2.23}$$

Because $V^I = V_i^I$ the Level I balance [Eq. (4.2.17)] becomes

$$\rho^{II}\left(V^{II} - V_i^{II}\right) = V^I\left(C_{A,i}^I - C_A^I\right), \tag{4.2.24}$$

where V_i^{II} is the initial volume of the solid salt. The quantities $\rho^{II}V_i^{II}$ and $C_{A,i}^I V^I$ are the initial mass of the solid and the initial mass of the salt dissolved in the water, respectively. In our case, the latter term is zero.

The interfacial area a is related to the solid volume V^{II} through the geometry of the tablets. The N tablets are assumed to be identical, so the volume per tablet is V^{II}/N. The area per tablet is then $\alpha(V^{II}/N)^{2/3}$, where α is a constant that depends on the tablet geometry. The value for this constant is presented in Table 4.6 for several common shapes. The salt tablets used in this experiment have an α value

Table 4.6. Surface-to-volume factors for some common shapes

Shape	Dimensions	Shape factor (α)
Cube	Length = Width = Height	6
Square cylinder	Diameter = Height	$\sqrt[3]{54\pi} \approx 5.536$
Sphere	Radius = r	$\sqrt[3]{36\pi} \approx 4.836$
Rectangular box	Length = a, Width = b, Height = c	$\dfrac{2}{(abc)^{2/3}}(ab + ac + bc)$
Cylinder	length/diameter = ε	$\sqrt[3]{2\pi\varepsilon}(2 + \varepsilon^{-1})$

close to that of a square cylinder. The total surface area is then N times the area per tablet:

$$a = N\alpha \left(\frac{V^{II}}{N}\right)^{2/3} = \alpha N^{1/3} \left(V^{II}\right)^{2/3}. \tag{4.2.25}$$

Equation (4.2.22) for the mass of undissolved salt becomes

$$\rho^{II}\frac{dV^{II}}{dt} = -K_m\alpha N^{1/3}C_{A,eq}^{I}\left(V^{II}\right)^{2/3}. \tag{4.2.26}$$

This can be separated, which, on integration, yields

$$V^{II} = \left[\left(V_i^{II}\right)^{1/3} - \frac{K_m\alpha N^{1/3}C_{A,eq}^{I}}{3\rho^{II}}t\right]^{3}. \tag{4.2.27}$$

This is substituted into (4.2.24) to obtain the aqueous-phase salt concentration

$$C_A^I = C_{Ai}^I + \frac{\rho^{II}}{V^I}\left\{V_i^{II} - \left[\left(V_i^{II}\right)^{1/3} - \frac{K_m\alpha N^{1/3}C_{A,eq}^{I}}{3\rho^{II}}t\right]^{3}\right\}. \tag{4.2.28}$$

This solution can be cast into a form that is linear in time so as to facilitate fitting to data:

$$(1 - \hat{c})^{1/3} = 1 - \omega t, \tag{4.2.29}$$

where

$$\hat{c} = \frac{C_A^I V^I}{\rho^{II}V_i^{II}} = \frac{\text{(salt in solution)}}{\text{(total salt added)}}, \qquad \omega = \frac{K_m\alpha N^{1/3}C_{A,eq}^{I}}{3\rho^{II}(V_i^{II})^{1/3}}.$$

The data and the fit of the model for this system are plotted according to (4.2.29) in Figure 4.3. As required by the model, the data follow a straight line passing through unity when plotted in this form. Using the best-fit value of the slope, $\omega = 1.46 \times 10^{-3}$ s^{-1} and the data in Table 4.5, we calculate K_m to be 3.3×10^{-5} m/s. The excellent agreement between experimental data and model behavior supports the assumptions underlying the rate law and model development. Although we shall not show any further experimental data, it is found that K_m in two-phase liquid tank-type systems is nearly always within an order of magnitude of the value computed here. Because K_m is often constrained to a relatively narrow range of values, the significant problem in mass contactor design is the determination of the interfacial area.

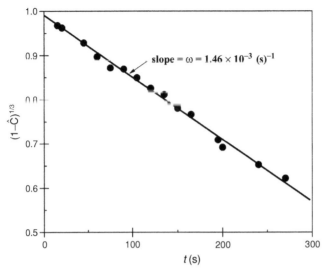

Figure 4.3. Experimental data for solid–liquid mass transfer experiment.

In this experiment, the interfacial area determination is trivial because of the clearly defined interface between the solid pellets and the liquid phase. However, in liquid–liquid and liquid–gas systems, determination of the interface becomes a much more complex problem involving multiphase fluid mechanics. This problem has received considerable attention in the literature, and a number of correlations have been developed relating the interfacial area to power input, vessel geometry, and physical properties. A more thorough discussion of interfacial area estimation is presented in Chapter 7.

4.2.2 Approach to Equilibrium

We remarked previously that batch experiments for mass transfer rate data are generally too rapid to obtain useful data. Now that we have a reasonable value for K_m, we can easily demonstrate this by using the model equations. We assume that we have two immiscible, liquid phases that are initially of equal volume and that the total amount of mass transferred between phases is so small that changes in the volume and density of each individual phase are negligible throughout the course of this experiment. The overall mass balance equations are therefore not needed in the analysis. The component balances for the two phases can be written as

$$V \frac{dC_A^I}{dt} = -K_m a \left(C_A^I - M C_A^{II} \right),$$

$$V \frac{dC_A^{II}}{dt} = K_m a \left(C_A^I - M C_A^{II} \right). \tag{4.2.30}$$

We have not included superscripts I and II for the volumes, as they are equal and constant. M will be taken as a constant.

Addition of the preceding equations gives

$$V\frac{dC_A^I}{dt} + V\frac{dC_A^{II}}{dt} = 0,\tag{4.2.31}$$

or, on integration,

$$\left(C_A^I - C_{Ai}^I\right) + \left(C_A^{II} - C_{Ai}^{II}\right) = 0.\tag{4.2.32}$$

Equation (4.2.32) is the Level I balance for this system, where C_{Ai}^I and C_{Ai}^{II} are the initial values of C_A^I and C_A^{II}, respectively. Substitution for C_A^{II} in (4.2.30) then gives

$$\frac{dC_A^I}{dt} = -\left(\frac{K_m a}{V}\right)(1 + M)\,C_A^I + M\left(\frac{K_m a}{V}\right)\left(C_{Ai}^I + C_{Ai}^{II}\right).\tag{4.2.33}$$

This is a linear ordinary differential equation of first order. It has a straightforward analytical solution, which, after some tedious algebra, is

$$C_A^I = C_{A,eq}^I - \left(C_{A,eq}^I - C_{Ai}^I\right)e^{-\beta t},\tag{4.2.34}$$

where the equilibrium concentration $C_{A,eq}^I$ is computed from Eq. (4.1.10),

$$C_{A,eq}^I = M\left(\frac{C_{Ai}^I + C_{Ai}^{II}}{1 + M}\right),$$

and the exponential quantity, β, is defined as $\beta = (K_m a/V)(1 + M)$.

Equation (4.2.34) shows us that the system will be at equilibrium when the exponential term vanishes. For the purpose of obtaining an estimate of the time, we assume that equilibrium is reached in a practical sense when the exponent has a value of about –3 (95% removal of A from one phase). The total time for the experiment is determined approximately by

$$\left(\frac{K_m a}{V}\right)(1 + M)\,t_{eq} \approx 3.\tag{4.2.35}$$

For the data shown in Figure 4.2, $M \approx 2$, and we have seen that K_m can be expected to be within an order of magnitude of 3×10^{-5} m/s. Thus if we consider this a wide range we obtain

$$\frac{10^4\,V}{3a} \le t_{eq} \le \frac{10^6\,V}{3a},$$

where lengths are in meters and time is in seconds. Now, if the agitation is such that one phase is effectively dispersed into N droplets then, according to (4.2.25), $a = \alpha N^{1/3} V^{2/3}$, where α is approximately 5. Thus,

$$10^3\left(\frac{V}{N}\right)^{1/3} \le t_{eq} \le 10^5\left(\frac{V}{N}\right)^{1/3}.$$

The term in parentheses represents the volume of a typical droplet. If we assume a spherical drop, $V/N = \pi d^3/6$, and t_{eq} will range between 538d and 53 800d.

Droplets with a diameter of 0.001 m are easily obtained in most low-viscosity systems. With this size drop t_{eq} will be between 0.54 and 54 s. This makes it very difficult to obtain concentrations of component A as functions of time, but it clearly shows that equilibrium can be reached in liquid–liquid systems very rapidly. If an experiment can be carried out with a drop diameter of 0.005 m, then t_{eq} will range between 2.7 and 270 s (4.5 min) and concentration–time data could be collected to obtain $K_m a$. When designing liquid–liquid experiments to obtain rate data, the drop size must be controlled so that equilibrium is not achieved too quickly. Alternatively, continuous-flow equipment can be used, as will be discussed in the next section.

Drop and bubble sizes can be estimated from process operating conditions and the physical properties of the fluids, as will be shown in Chapter 7. Often there is a trade-off involving area generation, energy input, and ease of separation such that typical sizes range from 5×10^{-4} to 10^{-2} m.

The execution and interpretation of the batch experiment for mass contactors have some critical issues that we did not have to consider when analyzing batch reactors and heat exchangers. Mass contactor experiments are more difficult to execute because of the rapid approach to equilibrium and because it is more difficult to sample two phases.

Batch reactor experiments can produce a verified rate expression that for many reactions can be used with confidence in designing a larger-scale reactor. Batch heat exchanger experiments need additional information on how the heat transfer coefficient varies with the microscale fluid mechanics, but this can be performed with some extra effort. Mass contactor batch experiments need to determine both K_m and the interfacial area a, and most experiments can determine only the combined quantity $K_m a$. Because both of these quantities depend on the microscale fluid mechanics and because a also depends on the two-phase macroscale fluid motions, we have a much more complicated problem than those we faced with reactors and heat exchangers.

4.3   Tank-Type Two-Phase Mass Contactors

There are three basic types of operation possible for tank-type mass contactors (Table 4.1). Both phases can be operated in batch mode—neither is added or removed from the system continuously. This is a batch mass contactor operation. If one of the phases is continuously added to and/or removed from the vessel, this is classified as a semibatch process. If both phases are added to and removed from the process continuously we have a continuous process. In this section, we investigate these three types of mass contactors.

Batch and semibatch contactors are most commonly employed for solid–fluid, liquid–liquid, and gas–liquid systems for which there is sufficient agitation to ensure that both phases are well mixed. For batch systems, the agitation must be stopped so that the phases can separate before the processing step is complete. The separation is trivial for solid–fluid systems and gas–liquid systems because of the large difference in density between the phases. Effective liquid–liquid separations are more difficult to achieve. Because the time required for separation is a function of the drop size generated by the agitation in the batch contactor, optimization of the design and operation of the separation and the mass contacting steps are coupled. Smaller droplets allow for more rapid mass transfer, but they increase the holding time required for separation. Proper optimization of a liquid—liquid mass transfer operation, either batch or continuous, requires a thorough analysis of each step as a function of droplet size.

A semibatch mass transfer process involves one batch phase held in the vessel throughout the mass transfer process and another phase that is flowing continuously through the system. Although we have already used the term "semibatch" to characterize tank-type chemical reactors and heat exchangers, there is an important distinction to be made for the mass contactors and heat exchangers. In a reactor, a reactant is continuously added to, but usually not removed from, the reactor. In a semibatch heat exchanger the batch process fluid is separated from the continuously flowing utility fluid by the material comprising the heat transfer area. In semibatch mass contactors the two phases are most frequently in direct contact with each other. In solid–fluid contactors the solid phase is the batch phase and either a gas or liquid is passed continuously through it. In gas–liquid semibatch contactors the liquid phase is the batch phase and the gas passes through the contactor in which the separation also occurs. In a typical process, the gas is introduced at the bottom of the contactor through spargers and rises to the surface of the liquid where it discharges into the vapor space and is removed from the contactor. The energy required for generating interfacial area and mixing the tank often comes from the pressure drop of the gas entering the sparger. Semibatch processes are used in the manufacture of specialty chemicals and pharmaceuticals and in environmental remediation. Laboratory-scale semibatch mass contactors–reactors are often employed to collect experimental rate data for reactions in which one of the reactants is delivered as a gas.

Two-phase tank-type mass contactors that operate continuously must have a separator, which also operates continuously. In gas–liquid systems, the contactor and separator make up one vessel. Such systems can have mixed–mixed fluid motions or plug–mixed fluid motions in which the gas rises through the liquid in plug flow. Liquid–liquid systems require a separator operating continuously, as do solid–fluid systems. However, in the latter case the separation is readily achieved.

4.3.1 Batch Mass Contactors

The general balance equations for a batch mass contactor were first presented in Section 4.1.

With an expression for the rate law already established, along with knowledge of equilibrium behavior for the phases in question, the model equations for batch contactors can be obtained for different two-phase systems. The two most common systems encountered are solid–liquid and liquid–liquid mixtures. Liquid–gas and solid–gas mixtures are very difficult to mix in batch systems of any size, although small, laboratory-scale equipment can sometimes achieve the required mixing.

Solid–Liquid Systems: A specific case involving salt pellets was discussed in the previous section.

Liquid–Liquid Systems: The analysis of a liquid-liquid batch contactor relies on mixed–mixed fluid motions within the contactor to effectively carry out the transfer of a material between the two liquids. Mixed–mixed fluid motion is required for analyzing experiments or for designing laboratory-, pilot-, or commercial-scale equipment. In the case in which the total mass transferred is small compared with the mass of each phase, the Level I equations reduce to

$$V^I \left(C_A^I - C_{Ai}^I\right) + V^{II} \left(C_A^{II} - C_{Ai}^{II}\right) = 0,$$

$$C_A^{II} = C_{Ai}^{II} - \frac{V^I}{V^{II}} \left(C_A^I - C_{Ai}^I\right). \tag{4.3.1}$$

With the batch heat exchanger there is a constraint on the Level I analysis imposed by the second law of thermodynamics that we quantified by a Level II analysis. With batch mass contactors there is a constraint imposed on the Level I analysis by Level II phase equilibrium considerations. For the batch system, assuming a simple equilibrium relation as discussed in Section 4.1, Eq. (4.1.9) applies:

$$C_{A,eq}^{II} = \frac{V^I C_{Ai}^I + V^{II} C_{Ai}^{II}}{V^I M + V^{II}}.$$

The Level IV analysis follows readily on application of the word statement to the control volume of the batch vessel. Using rate expression (4.2.10) we find

$$V^I \frac{dC_A^I}{dt} = -K_m a \left(C_A^I - M C_A^{II}\right). \tag{4.3.2}$$

The concentration in Phase II is given by a Level I balance as

$$C_A^{II} = C_{Ai}^{II} - \frac{V^I}{V^{II}} \left(C_A^I - C_{Ai}^I\right). \tag{4.3.3}$$

These component mass balance model equations for the batch mass contactor are similar in form to those for the batch reactor and batch heat exchanger. Table 1.3,

Table 1.3. Simplified model equations for batch heat exchangers and mass transfer contactors

Batch reactor $(A + B \rightarrow D)$ (irreversible, first order in A and B):

$$\frac{dC_A}{dt} = -kC_A C_B$$

$$\frac{dC_B}{dt} = -kC_A C_B$$

$$\frac{dC_D}{dt} = kC_A C_B$$

Batch heat exchanger (average thermal properties):

$$\rho_1 V_1 \hat{C}_{p1} \frac{dT_1}{dt} = -Ua (T_1 - T_2)$$

$$\rho_2 V_2 \hat{C}_{p2} \frac{dT_2}{dt} = Ua (T_1 - T_2)$$

Batch mass contactor (dilute in A):

$$\frac{V^I dC_A^I}{dt} = -K_m a \left(C_A^I - M C_A^{II}\right)$$

$$\frac{V^{II} dC_A^{II}}{dt} = K_m a \left(C_A^I - M C_A^{II}\right)$$

which is reproduced here for convenience, lists them all in one place for easy comparison.

To obtain C_A^I as a function of time, C_A^{II}, obtained from the Level I analysis, is substituted into the component balance equation for C_A^I and integrated:

$$C_A^I(t) = \frac{M}{M + \Lambda} \left(C_{Ai}^I + \Lambda C_{Ai}^{II}\right) + \frac{\Lambda}{M + \Lambda} \left(C_{Ai}^I - M C_{Ai}^{II}\right) e^{-\beta t},$$

$$C_{Ai}^{II} = \frac{V^I}{V^{II}} \left(C_A^I - C_{Ai}^{II}\right), \tag{4.3.4}$$

where

$$\beta = \frac{K_m a}{V^I} \left(1 + \frac{M}{\Lambda}\right), \quad \Lambda = \frac{V^{II}}{V^I}.$$

The equilibrium concentrations can be expressed in terms of Λ and M as follows:

$$C_{A,eq}^I = M \left(\frac{C_{Ai}^I + \Lambda C_{Ai}^{II}}{\Lambda + M}\right), \quad C_{A,eq}^{II} = \frac{C_{Ai}^I + \Lambda C_{Ai}^{II}}{\Lambda + M}. \tag{4.3.5}$$

Equations (4.3.4) can be rewritten in terms of the equilibrium concentrations, after some algebra, as

$$\ln \left(\frac{C_A^I(t) - C_{A,eq}^I}{C_{Ai}^I - C_{A,eq}^I}\right) = -\beta t. \tag{4.3.6}$$

This form is similar to that derived for batch heat exchangers, Eq. (3.2.15), emphasizing the analogy between mass contactors with negligible changes in phase volume and heat exchangers.

EXAMPLE 4.1 CONTINUED. Equations (4.3.4) can be used to calculate the amount of acetone that can be removed from the water for a given batch process time, provided we know M, K_m, and the interfacial area a for the batch contactor.

M for this system was obtained from a laboratory experiment and presented in Section 4.1. Both K_m and a depend on the microscale fluid mechanics in the mass contactor. This gives us a much more difficult problem than with the batch heat exchanger or the batch reactor, in which we had to determine only the heat transfer coefficient U or the specific reaction-rate constant k.

If we could measure the concentration of acetone in the water C_A^I as a function of time in an existing batch contactor, we could use Eqs. (4.3.4) to determine $K_m a$. This is very difficult to do because, as we have pointed out, equilibrium is achieved very quickly and it is experimentally quite difficult to obtain good concentration–time data for most liquid–liquid systems.

4.3.2 Semibatch Mass Contactors

A semibatch or semiflow mass contactor is either a gas–liquid system in which the liquid is well mixed and the gas flows through as the continuous phase or a fluid–solid system in which the dispersed solid phase is well mixed and the fluid phase flows through the system as the continuous phase. The mass contactors for gas–liquid and solid–fluid systems also serve as separators. Mixed–mixed fluid motions in gas–liquid systems are difficult to achieve. If the gas phase is assumed to be in plug flow, the model equations will more accurately represent the physical situation. In parallel with the heat exchanger model development, we develop both of these models in this section, although the mixed–mixed case is most applicable to solid–fluid mixtures.

As with the semibatch cases for heat-exchangers, semibatch mass contactor equations are time dependent, and numerical solutions are often necessary for accurate modeling. In some instances, it may be possible to neglect the explicit time variation in the model equation for the batch phase without introducing much error, but the subsequent models will always fail to capture the initial behavior of the system. This situation is similar to that discussed in Subsection 3.3.2 in the analysis of the semibatch heat exchanger.

4.3.2.1 Mixed–Mixed Fluid Motions

The Level I analysis proceeds as for the semibatch mixed–mixed heat exchanger:

$$\left(V^I C_A^I + V^{II} C_A^{II}\right)_{t+\Delta t} = \left(V^I C_A^I + V^{II} C_A^{II}\right)_t + q_F^I C_{AF}^I \Delta t - q^I C_A^I \Delta t.$$

Taking the limit and integrating yields

$$\frac{dV^I C_A^I}{dt} + \frac{dV^{II} C_A^{II}}{dt} = q_F^I C_{AF}^I - q^I C_A^I. \tag{4.3.7}$$

As with the semibatch heat exchanges we cannot obtain a simple algebraic expression for C_A^{II} in terms of C_A^I and the density ρ^I and flow rate q^I.

The Level II analysis requires that $C_{A,eq}^I = MC_{A,eq}^{II}$, a condition that has little practical application.

The dispersed solid-phase component balance equation is identical to the continuous-phase equation discussed in the batch operation of a liquid–liquid mass contactor. Because it is the dispersed phase, we designate it as Phase II in accordance with our previous convention. We again assume the quantity of species A transferred does not significantly affect the volumetric properties of either stream:

$$V^{II} \frac{dC_A^{II}}{dt} = K_m a \left(C_A^I - MC_A^{II}\right). \tag{4.3.8}$$

This case requires that the fluid phase is well mixed, and so the concentration of species A is uniform throughout this phase. The species mass balance is

$$V^I \frac{dC_A^I}{dt} = q\left(C_{AF}^I - C_A^I\right) - K_m a\left(C_A^I - MC_A^{II}\right). \tag{4.3.9}$$

There are two cases of interest with this solid–liquid mixed–mixed semibatch mass contactor. If the solid being transferred to the liquid is pure, $C_A^{II} = \rho^{II}$, and Eq. (4.2.27) shows how the volume of solid changes with time as it dissolves in the liquid that is flowing through the contactor. We showed how the model behavior for this situation in a batch contactor compared with experiment in Subsection 4.2.1, where the salt-tablet experiment is described.

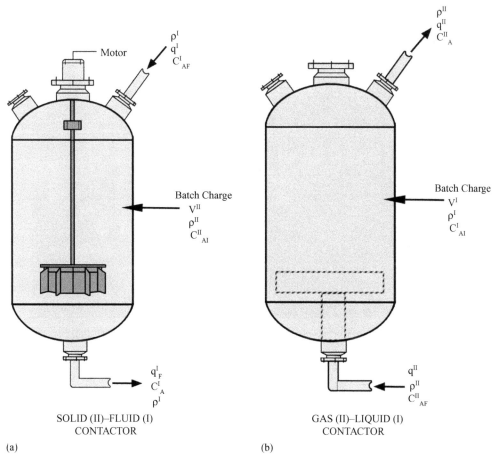

Figure 4.4. Semibatch mass contactors [note (I) is the continuous phase and (II) is the dispersed phase].

If component A is transferring from the liquid and adsorbing on the solid then C_A^{II} represents the concentration of A in the solid phase, and we need to solve the coupled set of Eqs. (4.3.8) and (4.3.9) numerically.

4.3.2.2 Mixed–Plug Fluid Motions

To achieve mixed–plug fluid motions one must continually flow a gas into the batch liquid through a well-designed sparger placed close to the bottom of the mass contactor, as sketched in Figure 4.4. Semibatch operations are labor intensive as the liquid phase must be loaded, contacted, and unloaded for each batch operation. They have the advantage that a bad batch can be discarded if the operation for some reason produces an off-spec product. Partly for this reason they are frequently employed in the pharmaceutical and fine chemicals industry. A photograph of a 10 000 L fermentor used by PDL BioPharma is shown in Figure 4.5.

Penicillin is made in semibatch mixed–plug vessels with either pure oxygen or sterilized air being sparged into the fermentor. In fact, it was the development of this type of mixed–plug fermentor that first allowed penicillin to be mass produced

Figure 4.5. Photograph of semibatch 10 000 L fermentor, courtesy of PDL BioPharma (Dr. Weichang Zhou).

in the early 1940s. This critical development was carried out in the United States in a very effective cooperative effort among industrial concerns, universities, and federal government agencies to provide enough penicillin to prevent serious infections from war wounds. This mixed–plug mass contactor separator replaced the "bottle plants" that processed some 100 000 batch flasks to produce the penicillin. A photograph depicting what such an operation would look like is shown in Figure 4.6.

Solids can be precipitated or crystallized from solution. For example, precipitated calcium carbonate is made by sparging carbon dioxide gas into lime water (CaO) in a semibatch operation.

The Level I analysis parallels that for the mixed–mixed case. The Level II analysis also follows from the mixed–mixed case. For the gas–liquid case the gas bubbles flowing through the mass contactor are the dispersed phase. The Level I balance becomes

$$\frac{d V^I C_A^I}{dt} + \frac{d V^{II} C_A^{II}}{dt} = q_F^{II} C_{AF}^{II} - q^{II} C_A^{II}. \qquad (4.3.10)$$

The species A component mass balance for batch Phase I is developed as we have done before for the semibatch heat exchanger in Subsection 3.3.2:

$$V^I \frac{d C_A^I}{dt} = -K_m a \left(C_A^I - M \langle C_A^{II} \rangle \right), \qquad (4.3.11)$$

where $\langle C_A^{II} \rangle$ is the average concentration of species A in the plug-flow phase (II). We can derive the species mass balance for the plug-flow gas phase by taking a

Figure 4.6. A simulated "bottle plant" operation.

control volume element of thickness Δz along the flow direction z with the tank's cross-sectional area A_c:

$$\frac{\partial}{\partial t}\left(\psi^{II} A_c C_A^{II}\right) = -\frac{\partial}{\partial z}\left(q^{II} C_A^{II}\right) + K_m a_v A_c\left(C_A^I - M C_A^{II}\right). \tag{4.3.12}$$

In this equation, ψ^{II} is the average fraction of the cross section occupied by the gas phase, and a_v is the interfacial area *per unit volume of the vessel*. These two equations are a system of coupled partial differential equations, which require numerical solution.

Because the change in C_A^{II} in the time it spends in the mass contactor is most often small compared with the time over which the concentration of A in the continuous phase changes substantially, we can assume that the dependence of C_A^{II} with time to be zero. The gas will rise through the tank in a time of the order of seconds. This occurs because the rising gas plume causes a very significant liquid velocity that carries the gas bubbles rapidly to the top of any vessel. Setting the time derivative for the phase in the plug-flow model equation to zero and assuming that volumetric properties of the phases will not change, we find that the model equations become

$$\frac{dC_A^I}{dt} = -\frac{K_m a}{V^I}\left(C_A^I - M \langle C_A^{II} \rangle\right), \tag{4.3.13}$$

$$\frac{dC_A^{II}}{dz} = \frac{K_m a_v A_c}{q^{II}}\left(C_A^I - M C_A^{II}\right). \tag{4.3.14}$$

The system has the following solution:

$$C_A^I(t) = MC_{AF}^{II} + \left(C_{A,i}^I - MC_{AF}^{II}\right)e^{-\beta t},\tag{4.3.15}$$

$$C_A^{II}(z, t) = C_{AF}^{II} + \frac{1}{M}\left(C_{A,i}^I - MC_{AF}^{II}\right)\left(1 - e^{\,\alpha z}\right)e^{-\beta t}\tag{4.3.16}$$

with

$$\alpha = \left(\frac{K_m a}{q^{II}}\right)\frac{1}{L}, \quad \beta = \frac{q^{II}}{V^I}\left(1 - e^{-\frac{K_m a}{q^{II}}}\right).$$

The expression for β becomes $K_m a/V^I$ at small values of $K_m a/q^{II}$. This is clearly seen by expanding the term in parentheses in a Taylor series.

If we consider the pragmatically important case in which oxygen is supplied to the mixed–plug mass contactor, the model equations simplify. C_A^{II}, the concentration of oxygen in the flowing gas stream, is essentially constant. If the oxygen is supplied by air, not enough oxygen is transferred to the liquid to change C_A^{II} with spatial position in the contactor. If pure oxygen is used then C_A^{II} is constant and does not change with time or spatial position. The driving force for mass transfer can also be expressed as $(C_A^I - C_{A,eq}^I)$ as presented in Subsection 4.1.2. $C_{A,eq}^I$ for water in contact with air is presented in Table 4.3. Equation (4.3.11) becomes

$$\frac{dC_A^I}{dt} = -\frac{K_m a}{V}\left(C_A^I - C_{A,eq}^I\right).\tag{4.3.17}$$

This equation has the following solution:

$$\ln\left(\frac{C_A^I - C_{A,eq}^I}{C_{Ai}^I - C_{A,eq}^I}\right) = \frac{-K_m a}{V^I}t.\tag{4.3.18}$$

Equations (4.3.15) and (4.3.16) are applicable to any mixed–plug mass contactor that has a flowing gas stream that obeys Henry's "Law." If the gas supplies one of the raw materials for a chemical reaction then an additional reaction-rate term must be added to Eq. (4.3.17). This then couples Eq. (4.3.17) to the component mass balances for the other reacting materials, a model of importance in fermentor analysis and design.

EXAMPLE 4.2. A cylindrical vessel with a diameter of 1.4 m and a height of 6.5 m is available for the study of oxygen transfer to water. There is one bubbling station in the center of the vessel. The rising plume of bubbles carries with it liquid that flows up in the two-phase plume until it reaches the surface of the liquid where it flows radially outward turning down at the wall. This liquid circulation has the effect of making the liquid in the vessel behave as well mixed because the liquid velocities in the plume are of the order of a meter per second. The power to achieve this well-mixed liquid comes from the flowing gas that in most systems of practical interest is several hundred watts per cubic meter (Otero, 1983).

The vessel is filled with water to a liquid height of 6.0 m. This water has an oxygen concentration of 1.7×10^{-3} kg/m^3. An experiment is carried out at 11.9 °C with an air flow rate to the orifice of 3.9×10^{-3} m^3/s. The oxygen concentrations in Table E.4.1 were measured in the water. Estimate $K_m a$ for this process.

Table E4.1. *Measured oxygen concentrations*

Time (min)	$C_A^I \times 10^3$ (kg/m^3)
0	1.7
10	4.8
20	7.5

SOLUTION. We can obtain an estimate of $K_m a$ from this data by using Eq. (4.3.15). At 11.9 °C the equilibrium concentration of oxygen in this water, $C_{A,eq}^I$, is 11.5 $\times$ 10^{-3} kg/m^3. Because the oxygen levels in the air are not significantly depleted during the operation, Eq. (4.3.17) applies:

$$\ln\left(\frac{4.8 - 11.5}{1.7 - 11.5}\right) = \frac{-K_m a}{V^I}600.$$

At 10 min,

$$\ln 0.68 = -0.38 = \frac{-K_m a}{V^I}600,$$

$$\frac{K_m a}{V^I} = 6.3 \times 10^{-4},$$

$$\ln\left(\frac{7.5 - 11.5}{1.7 - 11.5}\right) = \frac{-K_m a}{V^I}1200.$$

At 20 min,

$$\frac{K_m a}{V^I} = 7 \times 10^{-4}.$$

To determine K_m we must know how to estimate a, the bubble interfacial area. This problem in two-phase fluid mechanics is illustrated in Chapter 7. The area depends strongly on the gas flow rate, so if this vessel is to be used with a different air sparging rate, additional experiments would have to be carried out.

4.3.3 Continuous-Flow Two-Phase Mass Contactors

Continuous-flow two-phase systems are ones in which both phases are continuously fed into and removed from the system. They are widely employed for the variety of unit operations described in the introduction and may be carried out in both tank-type and tubular systems. They are also used in the laboratory to collect experimental data. An example of a continuous-flow liquid–liquid tank-type contactor is shown in Figure 4.7, where the process is seen to consist of a mass contactor and a separator to separate the phases after mass transfer has occurred.

Our goal in this section is to gain an appreciation of the problems of two-phase design and operation by examining a simple well-stirred continuous-flow tank-type mass contactor operation in detail. This will give us an introductory understanding of the more complex problems and some further practice with model development and manipulation for technically feasible analysis and design purposes.

Although only the mass contactor is discussed in this subsection, an additional piece of equipment, a separator, is required for continuously separating liquid–liquid

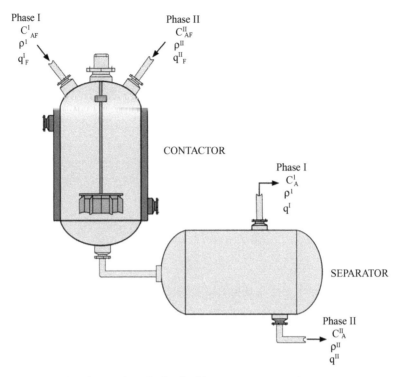

Figure 4.7. Continuous-flow liquid–liquid mass contactor and separator.

mixtures. We assume that all mass transfer takes place in the contactor and that the sole function of the separator is to separate the phases that are mixed in the contactor. The concentrations in the exit streams of the separator are assumed to be the same as those we would find if we sampled the individual phases in the contactor. This is reasonable because, without agitation, there is considerably less interfacial area available in the separator. We do not consider the separator operation in detail. It will often be no more than a holding tank large enough to allow separation by gravity. In some gas–liquid systems the separation takes place in the contactor and a distinct separation device is not needed. The contactor and the separator are often collectively referred to as a *stage*.

Continuous-flow systems operate at steady state so all time derivatives are zero, simplifying the analysis. The transient behavior of a process is important for problems concerning start-up, shutdown, and control, but technically feasible analysis and technically feasible design problems are solved with the steady-state model equations.

4.3.3.1 Mixed–Mixed Fluid Motions

Our control volumes will be the same as those designated for the batch two-phase systems, and we can develop the model equations by applying the law of conservation of mass in the same manner we did in Subsection 4.3.1. A mixed–mixed contactor is shown in Figure 4.7.

The Level I analysis is readily obtained:

$$C_A^I = C_{AF}^I + \lambda \left(C_{AF}^{II} - C_A^{II} \right),$$

$$\lambda = \frac{q^{II}}{q^I}. \tag{4.3.19}$$

The Level II analysis depends on the experimental determination of the concentration of component A in both phases when equilibrium has been reached in a batch experiment. For a liquid–liquid system, we can use the form we have used before:

$$C_{A,eq}^I = M C_{A,eq}^{II}$$

$$C_{A,eq}^{II} = \left(\frac{C_{AF}^I + \lambda C_{AF}^{II}}{\lambda + M} \right). \tag{4.3.20}$$

The component mass balance equations are

$$\frac{d}{dt} \left(C_A^I V^I \right) = q_F^I C_{AF}^I - q^I C_A^I - K_m a \left(C_A^I - M C_A^{II} \right), \tag{4.3.21}$$

$$\frac{d}{dt} \left(C_A^{II} V^{II} \right) = q_F^{II} C_{AF}^{II} - q^{II} C_A^{II} + K_m a \left(C_A^I - M C_A^{II} \right). \tag{4.3.22}$$

At steady state, this system reduces simply to two algebraic equations:

$$0 = q_F^I C_{AF}^I - q^I C_A^I - K_m a \left(C_A^I - M C_A^{II} \right), \tag{4.3.23}$$

$$0 = q_F^{II} C_{AF}^{II} - q^{II} C_A^{II} + K_m a \left(C_A^I - M C_A^{II} \right). \tag{4.3.24}$$

These equations are similar in form to those developed for continuous-flow mixed–mixed heat exchanger equations (3.3.22) and (3.3.23).

Together, these equations contain nine quantities: q^I, q^{II}, C_{AF}^I, C_{AF}^{II}, C_A^I, C_A^{II}, M, K_m, and a. Specification of any seven independent quantities allows us to determine the remaining two by solving the two algebraic equations. The situation is mathematically equivalent to the continuous-flow tank reactor problems we discussed in Chapter 2.

The solution for the effluent concentrations yields the following model equations:

$$C_A^I = \frac{(M + \eta) C_{AF}^I + M \lambda C_{AF}^{II}}{M + \lambda + \eta}, \tag{4.3.25}$$

$$C_A^{II} = \frac{C_{AF}^I + (\lambda + \eta) C_{AF}^{II}}{M + \lambda + \eta}, \tag{4.3.26}$$

$$\lambda = \frac{q^{II}}{q^I}, \quad \eta = \frac{q^{II}}{K_m a}.$$

Looking more closely at the parameter η, we see that it can be redefined as follows:

$$\eta^{II} = \frac{q^{II}}{K_m a} = \frac{1}{\tau^{II}} \left(\frac{V^{II}}{K_m a} \right), \tag{4.3.27}$$

where $\tau^{II} = V^{II}/q^{II}$ is the residence time of the dispersed phase. The equilibrium constant M will take on a value of the order of unity. We previously demonstrated that $V^{II}/K_m a$ will be of the order of several seconds if the agitation produces a dispersed phase with droplets of about a millimeter in diameter. If the system is designed for a holdup of the order of minutes, $\eta \ll M$, $\eta \ll \lambda$, and the equations reduce to

$$C_A^I = M \left(\frac{C_{AF}^I + \lambda C_{AF}^{II}}{\lambda + M} \right), \tag{4.3.28}$$

$$C_A^{II} = \left(\frac{C_{AF}^I + \lambda C_{AF}^{II}}{\lambda + M} \right). \tag{4.3.29}$$

This is the formal result of taking the limit as the product $K_m a$ approaches infinity. In other words, the holdup in the tank is significantly longer than the time required for mass transfer to take place. These equations are the natural result of the Level I and II analyses of the system for a liquid–liquid contactor. This allows us to refer to the contactor as an "equilibrium stage." The concept of the equilibrium stage is used extensively in design computations for the unit operations touched on in the introduction. In almost all chemical engineering programs there is one course devoted entirely to equilibrium stage design. Determining the number of stages to carry out a required separation is the goal of such courses, and it is an important part of mass contactor design. To complete the design of a stage its size and geometry must be specified. In the next subsection we present a method to estimate contactor volume.

4.3.3.2 Design of a Continuous Mixed–Mixed Mass Contactor

Given the importance of continuous-flow two-phase mass contactors in the chemical and allied industries, an example of how these model equations are used in the design process to estimate a mass contactor volume is subsequently illustrated.

A feasible design for the preceding contacting process requires the specification of three quantities:

- stream properties (flow rates, concentrations, etc.),
- power input (to yield the area for mass transfer),
- phase contactor vessel dimensions (which is related to the holdup).

We can estimate the required volume of a mass contactor to accomplish a given separation in a similar, but more complicated, procedure than that used to obtain a reactor volume in Chapter 2.

Equations (4.3.23) and (4.3.24) can be rearranged to obtain an expression for the interfacial area in terms of the outlet concentration of either phase:

$$a = \frac{q^I \left(C_{AF}^I - C_A^I \right)}{K_m \left(C_A^I - M C_A^{II} \right)} = \frac{q^{II} \left(C_A^{II} - C_{AF}^{II} \right)}{K_M \left(C_A^I - M C_A^{II} \right)}. \tag{4.3.30}$$

The preceding expression can be combined with Level I analysis [Eq. (4.3.19)] to obtain a as a function of the concentrations required in a particular phase. For example, to reduce the concentration of solute A in stream II to a given value, C_A^{II}, the required area for mass transfer is

$$a = \frac{q^{II} \left(C_A^{II} - C_{AF}^{II} \right)}{K_M \left(C_{AF}^I + \lambda \left(C_{AF}^{II} - C_A^{II} \right) - M C_A^{II} \right)}. \tag{4.3.31}$$

Equation (4.3.31) can be used to obtain an estimate of area a. If equilibrium is assumed the value of C_A^{II} can be obtained from a Level II analysis. If we then select a value of C_A^{II} close to $C_{A,eq}^{II}$ the area can be estimated and a contactor volume obtained.

Equation (4.3.28) can be expressed in terms of a stage efficiency. The Level I analysis [Eq. (4.3.19)] and the equilibrium relation $C_A^I = MC_A^{II}$ can be combined to show this.

The denominator in Eq. (4.3.30) can be rewritten by incorporating $C_{A,eq}^I$ and $C_{A,eq}^{II}$:

$$C_A^I - MC_A^{II} = C_A^I - C_{A,eq}^I - M\left(C_A^{II} - C_{A,eq}^{II}\right).$$

The Level I analysis, Eq. (4.3.19), becomes

$$C_A^{II} = C_{AF}^{II} - \frac{1}{\lambda}\left(C_A^I - C_{AF}^I\right),$$

$$C_{A,eq}^{II} = C_{AF}^{II} - \frac{1}{\lambda}\left(C_{A,eq}^I - C_{AF}^I\right);$$

subtraction gives

$$C_A^{II} - C_{A,eq}^{II} = -\frac{1}{\lambda}\left(C_A^I - C_{A,eq}^I\right),$$

$$\therefore C_A^I - MC_A^{II} = C_A^I - C_{A,eq}^I + \frac{M}{\lambda}\left(C_A^I - C_{A,eq}^I\right)$$

$$= \left(1 + \frac{M}{\lambda}\right)\left(C_A^I - C_{A,eq}^I\right).$$

Consequently we find the following equations for the area required for mass transfer:

$$a = \frac{q^{II}}{K_m\left(\lambda + M\right)}\left(\frac{C_A^I - C_{AF}^I}{C_{A,eq}^I - C_A^I}\right), \qquad (4.3.32)$$

$$a = \frac{q^{II}}{K_m\left(\lambda + M\right)}\left(\frac{C_A^{II} - C_{AF}^{II}}{C_{A,eq}^{II} - C_A^{II}}\right). \qquad (4.3.33)$$

In both the preceding equations, the terms in parentheses can be rewritten in terms of the *efficiency of separation*, defined as

$$\chi = \left(\frac{C_A^I - C_{AF}^I}{C_{A,eq}^I - C_{AF}^I}\right) = \left(\frac{C_{AF}^{II} - C_A^{II}}{C_{AF}^{II} - C_{A,eq}^{II}}\right). \qquad (4.3.34)$$

The efficiency is a ratio of the concentration change in the device to the maximum concentration change possible in a full equilibrium stage—it is just the fraction of complete equilibrium achieved. Equation (4.3.34) can be rearranged to give the interfacial area in terms of the efficiency of the stage:

$$a = \frac{q^{II}}{K_m\left(\lambda + M\right)}\left(\frac{\chi}{1 - \chi}\right) \qquad (4.3.35)$$

Because energy in the form of agitation must be put into two-phase systems to generate adequate interfacial area for mass transfer, it is evident that there will be situations for which equilibrium will not be reached in a stage. This has led to attempts to correlate $K_m a$ with factors like the power per kilogram in order to compute the effluent composition. One quantity often used is the stage efficiency defined in Eq. (4.3.34). When the stage efficiency approaches one, the effluent concentration approaches that given by the Level II analysis, the equilibrium concentration.

Manipulation of Eq. (4.3.35) gives the following relationship between the stage efficiency and the mass transfer coefficient:

$$\chi = \frac{M + \lambda}{M + \lambda + q^{II}/K_m a}. \tag{4.3.36}$$

Notice that, in assigning the efficiency, and consequently the interfacial area, the outlet concentrations of each phase are defined. Because the flow rates and density of each phase were assumed constant, all the stream properties are set at this point.

Defining the area does not specify the vessel size or the power input directly. The interfacial area can be written in terms of the number of drops (which are approximately spherical) and the effective drop size of the dispersed phase:

$$a = N a_{drop} \approx N \pi d^2. \tag{4.3.37}$$

The total volume occupied by the dispersed phase, meanwhile, can be written similarly as

$$V^{II} = N V_{drop}^{II} \approx N \frac{\pi}{6} d^3. \tag{4.3.38}$$

Combining (4.3.37) and (4.3.38), we obtain

$$V^{II} = \frac{1}{6}(a)(d). \tag{4.3.39}$$

The mean drop size can be estimated from any of a number of correlations available in literature, such as discussed in Chapter 7, and generally depends on the energy per volume imparted to the system. For gas bubbles in water, the diameter of the bubbles approaches a limiting value of about 4.5 mm.

The total volume of the tank is the sum of the volume of each phase. Given the assumption of homogeneity in a mixed–mixed tank, both phases will have the same residence time in the tank. Consequently, the holding times of each phase are equal, which provides the necessary relationship between the phase volumes and the flow rates, as

$$\tau^I = \frac{V^I}{q^I} = \frac{V^{II}}{q^{II}} = \tau^{II}. \tag{4.3.40}$$

Rearrangement gives $V^I = \lambda^{-1} V^{II}$, and the total tank volume is then

$$V = (V^I + V^{II}) = V^{II}(1 + \lambda^{-1}). \tag{4.3.41}$$

In practice, it is wise to overdesign the tank to account for uncertainties and to ensure that process safety and control issues can be addressed. Once the volume of the contactor has been determined, the geometry of the mixing vessel must be considered to determine the tank dimensions. Cylindrical tanks are the most common choice, and consideration of proper mixing generally restricts the height of the tank to between one and three times the diameter of the tank. This completes the specification of the technically feasible design.

EXAMPLE 4.3 A process unit has produced a 100-L/min water stream containing 20 g/L of acetone. The acetone concentration must be reduced to 2 g/L to be acceptable for the EPA to approve it for discharge into natural waters. We can assume that it is not feasible to modify the process producing the water contaminated with acetone to an acetone level below 20 g/L. There are 100 L/min of TCE available. This must be carefully handled so there is no operator contact. There is no acetone in the TCE so $C_{AF}^I = 0$.

SOLUTION. Laboratory studies have provided the equilibrium data shown in Figure 4.2. M is essentially 2 so $C_A^I = 2C_A^{II}$. The laboratory experiment indicates that the water–acetone phase is dispersed (Phase II) and the TCE is the continuous phase (Phase I).

A Level I analysis provides a relationship between C_A^I and C_A^{II}:

$$C_A^I = C_{AF}^I + \frac{q^{II}}{q^I} \left(C_{AF}^{II} - C_A^{II} \right),$$

therefore,

$$C_A^I = \left(20 - C_A^{II} \right).$$

Following the analysis presented before, we assume that the amount of acetone transferred will not affect the flow rates q^I and q^{II}.

A Level II analysis determines the value of acetone in the water that can be achieved in a single equilibrium stage, as shown in Figure E4.1.

Recall that the stage consists of both the mass contactor and a separator. The Level II model equation yields the maximum extraction possible:

$$C_{A,eq}^{II} = \frac{C_{AF}^I + \lambda C_{AF}^{II}}{\lambda + M}, \quad C_{AF}^I = 0, \quad \lambda = 1,$$

$$= \frac{20}{1 + 2} = 6.7 \text{ g/L}.$$

A single stage with a TCE flow rate of 100 L/min will not meet the required value of $C_A^{II} = 2.0$ g/L. If we double the flow rate of TCE to 200 L/min, then,

$$\lambda = \frac{q^{II}}{q^I} = \frac{100}{200} = \frac{1}{2},$$

$$C_A^{II} = \frac{\lambda C_{AF}^{II}}{\lambda + M} = \frac{1/2 \times 20}{1/2 + 2}$$

$$= 4.0 \text{ g/L}.$$

Doubling the flow rate of TCE reduces the acetone in the water only from 6.7 to 4 g/L.

The analysis in Example 4.3 illustrates that we need to consider the use of more stages if we are to reduce C_A^{II} to its acceptable value of 2.0 g/L. For example, if we consider three stages (Figure E4.2) with a flow of 100 L/min of TCE to each stage,

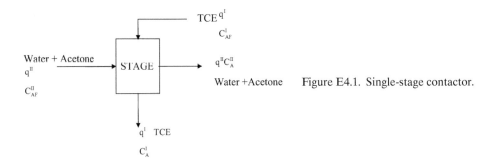

Figure E4.1. Single-stage contactor.

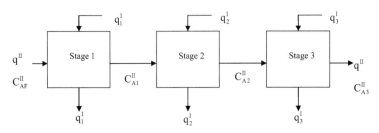

Figure E4.2. Three-stage cross-flow contactors.

$$C_{A2}^{II} = \frac{C_{A1}^{II}}{\lambda + M} = \frac{6.7}{1 + 2}$$

$$= 2.2 \text{ g/L},$$

$$C_{A3}^{II} = \frac{C_{A2}^{II}}{\lambda + M}$$

$$= \frac{2.2}{3}$$

$$= 0.74 \text{ g/L}.$$

The three-stage unit just shown is often described as a cross-flow extraction process. There are other ways of contacting the TCE and water–acetone solution, but before examining them it is useful to get an estimate of the volume of a single stage.

A Level IV analysis, to obtain a technically feasible design, requires a selection of a C_A^{II} value greater than the equilibrium value obtained from the Level II analysis. Taking 90% of the equilibrium stage concentration, $C_{A1}^{II} = 6.7/0.90 = 7.4$ g/L. Substituting into Eq. (4.3.31) yields

$$a = \frac{\left(\dfrac{100}{60 \times 1000} \dfrac{\text{m}^3}{\text{s}}\right)(7.4 - 20 \text{ kg/m}^3)}{(K_m \text{ m/s})[0 + 1(-7.4 + 20) - 2(7.4) \text{ kg/m}^3]}.$$

We can assume as a first approximation that $K_m \approx 3 \times 10^{-5}$ m/s, the value obtained in the salt-tablet experiment; therefore a = 318 m².

The volume of the mass contactor part of the separation stage in terms of drop diameter d is obtained with Eq. (4.3.39), which yields $V^{II} = 0.167(318)(d) = 53d$.

The following table shows V^{II} and V as functions of drop diameter.

d	V^{II}	V	$\dfrac{V^{II}}{q^{II}} = \dfrac{V^{I}}{q^{I}}$
1 mm	53 L	106 L	31 s
1 cm	530 L	1060 L	310 s

We must determine the optimum drop size by considering the separator design. It is easier to separate large drops of water from TCE than to separate small drops. There thus is an economic balance: Small drops lead to a small-volume contactor but a large-volume separator. Large drops increase the contactor volume but decrease the separator volume.

EXAMPLE 4.3 CONTINUED. If 200 L/min of TCE is used, it is clear that we obtain a greater reduction of acetone by using two stages and splitting the TCE to each stage rather than by using one stage:

$$C_{A1}^{II} = 6.7 \ \text{g/L} \quad \text{(1 stage)},$$
$$C_{A2}^{II} = 2.2 \ \text{g/L} \quad \text{(2 stages)}.$$

An optimum design for this acetone reduction mass requires a good deal more information as to both capital and operating costs, but the technically feasible design is a valuable first step. Calculate the acetone removal for two stages in a countercurrent arrangement.

SOLUTION. The amount of acetone in the water can be reduced even further if we use the TCE from the second stage in the first stage. This is a countercurrent arrangement of the stages, as shown in Figure E4.3.

The Level II analysis is readily developed with Eq. (4.3.20) used for the single stage and adapting it for the countercurrent stage:

$$C_A^{II} = \frac{C_{AF}^{I} + \lambda C_{AF}^{II}}{\lambda + M}.$$

Stage 1 countercurrent:

$$C_{A1}^{II} = \frac{C_{A2}^{I} + \lambda C_{AF}^{II}}{\lambda + M}.$$

Stage 2 countercurrent:

$$C_{A2}^{II} = \frac{C_{AF}^{I} + \lambda C_{A1}^{II}}{\lambda + M},$$

$$C_{A2}^{II} = \frac{C_{AF}^{I}}{\lambda + M} + \frac{\lambda}{(\lambda + M)^2} \left(C_{A2}^{I} + \lambda C_{AF}^{II}\right),$$

$$C_{A2}^{I} = M C_{A2}^{II}.$$

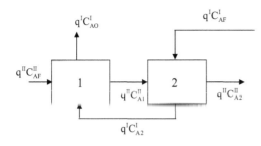

Figure E4.3. Two-stage countercurrent operation.

Equilibrium in stage 2 occurs when the exit concentration of acetone in the TCE, C_{A2}^I, is equal to the exit concentrations of acetone in the water, C_{A2}^{II}:

$$C_{A2}^{II} = \frac{C_{AF}^I}{\lambda + M} + \frac{\lambda M}{(\lambda + M)^2} C_{A2}^{II} + \frac{\lambda^2}{(\lambda + M)^2} C_{AF}^{II}.$$

In our case $C_{AF}^I = 0$; therefore,

$$C_{A2}^{II} \left[1 - \frac{\lambda M}{(\lambda + M)^2} \right] = \frac{\lambda^2}{(\lambda + M)^2} C_{AF}^{II},$$

$\lambda = 0.5$, $M = 2$, and $C_{AF}^{II} = 20$, and we find

$$C_{A2}^{II} = 0.95 \ \text{g/L}.$$

Using 200 L/min of TCE in two stages with a countercurrent piping arrangement, as shown in Figure E4.3, reduces the acetone in the water to less than half of that obtained in the two-stage cross flow.

The use of countercurrent staging of mass contactors provides advantages in efficiency, as shown by the previous example, and so this piping arrangement is often desired. Figure 4.8 illustrates an operating diagram for the two-equilibrium-stage countercurrent piping arrangement discussed in Example 4.3. The operating lines that define the stage operations can be drawn on a plot of concentration in Phase II versus that in Phase I, which includes the equilibrium line, as sketched in Figure 4.8. If each state reaches equilibrium, the exit conditions will be on the locus defined by the equilibrium line. Because mass transfer ceases when equilibrium is reached, the operating lines cannot cross the equilibrium line. The inlet to stage 1 is indicted by the solid square, where the feed of the process stream (II) is rich in component A. The operating lines for each stage are given by the Level I balance for each stage, where the slope is the ratio of flow rates. The exit of the second stage is indicated by the dot, which yields the maximum reduction in component A possible with two stages and the given flow rates. Such operating diagrams can be useful for understanding the operation of multistage mass contacting equipment.

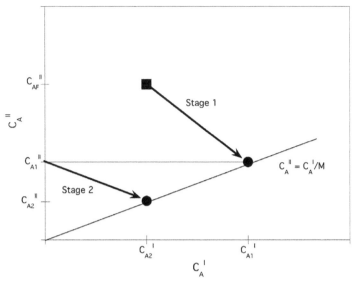

Figure 4.8. Operating diagram for two-equilibrium-stage countercurrent mass contactors. (The dots indicate the stage outlets, the filled square indicates the process inlet, and the diagonal line is the equilibrium line.)

4.3.3.3 Mixed–Plug Fluid Motions

Plug–mixed fluid motions are encountered only with gas–liquid systems both in semi-batch and continuous-flow mass contactors. Very often they are employed when one of the raw materials for a chemical reaction is a gas at the temperature and pressure required for the reactor. Such "reactors" are almost always mass transfer controlled and can be analyzed as a mass contactor, although model equations can be readily developed to account for both transfer into the liquid phase and reaction in the liquid phase. A typical mass contactor–reactor is shown in Figure 4.9.

A very common application is aeration of both domestic and industrial waste treatment ponds to which oxygen is supplied to ensure effective microorganism growth so that the organic waste can be oxidized. Figures 4.10(a) and 4.10(b) show the basic fluid motions in such systems. As can be seen from the photographs, the plug–mixed unit consists of a rising plume of gas bubbles with a substantial liquid flow surrounding the plume. At the surface the gas, in this case air, separates and flows into the air space above the surface. The liquid flows radially until it meets the liquid flowing from adjacent plumes when it turns and flows downward. There is thus liquid circulating in each plume with a flow into and out of each plume to accommodate the flow in and out of the treatment pond.

Industrially important gas–liquid reactions include chlorination of various hydrocarbons such as benzene. Chlorine gas is also widely used to purify drinking water supplies.

A Level I analysis can be derived as follows:

$$q_F^I C_{AF}^I + q_F^{II} C_{AF}^{II} = q^I C_A^I + q^{II} C_A^{II}.$$

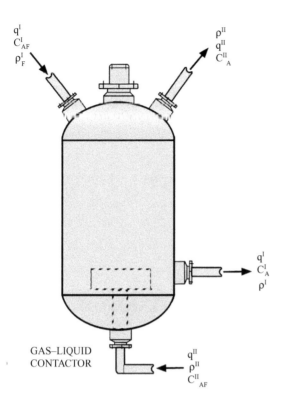

Figure 4.9. Continuous-flow gas–liquid mass contactor.

GAS–LIQUID
CONTACTOR

A Level II analysis follows when we assume that $C_A^I = MC_A^{II}$:

$$C_A^I = \frac{MC_{AF}^I + \lambda MC_{AF}^{II}}{M + \lambda} \quad C_A^{II} = \frac{C_{AF}^I + \lambda C_{AF}^{II}}{\lambda + M}.$$

The Level IV component balance for the liquid phase at steady state can be written as

$$0 = q_F^I C_{AF}^I - q^I C_A^I - K_m a \left(C_A^I - M \langle C_A^{II} \rangle \right). \tag{4.3.42}$$

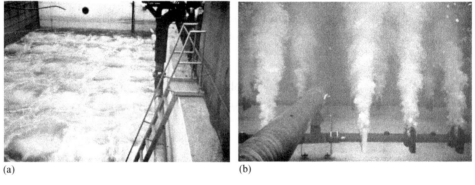

(a)   (b)

Figure 4.10. (a) Surface of industrial-scale air oxidation unit (Otera, 1983). (b) Underwater photograph showing gas–liquid plumes. (From Otero Z, Tilton JN, Russell TWF. Some observations of flow patterns. Int. J. Multiphase Flow 1985; 11: 584. Reproduced with permission.)

For the dispersed gas phase, the steady-state balance is

$$\frac{\partial}{\partial z}\left(q^{II}\,C_A^{II}\right) = K_m\,a_v\,A_c\left(C_A^I - MC_A^{II}\right).\tag{4.3.43}$$

These equations should not be unfamiliar; Eq. (4.3.43) is derived from Eq. (4.3.12) as they have been used in previous systems. The liquid-phase balance is only a slight variation on what we have seen thus far—it combines elements from earlier in the chapter. Again, $\langle C_A^{II}\rangle$ is just the spatial average value of C_A^{II}. The balance for the gas phase comes from the semibatch process discussed in Subsection 4.3.2.

Because both phases are operating at steady state and the continuous phase is well mixed, the concentration of species A in that phase is independent of time or space. Thus Eq. (4.3.43) is readily separable. The result can then be substituted back into (4.3.42) and solved for the other effluent concentration. The model equations for this system are

$$C_A^I = M\left[\frac{C_{AF}^I + \lambda(1 - e^{-\beta L})C_{AF}^{II}}{M + \lambda(1 - e^{-\beta L})}\right],$$

$$C_A^{II}(z) = \left[\frac{C_{AF}^I + \lambda(1 - e^{-\beta L})C_{AF}^{II}}{M + \lambda(1 - e^{-\beta L})}\right] - \left[\frac{C_{A,F}^I - MC_{AF}^{II}}{M + \lambda(1 - e^{-\beta L})}\right]e^{-\beta z},\tag{4.3.44}$$

where $\beta = M\frac{K_m a}{q^{II}}\frac{1}{L}$ and $\lambda = \frac{q^{II}}{q^I}$.

For the case in which there is a negligible change in C_A^{II} as the gas passes through the liquid, Eq. (4.3.42) becomes

$$0 = q_F^I\,C_{AF}^I - q^I C_A^I - K_m a\left(C_A^I - MC_A^{II}\right).\tag{4.3.45}$$

In the limit of large $K_m a$, the model equations reduce to the equilibrium stage results obtained earlier, Eq. (4.3.20). It is reassuring to verify that the fluid motions of the dispersed phase (well-mixed or plug flow) do not affect the equilibrium result, as the Level II analysis that gives the equilibrium stage result does not distinguish between these two schemes.

If we have the commonly encountered situation that we discussed in the semibatch mixed–plug case of a slightly soluble gas, then Eq. (4.3.42) simplifies to

$$0 = q_F^I\,C_{AF}^I - q^I C_A^I - K_m a\left(C_A^I - C_{A,eq}^I\right),\tag{4.3.46}$$

$$C_A^I = \frac{q_F^I C_{AF}^I}{q^I + K_m a} + \frac{K_m a C_{A,eq}^I}{q^I + K_m a},\tag{4.3.47}$$

For $q_F^I = q^I$,

$$C_A^I = \frac{C_{AF}^I}{1 + \dfrac{K_m a}{q^I}} + \frac{C_{A,eq}^I}{K_m a - \dfrac{1}{q^I}}.$$

EXAMPLE 4.2 CONTINUED. The vessel described in Example 4.2 is to be operated continuously to handle a flow of 340 L/min of water that contains no oxygen, $C_{AF}^I = 0$. The air sparge rate is as it was in the semibatch experiment, $q^{II} = 3.9 \times 10^{-3}$ m^3/s.

How much oxygen can be transferred to this water stream if its temperature is 20 °C?

SOLUTION. Equation (4.3.47) applies. Table 4.3 shows that $C_{A,eq} = 8 \times 10^{-3}$ kg/m^3. We find

$$C_A^I = 0 + \frac{(K_m a)\, 8.9 \times 10^{-3}\ \text{kg/m}^3}{0.0057\ \text{m}^3/\text{s} + K_m a}.$$

From the batch experiment we found

$$\frac{K_m a}{V^I} = 0.0007.$$

Because $V_I = \pi(6.0\,\text{m})[\frac{1.4\,\text{m}}{2}]^2 = 9.24\,\text{m}^3$, we calculate $K_m a = 0.0065$ and

$$C_A^I = \frac{0.0065}{0.0122}\,(8.9 \times 10^{-3})\ \text{kg/m}^3$$

$$= 4.8 \times 10^{-3}\ \text{kg/m}^3.$$

Note that the stage efficiency of this design is given by Eq. (4.3.34) as $\chi = 0.6$, indicating that there is room for improvement.

4.4   Tubular Two-Phase Mass Contactors

Tubular contactors come in two general classes, those that use membranes to separate the two fluids and control the area of contact, or those that directly contact two immiscible fluids. Membrane contactors can be analyzed with the model equations considered here, but additional issues arise because of mass transfer across the membrane used to separate the phases. Hence these are considered later in the book. Of particular interest here are contactors in which two immiscible phases are brought into direct contact as they travel through a pipe. These contactors are the mass transfer analog of the plug-flow reactor for chemical reaction and the shell-and-tube heat exchanger. Tubular mass contactors can be used with solid–fluid systems, liquid–liquid systems, and gas–liquid systems. Two-phase fluid mechanics can be quite complex and is an active research area today. A guide to the complex fluid mechanics of such flows is provided in the book *Flow of Complex Mixtures in Pipes* by Govier and Azis (1972). Both horizontal and vertical tubular systems can be used to transfer a species from one phase to another, but vertical countercurrent tubular towers for gas–liquid systems are the most common tubular system in use. Such towers can have "internals," such as trays, or they may be filled with different proprietary packing. Their design and operation have been the subject of considerable research that continues to this day.

In parallel with our organization of heat exchangers we first discuss cocurrent mass contactors. They can be either horizontal or vertical, and fluids can be processed in a pipe with or without internals. They are not commonly used for solid–fluid systems but can have definite advantages for liquid–liquid or gas–liquid systems.

Solid–fluid tubular cocurrent mass contactors require substantial fluid velocities to keep solids in suspension. This can be a severe disadvantage if a large resident time is required for the transfer to take place. It is usually more practical to carry out solid–fluid contacting in a tank-type contactor.

The drop size in liquid–liquid systems can be controlled by the fluid velocities in both vertical and horizontal configurations. This allows for some clever design and operation if the contactor and separator can be combined in one pipe system of varying diameters: small pipe diameters for area creation and larger pipe diameters for separation. As shown in Subsection 4.2.2, one can achieve mass transfer close to phase equilibrium in rather short times so it is not necessary to have very small drop sizes (such as those smaller than 0.5 mm). Drop size can be controlled by the power input to the tubular contactor, and a method for estimating drop size and thus interfacial area will be presented in Chapter 7.

Gas–liquid flows in pipes also produce a variety of flow patterns. A bubble flow pattern is usually the most efficient for mass transfer from the gas to the liquid phase, and this can generally be achieved with liquid velocities of about 3 m/s for low gas-to-liquid ratios and low-viscosity liquids. At high gas velocities, over 100 m/s, a flow pattern of liquid drops in the gas stream can be achieved. In between these extremes there can be stratified, slug, and annular flows in horizontal pipes. Typical flow pattern maps are shown in Govier and Azis (1972). As with liquid–liquid systems a separation can be achieved by changing the contactor diameter.

Countercurrent flow can be achieved in tubular vertical mass contactors that act both as a separator and mass contactor. The vertical mass contactor almost always has trays or packing and is designed and operated for gas–liquid systems. If a liquid–liquid system has sufficient density differences it is possible to have a vertical tubular contactor. There are critical constraints on the flow of both fluids to maintain countercurrent flow and to achieve uniform distribution of the two phases, and these will be discussed in Chapter 8. Continuous solid–fluid tubular countercurrent vertical contactors are rare. It is not possible to have a horizontal countercurrent tubular mass contactor unless the two fluids are separated by a membrane.

As with plug-flow heat exchangers and reactors, the individual fluids are expected to be well mixed radially, with the concentration of solute in each phase varying in only the axial direction. For a number of multiphase flow patterns we can obtain useful results by assuming that both fluids approximate plug flow in developing the model equations. Predicting the interfacial area requires a more detailed understanding of two-phase fluid mechanics, and this will be covered in Chapter 7.

This section focuses on the development of the model equations describing both cocurrent and countercurrent flow in a tubular mass contactor. The properties of these equations are also discussed in the context of equilibrium, and the model behavior is shown. In keeping with our strategy, we assume that the area for mass contact and the mass transfer coefficient are given or can be obtained from a suitable

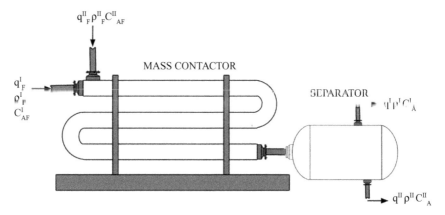

Figure 4.11. Tubular horizontal mass contactor and separator.

experiment. Considerations of area generation and estimation of the mass transfer coefficient will be discussed in Part II.

4.4.1 Cocurrent Flow

We start our analysis of the cocurrent contactor shown in Figure 4.11 at Level I by considering the inputs and outputs of the tubular mass contactor and separator:

$$q_F^I C_{AF}^I + q_F^{II} C_{AF}^{II} = q^I C_A^I + q^{II} C_A^{II}. \tag{4.4.1}$$

As the area available for mass transfer increases, the outlet concentrations approach a limiting value. A Level II analysis for cocurrent operation yields the following limiting values:

$$C_{A,\,eq}^I = M\left(\frac{C_{AF}^I + \lambda\, C_{AF}^{II}}{\lambda + M}\right), \quad C_{A,\,eq}^{II} = \left(\frac{C_{AF}^I + \lambda\, C_{AF}^{II}}{\lambda + M}\right). \tag{4.4.2}$$

The two streams approach equilibrium with each other as they travel down the length of the contactor such that $C_{A,eq}^I = M C_{A,eq}^{II}$. Again, this is the same equilibrium result as we derived for the previous two continuous-flow mass contactors.

The law of conservation of mass can be applied to the differential control volumes depicted in Figure 4.12. As we have done throughout this chapter, a mass balance is considered for both the total mass and the mass of species A in each phase.

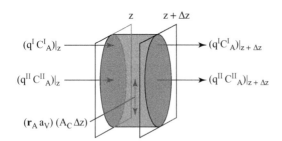

Figure 4.12. Cross-sectional slice of a cocurrent mass contactor.

For situations in which the mass transferred is small relative to the overall mass flows, the overall mass balances can be neglected, and the species balance yields

$$\frac{\partial}{\partial t}\left(\psi^{I} A_c C_A^{I}\right) = -\frac{\partial}{\partial z}\left(q^{I} C_A^{I}\right) - K_m a_v A_c \left(C_A^{I} - M C_A^{II}\right), \qquad (4.4.3)$$

$$\frac{\partial}{\partial t}\left(\psi^{II} A_c C_A^{II}\right) = -\frac{\partial}{\partial z}\left(q^{II} C_A^{II}\right) + K_m a_v A_c \left(C_A^{I} - M C_A^{II}\right), \qquad (4.4.4)$$

where ψ^{I} is the average fraction of the cross section occupied by Phase I, ψ^{II} is the average fraction of the cross section occupied by Phase II, and a_v is the interfacial area *per unit volume of the vessel*. If we also take the volumetric flow rates as constant, and restrict the analysis to *steady-state* operation, the component balances and their initial conditions reduce to

$$q^{I}\frac{dC_A^{I}}{dz} = -\frac{K_m a}{L}\left(C_A^{I} - M C_A^{II}\right) \qquad C_A^{I}(z=0) = C_{AF}^{I}, \qquad (4.4.5)$$

$$q^{II}\frac{dC_A^{II}}{dz} = \frac{K_m a}{L}\left(C_A^{I} - M C_A^{II}\right) \qquad C_A^{II}(z=0) = C_{AF}^{II}. \qquad (4.4.6)$$

The solutions of these equations, subsequently shown, are obtained in a straightforward manner outlined earlier for cocurrent heat exchangers:

$$C_A^{I} = \frac{M}{\lambda + M}\left(C_{AF}^{I} + \lambda C_{AF}^{II}\right) + \frac{\lambda}{\lambda + M}\left(C_{AF}^{I} - M C_{AF}^{II}\right)e^{-\beta z}, \qquad (4.4.7)$$

$$C_A^{II} = \frac{1}{\lambda + M}\left(C_{AF}^{I} + \lambda C_{AF}^{II}\right) - \frac{1}{\lambda + M}\left(C_{AF}^{I} - M C_{AF}^{II}\right)e^{-\beta z}, \qquad (4.4.8)$$

where

$$\beta = \frac{K_m a}{q^{II} L}(\lambda + M), \quad \lambda = \frac{q^{II}}{q^{I}},$$

$$\beta = \frac{K_m a}{q^{II}}\left(\frac{q^{II}}{q^{I} M} + 1\right)\frac{M}{L}.$$

4.4.2 Countercurrent Flow

Countercurrent flow in a mass contactor can occur only in vertical devices with phases that have significant density differences or in membrane contactors. Solid–fluid contactors operate with the solid dispersed phase falling through an upward-flowing liquid or gas. The fluid velocity must be kept low enough to allow the solid particles to move from the top of the device where they are introduced to the bottom where they are removed. A plug-flow contactor in which liquid drops are allowed to flow by gravity from top to bottom has the same limitations.

In a contactor that has the gas rising in bubbles through a downward-flowing liquid, the liquid velocity must be kept well below the rise velocity of the bubble swarm. The need to maintain both a low fluid velocity and uniform distribution of solids, drops, and bubbles can severely limit the capacity of countercurrent plug-flow mass contactors.

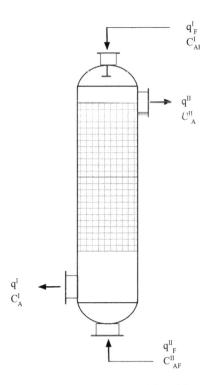

q_F^I
C_{AF}^I

q^{II}
C_A^{II}

q^I
C_A^I

q_F^{II}
C_{AF}^{II}

Figure 4.13. Tubular vertical mass contactor (packed tower).

Scrubbers, fluidized beds, spray towers, and wetted wall columns are the common terms for some countercurrent mass contactors. Design and operation to provide the appropriate area for mass transfer and avoid problems such as "slugging," "streaming," "entrainment," unsteady flow behavior, and other related problems are nontrivial.

Horizontal devices with countercurrent flow are feasible only if the two phases are separated by a membrane. Transport through membranes will be treated in Chapter 5, where the constitutive equation for the rate of mass transfer will be developed further. The model equations for a membrane mass contactor follow from the analysis provided here.

The Level I analysis for the device sketched in Figure 4.13 is

$$q_F^I C_{AF}^I + q_F^{II} C_{AF}^{II} = q^{II} C_A^{II} + q^I C_A^I.$$

Once again, the model equations are derived from the law of conservation of mass applied to species A (the solute) as per the differential control volumes in Figure 4.14.

The Level II analysis parallels that presented and discussed in detail for the countercurrent heat exchanger in Subsection 3.4.2. The most relevant situation for the mass contactor is, for a given process stream (stream I), to determine the minimum utility flow rate required for a desired amount of removal. For an infinitely long tower, $C_{AF}^I = M C_A^{II}$, and the minimum utility stream flow rate is given by the Level I analysis, given above, which can be rearranged to yield

$$q_{min}^{II} = q^I \frac{\left(C_{AF}^I - C_A^I\right)}{\left(M^{-I} C_{AF}^I - C_{AF}^{II}\right)}. \qquad (4.4.9)$$

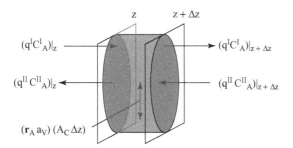

Figure 4.14. Cross-sectional slice of a counter-current mass contactor.

The other possible cases can be derived similarly:

$$\frac{\partial}{\partial t}\left(\psi^{I} A_{c} C_{A}^{I}\right) = -\frac{\partial}{\partial z}\left(q^{I} C_{A}^{I}\right) - K_{m}a_{v} A_{c}\left(C_{A}^{I} - MC_{A}^{II}\right), \qquad (4.4.10)$$

$$\frac{\partial}{\partial t}\left(\psi^{II} A_{c} C_{A}^{II}\right) = \frac{\partial}{\partial z}\left(q^{II} C_{A}^{II}\right) + K_{m}a_{v} A_{c}\left(C_{A}^{I} - MC_{A}^{II}\right). \qquad (4.4.11)$$

Making the same assumptions used in the cocurrent derivation (for low solute concentrations we can neglect the overall mass balances and assume the flow rates do not change appreciably), the steady-state differential component balances and their respective initial conditions become

$$q^{I}\frac{dC_{A}^{I}}{dz} = -\frac{K_{m}a}{L}\left(C_{A}^{I} - MC_{A}^{II}\right) \quad C_{A}^{I}(z=0) = C_{AF}^{I}, \qquad (4.4.12)$$

$$q^{II}\frac{dC_{A}^{II}}{dz} = -\frac{K_{m}a}{L}\left(C_{A}^{I} - MC_{A}^{II}\right) \quad C_{A}^{II}(z=L) = C_{AF}^{II}. \qquad (4.4.13)$$

These model equations are the same as those derived for the cocurrent flow except for the sign of the rate expression in the model equation for Phase II and a change in the boundary condition for Phase II [compare Eq. (4.4.6) with Eq. (4.4.13)]. This change in boundary conditions complicates the solution of the equations, as discussed previously with reference to countercurrent heat exchangers. The analytic solution (for the case $\lambda \neq M$) is

$$C_{A}^{I} = M\left(\frac{\lambda C_{AF}^{II} - e^{-\beta L}C_{AF}^{I}}{\lambda - Me^{-\beta L}}\right) + \lambda\left(\frac{C_{AF}^{I} - MC_{AF}^{II}}{\lambda - Me^{-\beta L}}\right)e^{-\beta z}, \qquad (4.4.14)$$

where

$$\beta = \frac{K_{m}a}{q^{II}L}\left(\lambda - M\right), \lambda = \frac{q^{II}}{q^{I}}, \text{ such that } \beta = \frac{K_{m}a}{q^{II}}\left(\frac{q^{II}}{q^{I}M} - 1\right)\frac{M}{L}.$$

C_{A}^{II} is given by the Level I balance written on a control volume from one end of the exchanger to location z.

Although the equilibrium analysis for tubular contactors is useful, it is also important to understand the behavior of the model equations when equilibrium conditions are not achieved in the contactor. The heat exchanger analysis in Chapter 3 is a good place to start, as the model equations for the heat exchangers are very similar to those of the mass contactors. If the equilibrium relation is taken to be $M = 1$, then the equations for mass contactors are identical in form to those describing the equivalent heat exchanger. Moreover, if the fluids on both sides of the heat exchanger are the same, the thermal capacity ratio term (Θ) discussed in Chapter 3 reduces to λ. However, the more general equilibrium relationship in mass contactors creates an additional level of complexity over heat exchangers. Consequently, some new features are possible in the mass contactor model equations. Specifically, the concentration profiles of the two phases may cross over one another depending on the values of λ and M; this can happen for either cocurrent or countercurrent flows. This behavior is shown in Example 4.4.

The final point to be made is that if the countercurrent operation is possible it will always be as good as and nearly always better than the equivalent cocurrent operation. This can be seen in the following example. It may, however, be impossible to achieve an equivalent countercurrent operation because of two-phase fluid mechanical limitations.

Cocurrent plug-flow mass contactors may generate more interfacial areas or be more practical to implement. It may also be more effective to consider tank-type contactors.

EXAMPLE 4.4. A tubular mass contactor operating with the properties subsequently listed is used to continuously remove acetone from water by contacting with TCE. Compare the concentration profiles and performance of a tubular mass contactor operating in cocurrent and countercurrent operation.

> Data: The flow rates of water (II) and TCE (I) are 100 L/min. The droplet size and contactor volume are such that the contact area is 50 m^2. Further, $K_m = 3.3 \times 10^{-5}$ m/s. The equilibrium relationship is M = 2 at the concentrations of interest (see Figure 4.2).

SOLUTION. A Level IV analysis for cocurrent operation is given in Eqs. (4.4.7) and (4.4.8). This analysis can be readily calculated to yield the concentration profiles shown in Figure E4.4(a).

Notice that the concentration profiles look similar to the temperature profiles observed in the cocurrent heat exchanger, shown in Figure E3.5. However, these concentration profiles cross in the contactor. This is a consequence of the distribution coefficient M = 2. To make a clearer analogy to heat transfer, the results are plotted as C_A^I and $2C_A^{II}$ in Figure E4.4(b). The driving force for mass transfer at any position (z/L) in the contactor is $(C_A^I - MC_A^{II})$, which is the vertical difference between the curves. As seen, the driving force for mass transfer approaches zero near the exit of the contactor.

Countercurrent operation, given by Eq. (4.4.14) and the Level I balance, results in the concentration profiles given in Figure E4.5(a). Note that the direction of stream I is held constant in this example, whereas process stream II is reversed in direction. The concentration profiles are similar to those observed in the countercurrent heat exchanger, Figure E3.5. The driving force for mass transfer is illustrated by the plots of C_A^I and $2C_A^{II}$ in Figure E4.5(b).

Comparison of the performance of the cocurrent and countercurrent mass contactors illustrates that, for a given contactor size, flow rates, and area for mass transfer,

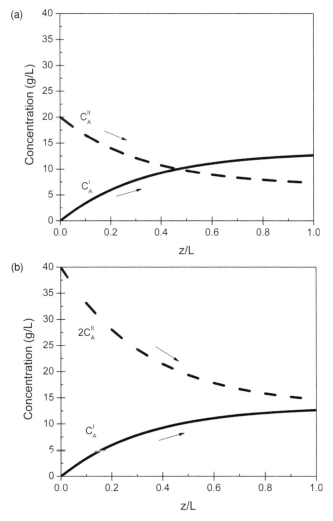

Figure E4.4. (a) Concentration profiles in the cocurrent mass contactor; (b) the profile accounting for M.

the countercurrent operation yields a larger average driving force for mass transfer and, hence, better performance. The amount of acetone in the exit water stream is calculated to be

	C_A^I (g/L)	C_A^{II} (g/L)	$(C_{AF}^{II} - C_A^{II})$ (g/L)
Cocurrent	12.6	7.4	12.6
Countercurrent	15.4	4.6	15.4

As shown, the countercurrent mass contactor removes nearly 25% more acetone. However, designing and operating a countercurrent direct-contact mass contactor for immiscible liquids will require additional considerations beyond those needed to design and operate a cocurrent mass contactor. The primary issues are maintaining the area for mass transfer and steady, countercurrent flow in the tubular contactor.

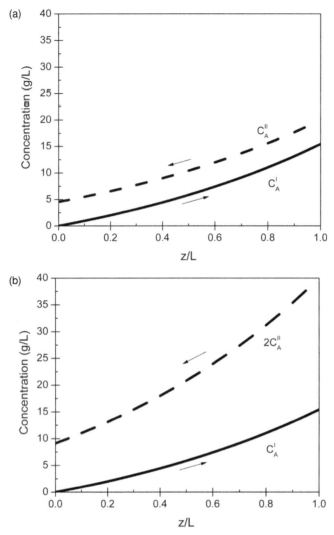

Figure E4.5. (a) Concentration profiles in the countercurrent mass contactor; (b) the profile accounting for M.

This will most certainly necessitate a vertical tower, for which the countercurrent mass contactor will also act as a separator. These issues are discussed in more detail in Chapters 7 and 8.

4.4.3 Gas–Liquid Countercurrent Contactors

Countercurrent operation yields the highest efficiencies for mass contacting. When the relative density of the fluids is sufficient to sustain a countercurrent flow by gravity, a tower-type contactor can be designed that is simpler to construct and operate. A schematic of such a device is shown in Figure 4.13. The development of the Level I, II, and IV balances appropriate for this device can be found in Section 4.4. Here, we translate that analysis and design into the language and symbols used most often in the chemical process industry for gas–liquid contacting.

To illustrate the analysis with a minimum of algebra, we consider the case of gas–liquid contacting where the gas phase (g) and the liquid phase (l) are considered immiscible. Further, we assume that the amount of species A transferred is small such that we can ignore changes in these rates through the column, such that Level III balances are used. Also, a simple equilibrium relationship such as Raoult's "Law" is assumed, namely,

$$p_A = y_A P = p_{A,vap} x_A,$$

$$\therefore \ y_A = m x_A \quad \text{where} \quad m = \frac{p_{A,vap}}{P}, \tag{4.4.15}$$

where P is the pressure in the gas side and y_A is the mole fraction of A in the gas phase. The mole fraction of A in the liquid is denoted as x_A. Of course, in a more general case there will be an equilibrium curve relating the mole fractions of A in each phase.

A Level I balance, Eq. (4.4.1), gives a relationship between the amount of species A in each phase. It is usual to specify an inlet concentration (say in the gas side for scrubbing operations) and the extent of removal desired (such as 90%). For purposes of illustration, assume that species A (a volatile organic) is to be removed from the gas stream (by absorption into a nonvolatile liquid) that is at a pressure of 1 bar and at 1% mole fraction. The inlet liquid stream is assumed to be pure oil. Then the Level I balance gives a relationship for the mass of species A leaving the contactor in the liquid side. It is usual to define G as the molar gas flow rate divided by the *empty* column cross-sectional area A_c and L as the molar liquid flow rate, again divided by the empty column cross section. The Level I component A mass balance written in these units (the molecular weight of A cancels from both sides of the equation) is

$$G A_c (y_{AF} - y_{Aout}) = -L A_c (x_{AF} - x_{Aout}). \tag{4.4.16}$$

The Level II balance follows by substitution of the equilibrium condition. For example, to determine the minimum liquid flow rate for a given gas flow rate, feed and desired exit gas mole fractions, the Level II balance becomes

$$L_{min} A_c = \frac{G A_c (y_{AF} - y_{Aout})}{y_{AF}/m}. \tag{4.4.17}$$

To determine the volume of this tubular mass contactor as we did for the mixed–mixed tank-type contactor in Subsection 4.3.3.2, both the tower height and the cross-sectional area A_c must be determined. The cross-sectional area is determined by an analysis of the gas flow and its pressure drop through the tower. The height of the tower requires a Level IV analysis (which simplifies to a Level III analysis as the amount of mass transferred is small). The analysis developed in Subsection 4.2.1, Eq. (4.2.10), shows that the rate of mass transfer can be rewritten in the terminology used here as

$$r_A a = G A_c (y_{AF} - y_{Aout}) = K_G a (y_A - m x_A). \tag{4.4.18}$$

Here, the notation for the overall mass transfer coefficient K_G is defined for a driving force for the rate of mass transfer based on the mole fraction of A in the gas phase, i.e., the (SI) units of K_G would be moles per square meter per seconds. The conversion from concentration to mole fractions is readily done in this problem if the gas phase can be modeled as an ideal gas. Then,

$$N_A = K_M \left(C_A^g - mC_A^l \right) = K_M \frac{P}{RT} \left(y_A - mx_A \right) = K_G \left(y_A - mx_A \right).$$

A similar derivation can be performed for the overall mass transfer coefficient defined on the liquid phase given the molar concentration of the liquid.

Using the differential slice defined in Figure 4.14 and steady-state Level III balances lead to the relations for the concentration of A anywhere in the contactor by use of a_v, the area for mass transfer per unit volume of contactor:

$$0 = A_c \frac{d}{dz} \left(G y_A \right) + K_G a_v A_c \left(y_A - mx_A \right),$$

$$0 = A_c \frac{d}{dz} \left(L x_A \right) + K_G a_v A_c \left(y_A - mx_A \right). \tag{4.4.19}$$

Note that these equations are simply Eqs. (4.4.12) and (4.4.13) written in terms of the new nomenclature.

For a tower of a given diameter, the height of the tower (Z_T) can be determined by integration of the equations. The first of Eqs. (4.4.19) can be rearranged and integrated (or the second when rewritten for a liquid side driving force) as

$$Z_T = \frac{V^g + V^l}{A_c} = \int_0^{Z_T} dz = \int_{y_{AF}}^{y_{Aout}} \frac{G}{K_G a_v} \left(\frac{1}{y_A - mx_A} \right) dy_A. \tag{4.4.20}$$

The gas and liquid flow rates can be approximated as constant throughout the tower if the amount of mass transfer is small, the gas and liquid are insoluble, and steady-state operation is maintained. Further, the product of mass transfer coefficient and area per unit volume is also assumed to be constant over the height of the column. With these assumptions, the height of the tower can be calculated as

$$Z_T \approx \frac{G}{K_G a_v} \int_{y_{AF}}^{y_{Aout}} \left(\frac{1}{y_A - mx_A} \right) dy_A. \tag{4.4.21}$$

For design purposes, this equation can be split into two parts defined as the *height of a transfer unit* H_T, and the *number of transfer units* N_T:

$$Z_T = H_T^g N_T^g = \frac{G}{K_G a_v} \int_{y_{A,F}}^{y_{Aout}} \left(\frac{1}{y_A - mx_A} \right) dy_A,$$

$$H_T^g = \frac{G}{K_G a_v}, \tag{4.4.22}$$

$$N_T^g = \int_{y_{A,F}}^{y_{Aout}} \left(\frac{1}{y_A - mx_A} \right) dy_A.$$

The superscript "g" is to remind us that the height and number of transfer units are based on a driving force written in terms of the gas mole fraction. A similar set of equations can be defined based on the liquid-phase mole fraction if desired. The number of transfer units is a measure of how difficult the separation is, and the height of a transfer unit is the length of tower required for achieving one transfer unit.

For a linear equilibrium relationship, the number of transfer units can be calculated by integration as the difference, $y_A - mx_A = C_1 y_A + C_2$, where the constants are determined from knowledge of the inlet and outlet conditions. The result is

$$N_T^g = \int_{y_{AF}}^{y_{Aout}} \left(\frac{1}{y_A - mx_A}\right) dy_A = \frac{y_{AF} - y_{Aout}}{\Delta y_{A,lm}},$$

$$\Delta y_{A,lm} = \langle \Delta y_A^{I-II} \rangle = \frac{(y_{Aout} - mx_{AF}) - (y_{AF} - mx_{Aout})}{\ln\left[\frac{(y_{Aout} - mx_{AF})}{(y_{AF} - mx_{Aout})}\right]}, \qquad (4.4.23)$$

where $\Delta y_{A,lm}$ is the *log-mean* mole fraction difference. This is discussed in more detail in Section 4.5. Students studying mass transfer operations will encounter the analysis and design equations for countercurrent gas–liquid mass contactors in this nomenclature. Next, we consider the general design of continuous mass contactors, but return to the use of these analysis and design equations in Chapter 8. Before doing so, however, we consider an example of the use of these equations in calculating the height of a gas–liquid contactor.

EXAMPLE 4.5. Determine the tower height required for performing the following operation. Consider a gas flow rate of 0.25 mol/min in a tower with a column cross-sectional area (empty) of $A = 1$ m^2 contacting a nonvolatile oil with a flow rate of 0.1 mol/min at a temperature of 25 °C. The gas contains an undesired organic contaminant at a mole fraction of 1% that must be reduced to 0.1% mole fraction. Experiments show that the volatile contaminant has a partition coefficient of $m = 0.3$. The choices of column diameter and liquid flow rate are made to maintain good gas–liquid contacting without flooding and will depend on the choice of packing and the properties of both streams, as well as the pressure drop, as will be discussed further in Chapter 8.

SOLUTION. First, we can check whether the specified oil flow rate is above the minimum required for the desired exit purity of the gas. The Level II balance indicates that the minimum flow rate is

$$L_{min} A_c = \frac{GA_c (y_{AF} - y_{Aout})}{y_{AF}/m} = \frac{(0.25 \text{ mol/min})(0.01 - 0.001)}{0.01/0.3} = 0.0675 \text{ mol/min}.$$

The specified oil flow rate exceeds this minimum. Now, the exit mole fraction of A in the oil can be determined from the Level I analysis as

$$x_{Aout} = \frac{GA_c}{LA_c}(y_{AF} - y_{Aout})$$

$$= \frac{0.25 \text{ mol/min m}^2}{0.1 \text{ mol/min m}^2}(0.01 - 0.001)$$

$$= 0.0225.$$

Next, the number of transfer units required for achieving the separation is given by Eq. (4.4.23) as

$$\Delta y_{A,lm} = \frac{(0.001 - 0) - (0.01 - 0.3 \times 0.0225)}{\ln\left[\frac{(0.001 - 0)}{(0.01 - 0.3 \times 0.0225)}\right]} = \frac{0.001 - 0.00325}{\ln\frac{0.001}{0.00325}} = 0.0019,$$

$$N_T^g = \frac{y_{AF} - y_{Aout}}{\Delta y_{A,lm}} = \frac{0.01 - 0.001}{0.0019} = 4.7.$$

The design is specified when the height of a transfer unit is calculated. This requires knowledge of both the mass transfer coefficient and the interfacial area for mass transfer per unit volume of contactor. Determination of these will be discussed in detail in Part II. However, for the problem at hand, laboratory data have been obtained to show that

$$K_G a_v = 2.25 \times 10^{-3} \text{ mol/m}^3 \text{ s},$$

$$\therefore H_T^g = \frac{G}{K_G a_v} = \frac{0.25 \text{ mol/m}^2 \text{ min}\frac{1 \text{ min}}{60 \text{ s}}}{0.00225 \text{ mol/m}^3 \text{ s}} = 1.85 \text{ m},$$

$$\therefore Z_T = H_T^g N_T^g = 1.85 \text{ m} \times 4.7 = 8.7 \text{ m}.$$

Therefore our design specifies that the tower be constructed to have a cross-sectional area of 1 m^2 and be nearly 9 m high.

This simplified analysis and example of a technically feasible design will be elaborated on in Part II of this text. A more complicated analysis, beyond the scope of this text, is required to handle cases in which the amount of mass transfer cannot be considered small, in which isothermal operation may not be a good assumption, and in which the equilibrium line is not a constant. The interested student is referred to the references given for this chapter.

A graphical analysis paralleling what was done for the two-stage countercurrent operation is shown in Figure 4.15. The equilibrium line follows Eq. (4.4.15), and the *operating line* is given by the Level I balance, Eq. (4.4.16). For this simple problem, both the operating line and the equilibrium line are shown to be linear. The number of transfer units is simply the total difference (in mole fraction) in the gas phase divided by the log-mean average driving force (Δy_{lm}). It has the physical interpretation that it is the number of equivalent "stages" required for performing the separation, in which each "stage" achieves the equivalent removal given by $GA_c\Delta y_{A,lm}$. The *height of a stage* is the height of column (which, when multiplied by the cross-sectional area yields the volume) required for achieving one, average "stage." Of course, there are no discrete stages in the countercurrent packed-bed mass contactor, but the utility of such an analysis has proven most useful for a packed-tower design.

4.5   Continuous-Flow Mass Contactor Design Summary

The previous sections concentrated on the development of models for a variety of different mass transfer equipment. Specifically, we showed the Level II and IV

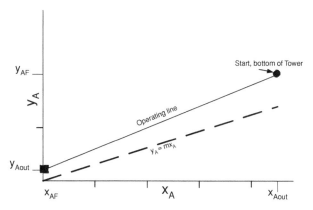

Figure 4.15. Operating diagram (X–Y representation) for a countercurrent gas–liquid mass contactor. (The dot shows the inlet at the bottom of the column, and the square the outlet. The operating line is solid and the dashed line is the equilibrium line.)

analyses for each system. At this point, we summarize the procedure with a step-by-step methodology for the sizing of the equipment involved in continuous mass transfer systems. *The goal of the procedure is to produce a technically feasible design.*

Step 1. Determination of Objectives. The first task is to determine the objectives and level of performance expected of the design. For a mass contactor, a typical objective is to reduce or increase the concentration of a species in a given stream to a set level. Hence a process stream temperature and flow rate, initial concentrations, and desired final concentration(s) will be known, in general. Other constraints may exist, but these are not critical for a feasible design. The optimization of a design is a further step that is beyond the scope of the analysis presented here and usually requires consideration of process economics.

Step 2. Calculation of Mass Transfer Load (Level I Analysis). From the objectives, we can directly calculate one quantity—the mass transfer load required. This is given by

$$m_{load} = \left(q^I C_A^I - q_F^I C_{AF}^I \right) = - \left(q^{II} C_A^{II} - q_F^{II} C_{AF}^{II} \right). \qquad (4.5.1)$$

Note that the load for mass transfer is defined to be positive if Phase I is enriched in species A by the process. This definition is arbitrary, but the choice of sign must be used consistently within a given calculation. The sign of the load for heat transfer is defined in a similar fashion. No additional material properties are required for this calculation. Often, the amount of solute transferred is small, such that the volumetric properties of the stream are not changed significantly and the volumetric flow rate can be approximated as constant. The rest of this discussion will make this simplifying assumption.

Step 3. Choose a Mass Transfer Agent. With the mass transfer load, we now turn our attention to the extraction–infusion agent. Often, to facilitate separation and prevent further contamination of the process stream, a fluid that is immiscible with the process stream will be chosen. For extraction, it is desirable to use a fluid (Phase II) that is the preferred medium for the transferred substance. In some cases, some fluids will allow for faster mass transfer than in other cases (i.e., gases and supercritical fluids over liquids). Thermodynamic property information, such as shown in Figure 4.2, and chemical safety and toxicity data are required for choosing a mass transfer agent.

Step 4. Flow Rate Determination (Level II Analysis). Next, the minimum flow rate of the chosen mass transfer agent required for accomplishing the goals of the process is calculated. This begins by applying the Level II analysis. This calculation is not directly used in the final design, but it enables an estimate (without mass transfer coefficients or interfacial area data) if the extraction–infusion agent is a reasonable choice for the particular process.

The Level II analysis yields the minimum flow rate of the utility phase required to effect the mass transfer as

$$q^{II}_{min} = q^{I} \left(\frac{C^{I}_{AF} - C^{I}_{A}}{C^{I}_{A} M^{-1} - C^{II}_{AF}} \right),$$

$$q^{I}_{min} = q^{II} \left(\frac{C^{II}_{AF} - C^{II}_{A}}{M C^{II}_{A} - C^{I}_{AF}} \right), \tag{4.5.2}$$

depending on which phase is the utility phase.

For the case of the countercurrent mass contactor, considered in Subsection 4.4.2, the Level II balance that yields the minimum utility flow rate is when the utility stream exits in equilibrium with the feed concentration of the process stream. This holds both for reduction in process stream concentration as well as for enrichment of the process stream.

The minimum utility flow rate is then given as

$$q^{II}_{min} = q^{I} \left(\frac{C^{I}_{AF} - C^{I}_{A}}{C^{I}_{AF} M^{-1} - C^{II}_{AF}} \right),$$

$$q^{I}_{min} = q^{II} \left(\frac{C^{II}_{AF} - C^{II}_{A}}{M C^{II}_{AF} - C^{I}_{AF}} \right). \tag{4.5.3}$$

Note that, although Eqs. (4.5.2) and (4.5.3) are similar in form, they give different results—namely that a countercurrent operation will have a lower minimum flow rate than that of other designs, all other things being equal. This is consistent with the observation that countercurrent designs are often the most efficient because they maintain the greatest average driving force for mass transfer.

Next, the engineer must size the equipment by specifying the volume(s) and contact area by specifying a new flow rate that is larger than the minimum for the actual design. It is important to reinforce the concept that identification of equilibrium does not enable designing the equipment, nor will, in general, the equipment achieve an equilibrium stage in operation. As can be seen by inspection of the model equations, which for the most part approach equilibrium exponentially slowly, in general it will take an infinite residence time to achieve true equilibrium.

In the technically feasible design of a heat exchanger, we determined a feasible utility flow rate by requiring that the minimum temperature difference be 10 °C. This ensured a sufficient driving force for heat transfer so that the area of the exchanger was not unnecessarily large, and yet the utility flow rate was not excessively above the minimum required. There is no equivalent "rule of thumb" for mass contactors that can be simply expressed in terms of concentrations. However, it is possible to use the concept of stage efficiency χ, as introduced in Section 4.3 [Eq. (4.3.34)]. For mass contactors other than countercurrent, choosing a stage efficiency of 90% generally yields a reasonable trade-off between utility stream flow rates and contact area (and hence process vessel size and power input). Using the Level I balance and the definition of the stage efficiency [Eq. (4.3.34)], we can determine a technically feasible utility flow rate as

$$q^{II} = \frac{q^I M}{\left[\chi \left(\frac{MC_{AF}^{II} - C_{AF}^I}{C_A^I - C_{AF}^I} \right) - 1 \right]} = \frac{q_{min}^{II}}{\left[\chi - (1-\chi) \frac{q_{min}^{II}}{q^I M} \right]}. \qquad (4.5.4)$$

A similar result can be obtained for q^I if Phase I is the utility stream. For the countercurrent exchanger, there is no equivalent definition of efficiency as equilibrium is not reached. However, similar reasoning suggests that using $q^{II} \approx q_{min}^{II}/0.9$ for the minimum utility flow is a reasonable starting estimate.

The relative magnitude of the utility flow rate relative to that of the process stream should be evaluated within the context of the proposed operation. If a mixed–mixed contactor is being designed, the two streams should have comparable flow rates, or the continuous phase should be in excess. On the other hand, typical gas-sparged contactors have gas volumetric flow rates an order of magnitude lower than liquid rates. Further considerations of feasible utility flow rates will be discussed in relation to specific contactor designs illustrated in Chapter 8.

Step 5. Mass Transfer Coefficient Estimation. With the stream properties of the design determined, the appropriate Level IV balance can be applied to determine the required $K_m a$. As we have emphasized throughout this chapter, estimates of K_m, the overall mass transfer coefficient, are possible. In some cases, this may be obtained from an experiment, though usually a rough estimate is obtained from existing data, handbooks, or theory. Any experimental determination of the mass transfer coefficient must uncouple the mass transfer coefficient from the interfacial area, which is a nontrivial process. Further guidance in this task will be presented in the Part II of this book.

Step 6. Interfacial Area Determination and Design (Level IV Analysis). At this point in the design, we know the outlet concentrations for both streams (or stream and batch) in the mass contactor. The required interfacial area can be determined from the appropriate model equations developed in a Level IV analysis in the same manner as performed for the heat exchanger design. The mass "load," in analogy to the heat load for heat exchangers, is the amount of solute to be transferred. The following expression defines the mass load of the contactor, which, as shown previously, results from rewriting the Level IV analysis equations in terms of the mass load:

$$m_{load} = K_m a \langle \Delta C_A^{I-II} \rangle. \qquad (4.5.5)$$

Table 4.7. Area determination for continuous-flow models

Continuous-flow model	Average concentration difference
Mixed–mixed tank	$-\left(C_A^I - MC_A^{II}\right)$
Plug–mixed tank	$-\dfrac{\left(C_A^I - MC_A^{II}\right) - \left(C_A^I - MC_{AF}^{II}\right)}{\ln\left(\dfrac{C_A^I - MC_A^{II}}{C_A^I - MC_{AF}^{II}}\right)}$
Cocurrent	$-\dfrac{\left(C_A^I - MC_A^{II}\right) - \left(C_{AF}^I - MC_{AF}^{II}\right)}{\ln\left(\dfrac{C_A^I - MC_A^{II}}{C_{AF}^I - MC_{AF}^{II}}\right)}$
Countercurrent ($\lambda \neq M$)	$-\dfrac{\left(C_A^I - MC_{AF}^{II}\right) - \left(C_{AF}^I - MC_A^{II}\right)}{\ln\left(\dfrac{C_A^I - MC_{AF}^{II}}{C_{AF}^I - MC_A^{II}}\right)}$
Countercurrent ($\lambda = M$)	$-\left(C_A^I - MC_{AF}^{II}\right)$

The concentration difference in the rate expression depends on the geometry of the contactor. It has a physical interpretation—it is the average driving force for mass transfer in the contactor. By rearranging the model equations derived in this chapter, this average concentration difference has been developed for each of the models described in this chapter: They are given in Table 4.7.

Although we can easily calculate the required interfacial area, relating that result to the size of the process vessel is a much more complicated task than in heat exchanger design. We will show in Chapter 7 that the interfacial area generated in most contactors is related to the power per unit volume imparted to the system. However, determining the power per volume is a very complicated task that requires information concerning the vessel size and geometry, flow rates, and fluid properties, as well as the fluid mechanics (which are not rigorously understood for multiphase systems).

Step 7. Design Evaluation and Iteration. The final values of flow rates, area, and vessel geometry must be evaluated within the context of the assumptions used to derive the model equations and the rate expressions. In general, the area and mass transfer coefficient will be shown to be functions of the flow rates and vessel geometry in one or both fluids, as well as the fluid properties. Therefore specification of the flow rates and vessel geometry may necessitate recalculation of the area and mass transfer coefficient and hence iteration. A further discussion of this is presented in Part II of this book.

Note that, unlike with heat exchanger design, the technically feasible design of a mass contactor is not complete until the method of area generation has been specified, which is highly equipment specific. Equations relating the drop diameter to the power input and fluid properties are required. In cases in which gas pressure is used to drive mixing, iterative procedures are required for achieving a technically feasible design, as will be illustrated in the next part of the book.

The procedure to develop a technically feasible design of a mass contactor is summarized in Table 4.8 and illustrated in the following example.

Table 4.8. Overview of mass contactor analysis and design procedure

Step	Level of analysis or action	Input parameters	Outputs	Comments
1	Determine objectives	Problem statement	Final concentration in one phase	Usually a process stream flow rate, initial solute, and target effluent concentrations are specified
2	**Level I**	Flow rate and concentrations of process stream	Mass load, type of contactor	Consider different contactor designs based on objectives and method of operation
3	Choose transfer fluid	Thermodynamic properties of fluids	Distribution coefficient	Consider miscibility, solubility, and mass transfer rates
4	**Level II**	M, flow rates, and concentrations in process stream	Minimum flow rate of transfer fluid for equilibrium stage operation and actual design flow rate (stage efficiency)	Evaluate choice of transfer fluid, bounds minimum flow rate
5	Estimate K_m	Contactor type and operation	K_m	May require experiments, correlations, and possible iteration
6	**Level IV**	K_m, M, flow rates and geometry	Area required for effecting mass transfer	Can evaluate the effect of different geometries on required area
7	Evaluation and validation	Design and underlying assumptions	Refined design based on possible modification of input parameters	Check assumptions and value of K_m given operation of device

---

EXAMPLE 4.6. A water stream flowing at 100 L/min. at 25 °C contaminated with acetone at 5 g/L must be reduced to an acetone concentration of 2 g/L. TCE at 25 °C is available for this process.

A technically feasible design is to be proposed following the design procedure previously outlined.

SOLUTION.

1. The process stream (water) is specified and the feed and desired outlet concentrations given. Given our previous experience with TCE–water systems, the water will be the dispersed phase (II) and TCE the continuous phase (I), and a mixed–mixed tank type contactor will be designed.
2. The load is calculated with Eq. (4.5.1) as

$$m_{load} = (-100 \text{ L/min})(2 - 5 \text{ g/L}) = 300 \text{ g/min}.$$

Note the load is positive as Phase I is to be enriched with acetone.

3. The solvent TCE is available and sufficient thermodynamic data are available (Figure 4.2). Furthermore, the concentrations are low enough such that M = 2. As noted in the earlier example, TCE has environmental and health effects such that it should be handled carefully. However, TCE is largely insoluble in water and has a stronger affinity for acetone, both being advantageous for separation.

4. The minimum TCE flow rate is given by the second of Eqs. (4.5.2) for a tank-type contactor as

$$q_{min}^I = (100 \text{ L/min}) \left(\frac{5-2}{2 \times 2 - 0} \text{ g/L} \right) = 75 \text{ L/min}.$$

We will assume a 90% stage efficiency, which yields the estimated flow rate of TCE for the technically feasible design to be

$$q^I = \frac{M^{-1}q^{II}}{\left[\chi \left(\frac{M^{-1}C_{AF}^I - C_{AF}^{II}}{C_A^{II} - C_{AF}^{II}} \right) \right]} = \frac{2^{-1} \times 100 \text{ L/min}}{\left[0.9 \left(\frac{0 \times 2^{-1} - 5}{2 - 5} \right) - 1 \right]} = 100 \text{ L/min}.$$

Note that this choice results in $\lambda = q^{II}/q^I = 1$, which is appropriate for the mixed–mixed mass contactor under design.

5. The mass transfer coefficient is, in the absence of other experimental data, taken to be $K_m \approx 3 \times 10^{-5}$ m/s. This estimate should be checked and its uncertainty recognized in the final recommendation.

6. From the Level I balance, we find the exit concentration of acetone in the TCE to be $C_A^I = 3$ g/L. The area is determined from Eq. (4.5.5), with the value of

$$\langle \Delta C^{I-II} \rangle = -(C_A^I - MC_A^{II}) = -(3 - 2 \times 2 \text{ g/L}) = 1 \text{ g/L},$$

$$\therefore \quad a = \frac{\dot{m}_{load}}{K_M \langle \Delta C^{I-II} \rangle} = \frac{(300 \text{ g/min})(1/60 \text{ min/s})}{(3 \times 10^{-5} \text{ m/s})(1 \text{ g/L})(10^3 \text{ L/m}^3)} = 167 \text{ m}^2.$$

7. The volume of the tank will depend on how the area is created. A reasonable bubble size is 2 mm. Therefore the volume of the dispersed phase in the tank can be calculated from the definition of the sauter mean drop size as

$$V^{II} = \frac{1}{6}(167 \text{ m}^2)(2 \times 10^{-3} \text{ m}) = 0.056 \text{ m}^3 = 56 \text{ L}.$$

Because the residence times in the tank are equal, the volume of the continuous phase is given by

$$\frac{V^I}{q^I} = \frac{V^{II}}{q^{II}} \quad \therefore \quad V^I = V^{II} \frac{q^I}{q^{II}} = 56 \text{ L}.$$

Hence, the total volume of the tank is 112 L, with a residence time of 33 s. A reasonable geometry would be a tank of 0.78 m in diameter and height.

To complete this design, the power required for generating the specified area must be specified and, based on the mixing in the tank, a better estimate of the mass transfer coefficient calculated. These topics will be covered in Part II of this text.

REFERENCES

Astarita G. *Mass Transfer with Chemical Reaction*. Amsterdam: Elsevier, 1967.

Bird RB, Stewart WE, Lightfoot EN. *Transport Phenomena* (2nd ed.). New York: Wiley, 2002.

Chemical Engineers' Handbook (7th ed.). New York: McGraw-Hill Professional, 1997, Sections 18 and 19.

Fair JR, Steinmeyer DE, Penny WR, Crocker BB. Gas absorption and gas-liquid system design. In *Chemical Engineers' Handbook* (7th ed.), Perry RH, Green DW, eds. New York: McGraw-Hill Professional, 1997, Chap. 14.

Govier GW, Aziz K. *The Flow of Complex Mixtures in Pipes*. New York: Van Nostrand Rienhold, 1972

Middleman S. *An Introduction to Mass and Heat Transfer*. New York: Wiley, 1998.

Otero Z. Liquid circulation and mass transfer in gas-liquid contactors. Ph.D. dissertation, University of Delaware, 1983.

Russell TWF, Denn MM. *Introduction to Chemical Engineering Analysis*. New York: Wiley, 1972.

Sandler SI. *Chemical Engineering Thermodynamics* (4th ed.). New York: Wiley, 2006.

Sherwood TK, Pigford RL. *Adsorption and Extraction*. New York: McGraw-Hill, 1952.

Treybal RE. *Mass Transfer Operations* (2nd ed.). New York: McGraw-Hill, 1968.

PROBLEMS

4.1 In Example 4.1 the initial concentration of acetone in the water was reduced from $C_{Ai}^{II} = 100$ g/L to 33 g/L by contacting it with 700 L of TCE in which there is no acetone, $C_{Ai}^{I} = 0$ The batch mass contactor vessel required to do this has a volume of 1500 L. A second vessel of 2200 L is available and 1400 L of TCE is available. You can use either vessel and you can assume that the acetone in the water has already been reduced to 33 kg/m^3. What is the lowest concentration of acetone in the water that can be achieved by carrying out a Level II analysis? Describe in detail the operations that need to be carried out.

The water–TCE mixture is immiscible. The density of pure TCE is about 1450 kg/m^3, and even if all the acetone is transferred to the TCE it reduces the density to only 1350 kg/m^3. The water density is about 1000 kg/m^3. Even with the all the acetone in the water the density does not essentially vary from that of water, 1000 kg/m^3.

4.2 The time required for carrying out the mass contacting step in Problem 4.1 needs to be determined. Develop the model equations that will enable this to be done. How would you simplify this model and what assumptions did you make? What quantities must you know to determine the time to carry out the extraction specified in Example 4.1?

4.3 It is suggested by your acting supervisor who is a business school graduate that you carry out the TCE extraction in a semibatch operation in which the water–acetone charge of 700 L is placed in the contactor and the 700 L of TCE is pumped through the water–acetone mixture at a rate of 10 L/min. Develop the required model equations that will describe the operation. Can such an operation be carried out? Outline your findings in a short memo to the acting supervisor.

4.4 The saturated concentration of NaCl in water, C_S^I, needs to measured. Describe the experiment required to do so in enough detail so it can be carried out by a grade school student. All that is available is a balance, table salt, water, and a 1-L beaker.

4.5 The following data were collected by Rushton, Nagata, and Rooney (*AIChE J.*, 1964; 10:298) in a batch experiment on the distribution of octanoic acid between an aqueous (corn syrup water) and organic xylene (phase.) 2000 mL of aqueous phase was initially placed in the tank with 200 mL of xylene. The mixer was started, and this system was brought to equilibrium with respect to drop size. When this completely mixed state was achieved, 250 mL of an aqueous solution of octanoic acid was added and the concentrations recorded.

Time t (secs)	$C_A^I \times 10^4$, Concentration of octanoic acid in water phase (g-mol/L)
0	2.75
10	2.13
20	1.72
30	1.45
40	1.23
60	1.03
80	0.94
120	0.83
∞	0.78

a. Develop the mathematical description for this particular experiment.
b. How should the data be plotted so that it will lie on a straight line?
c. Prepare the plot, determine the slope of the line, calculate $K_m a$, and compare your value with that reported, 75 cm³/s.

4.6 The bubbling of CO_2 through water in a small vessel results in a lower solution pH. This effect was achieved through the following reaction pathway:

$$CO_2 + H_2O \Leftrightarrow H_2CO_3, \qquad K_1 \ (T = 25\,°C) = 1.00,$$
$$H_2CO_3 + H_2O \Leftrightarrow HCO_3^- + H_3O^+, \qquad K_2 \ (T = 25\,°C) = 4.45 \times 10^{-7},$$
$$HCO_3^- + H_2O \Leftrightarrow CO_3^{-2} + H_3O^+, \qquad K_3 \ (T = 25\,°C) = 4.69 \times 10^{-11}.$$

a. If the experiment is done at a temperature of 25 °C and the CO_2 partition coefficient [defined as C_{CO_2} (Molar) $= MP_{CO_2}$ (atm)] is equal to 3.34×10^{-2} M/atm, determine the equilibrium pH of the aqueous solution. Specify the level of analysis used.
b. The time required for reaching 90% of equilibrium for the 2-L tank with a 20-hole sparger has been measured to be 30 s. Estimate the time required for reaching 90% of the equilibrium CO_2 concentration for the 2-L tank with a single hole in the sparger. The CO_2 flow rate is the same in both experiments. Explain your reasoning and *clearly state all assumptions.* Does your answer seem reasonable? *Hint: The residence time will depend on the bubble rise velocity, which can be estimated by balancing Stokes drag against the buoyancy force.*

4.7 Consider the absorption process in which methyl chloride (MeCl) is transferred from a N_2 stream into water. Write mass balance equations for the design of a mass contactor.

The ratio of MeCl concentration in the gaseous phase with respect to the aqueous phase at equilibrium is M^{-1}. Assume that the mass transfer coefficient K_m *based on concentration difference in the liquid phase* is also given. In all four cases the initial state or feed is either pure water or gaseous MeCl in N_2, and the gas is dispersed as bubbles. Further, you may NOT assume that the concentration of MeCl is small in either phase.

Case I: You have a well-stirred batch system initially containing bubbles of gaseous MeCl and N_2 in a tank of pure water. *Derive* a mass balance and species mass balance for the concentration of MeCl for each phase as a function of time.

Case II: You have a well-mixed tank type semibatch contactor charged at t = 0 with pure water. The MeCl/N_2 stream is continuously bubbled through the contactor at a particular flow rate. *Derive* a mass balance and a species mass balance for the concentration of MeCl for each phase.

Case III: You have a well-mixed tank-type contactor with the both phases continuously flowing through it. *Derive* a mass balance and species mass balance for the concentration of MeCl for each phase.

Case IV: You have a well-mixed tank-type contactor with both phases continuously flowing through it. This time, the MeCl is consumed in the aqueous phase by a reagent N. This reaction has been determined to follow the following constitutive equation:

$$\frac{dC_{MeCl}}{dt} = -k_R C_{MeCl} C_N$$

Derive a mass balance and species mass balance for the concentration of MeCl for each phase

Finally: Briefly explain how you would perform a technically feasible design, including what data would be required for Case IV. Be careful to indicate how the equations may or may not be coupled.

4.8 Joe Blue Hen has recently designed a controlled-release drug-delivery capsule. The capsule works by maintaining the internal concentration of the active drug ingredient constant (at 1×10^{-7} mol/cc) by a proprietary technique. Further, the capsule has a membranelike surface layer that controls the rate of release. For treatment purposes, the dosage of active ingredient should be 1×10^{-11} mol/s. Propose a technically feasible design for the capsule size and required mass transfer coefficient if the release should last for 8 h. Clearly state all assumptions and indicate the primary uncertainties in your design. You may assume that the active ingredient mixes quickly in the gut and is rapidly transported into the body.

4.9 A dilute reaction mixture was allowed to react under an inert atmosphere (no oxygen). Once the reaction is complete, the cylindrical vessel is exposed to a chamber with a normal oxygen concentration ($P_{O_2} = 0.21\, P_{air}$). The oxygen is absorbed by the water. The mass transfer coefficient is 2.1×10^{-3} cm/s, and Henry's "Law" constant is 1.26×10^{-3} mol/ (L atm). The radius of the cylinder is 10 cm, and the liquid height is 15 cm. The volume of the air chamber is 5 L. Plot the oxygen concentration in solution versus time for chamber pressures of 1, 5, and 10 atm.

4.10 An artificial lung is a countercurrent tubular mass contactor, in which blood flows through hydrophobic microporous polypropylene tubes, having a pore diameter of 0.1 μm. This allows the blood to come in direct contact with the gaseous oxygen flowing around the tube. The venous blood enters the artificial lung, otherwise known as a blood oxygenator, with an oxygen concentration of 40 mm Hg. *Note: The concentration in the liquid phase is often expressed in the medical profession in units of pressure.* The oxygenated blood must leave the oxygenator with an oxygen concentration of 95 mm Hg. If the partition coefficient of oxygen into blood is 1.53×10^{-6} (mol/cm^3)/mm Hg, and the overall mass transfer coefficient between the gaseous oxygen and the blood through the pore is 2.91×10^{-3} cm/s, and the blood flow rate is 200 mL/min, propose a feasible design for the unit. The design must include the length of the unit, the thickness of the tubes, the number of the tubes used, tube porosity, and the flow rate of oxygen. *Hint: You may assume that mass transfer through the polypropylene is negligible.*

4.11 You join a group of engineers working on the design of a coffee bean decaffeination process. It has been decided that the proposed design will include two mass contactors. The first will be used to extract caffeine from coffee beans by use of supercritical CO_2, whereas the second will be used to extract caffeine from supercritical CO_2 to water. Several mass contactor configurations are being considered. You are asked to propose models for the possible design scenarios (ρ^I is the density of the CO_2 phase in kg/m^3):

a. Write mass balance equations for the design of the first mass contactor. The ratio of caffeine concentration in the coffee phase with respect to the CO_2 phase at equilibrium is M. The mass transfer coefficient K_m is also given.

Case I: You have a batch system charged with coffee beans and supercritical CO_2 at time $t = 0$. Write mass balance for the concentration of caffeine in each phase as a function of time.

Case II: You have a tank-type semibatch contactor charged at $t = 0$ with the coffee beans. The CO_2 is continuously flown through the contactor with a particular flow rate. Write mass balance equations for the concentration of caffeine in each phase.

b. Write mass balance equations for the design of the second contactor, the one used to separate extract caffeine from the CO_2 phase into the water phase: Again you are given M and K_M for this particular situation.

Case I: You have a tank-type contactor with both phases continuously flowing through it. Write mass balance equations for the concentration of caffeine in each phase.

Case II: You have a tubular reactor with both phases flowing concurrently through this reactor. Here the concentration of caffeine in each phase will be changing as a function of the length of the tubular reactor. Write mass balance equations for the concentration of caffeine in each phase as a function of the length of the reactor.

APPENDIX A: "LOG-MEAN" CONCENTRATION DIFFERENCE

The starting point is the definition of the load, which is positive for mass transfer into Phase I from Phase II [Eq. (4.5.5)]:

$$m_{\text{load}} = q^I \left(C_A^I - C_{AF}^I \right) = K_m a \left(\Delta C_A^{I-II} \right). \tag{4.A1}$$

The goal is to derive the value of the average concentration difference $\langle \Delta C_A^{I-II} \rangle$ for specific geometries. This is performed by use of the Level IV analysis relating the concentrations to the area (and mass transfer coefficient, flow rates, and distribution coefficient).

As an example, consider the simple case of a continuous-flow tank-type device with a mixed–mixed flow pattern. The Level II balance yields

$$q_F^I \left(C_A^I - C_{AF}^I \right) = -K_m a \left(C_A^I - M C_A^{II} \right),$$

$$\therefore \left\langle \Delta C_A^{I \to II} \right\rangle = - \left(C_A^I - M C_A^{II} \right). \tag{4.A2}$$

For a more complex device, the derivation follows the same pathway. For example, for a cocurrent tubular exchanger with plug–plug fluid motion, the Level III balance equations (again, for small levels of mass transfer) become

$$q^I \frac{dC_A^I}{dz} = -K_m \frac{a}{L} \left(C_A^I - M C_A^{II} \right),$$

$$q^{II} \frac{dC_A^{II}}{dz} = K_m \frac{a}{L} \left(C_A^I - M C_A^{II} \right). \tag{4.A3}$$

These equations can be rearranged and added to yield the following:

$$\frac{dC_A^I}{dz} = -K_m \frac{a}{L q^I} \left(C_A^I - M C_A^{II} \right),$$

$$-M \frac{dC_A^{II}}{dz} = -M K_m \frac{a}{L q^{II}} \left(C_A^I - M C_A^{II} \right), \tag{4.A4}$$

$$\frac{d \left(C_A^I - M C_A^{II} \right)}{dz} = -K_m \frac{a}{L} \left(\frac{1}{q^I} + \frac{M}{q^{II}} \right) \left(C_A^I - M C_A^{II} \right).$$

This can be separated and integrated from one end $(z = 0)$ to the other end $(z = L)$ of the tubular exchanger, to yield

$$\int_{C_A^I - M C_A^{II}}^{C_{AF}^I - M C_{AF}^{II}} d \ln \left(C_A^I - M C_A^{II} \right) = -K_m \frac{a}{L} \left(\frac{1}{q^I} + \frac{M}{q^{II}} \right) \int_0^z dz,$$

$$\ln \left(\frac{C_{AF}^I - M C_{AF}^{II}}{C_A^I - M C_A^{II}} \right) = -K_m a \left(\frac{1}{q^I} + \frac{M}{q^{II}} \right),$$

$$\therefore K_m a = - \ln \left(\frac{C_{AF}^I - M C_{AF}^{II}}{C_A^I - M C_A^{II}} \right) \Big/ \left(\frac{1}{q^I} + \frac{M}{q^{II}} \right), \tag{4.A5}$$

$$\therefore q^I \left(C_A^I - C_{AF}^I \right) = \frac{- \ln \left(\dfrac{C_{AF}^I - M C_{AF}^{II}}{C_A^I - M C_A^{II}} \right)}{\left(\dfrac{1}{q^I} + \dfrac{M}{q^{II}} \right)} \left\langle \Delta C_A^{I-II} \right\rangle.$$

Rearranging yields the relationship

$$\left(\Delta C_A^{I-II}\right) = -\frac{q^I\left(C_A^I - C_{AF}^I\right)\left(\frac{1}{q^I} + \frac{M}{q^{II}}\right)}{\ln\left(\frac{C_{AF}^I - MC_{AF}^{II}}{C_A^I - MC_A^{II}}\right)} = -\frac{\left(C_A^I - C_{AF}^I\right)\left(1 + \frac{q^I M}{q^{II}}\right)}{\ln\left(\frac{C_{AF}^I - MC_{AF}^{II}}{C_A^I - MC_A^{II}}\right)}. \quad (4.A6)$$

A Level I balance is used to relate the flow rates as

$$q^I\left(C_A^I - C_{AF}^I\right) = q^{II}\left(C_{AF}^{II} - C_A^{II}\right),$$

$$\frac{q^I\left(C_A^I - C_{AF}^I\right)}{q^{II}} = \left(C_{AF}^{II} - C_A^{II}\right). \quad (4.A7)$$

Thus, substitution yields

$$\left(\Delta C_A^{I-II}\right) = -\frac{\left(C_A^I - C_{AF}^I\right) + M\left(C_{AF}^{II} - C_A^{II}\right)}{\ln\left(\frac{C_{AF}^I - MC_{AF}^{II}}{C_A^I - MC_A^{II}}\right)}$$

$$= -\frac{\left(C_A^I - MC_A^{II}\right) - \left(C_{AF}^I - MC_{AF}^{II}\right)}{\ln\left(\frac{C_{AF}^I - MC_{AF}^{II}}{C_A^I - MC_A^{II}}\right)} \quad (4.A8)$$

Similarly derivations can be performed for the other cases listed in Table 4.7.

APPENDIX B: EQUIVALENCE BETWEEN HEAT AND MASS TRANSFER MODEL EQUATIONS

As repeatedly emphasized throughout this chapter, the equations for mass contactors become equivalent to those describing heat exchangers with similar fluid motions when the amount of mass transfer is small such that the phase volumes remain unchanged. In this limit, the equivalences in Table B4.1 facilitate translation of the model equations from one transport mode to the other.

Table B4.1. Equivalences in comparing heat exchanger and mass contacting model equations

	Mass contactor	Heat exchanger
Property (per volume) being transported	C_A	$\rho \hat{C}p T$
Transport coefficient	K_m	U
Area	a	a
Distribution coefficient	M	$M = 1$
Capacity ratio	$\lambda = \frac{q^{II}}{q^I}$	$\theta = \frac{\rho_2 q_2 \hat{C}_{p2}}{\rho_1 q_1 \hat{C}_{p1}}$

Nomenclature for Part I

a	Interfacial area (m^3)
A_c	Cross-sectional area (m^3)
C	Concentration (kg/m^3 or kg-mol/m^3)
$\hat{C}_p$	Heat capacity (kJ/kg K)
D	Diameter (m)
D_{AB}	Diffusion (m^2/s)
E_A	Energy of activation (cal/g-mol)
h	Height (m)
h	Heat transfer coefficient [kJ/(m^2 s K), kW/m^2 K]
H	Enthalpy (kJ)
ΔH_{rxn}	Heat of reaction (kJ/kg-mol)
k	Conductivity [J/(m s K)]
k	Reaction rate constant (various)
K_e	Equilibrium constant (various)
k_m	Mass transfer coefficient (m/s)
K_m	Overall mass transfer coefficient (m/s)
m	mass (kg)
M	Distribution coefficient (dimensionless)
p	Partial pressure (N/m^2, kg/m s^2)
q	Volumetric flow rate (m^3/s)
Q	Heat transfer rate (kJ/s)
r	Reaction rate (kg-mol/m^3 s)
r	Mass transfer rate (kg/m^2 s)
T	Temperature (K)
t	Time (s)
U	Overall heat transfer coefficient [kJ/(m^2 s K), kW/m^2 K]
v_i	Velocity (m/s)
V	Volume (m^3)
x	Mass fraction
z	Length (m)

Greek

Θ	Holding time (s)
μ	Viscosity (Pa-s, kg/m s)

ρ	Density (kg/m^3)
τ	Residence time
τ_{ij}	Stress, ij component (Pa)

Subscripts

A	Species A
i	Initial
eq	Equilibrium
sat	Saturation
min	Minimum
∞	Infinite time
w	Wall
1, 2	Fluid 1, 2

Superscripts

I	Control Volume I, Phase I
II	Control Volume II, Phase II

PART II

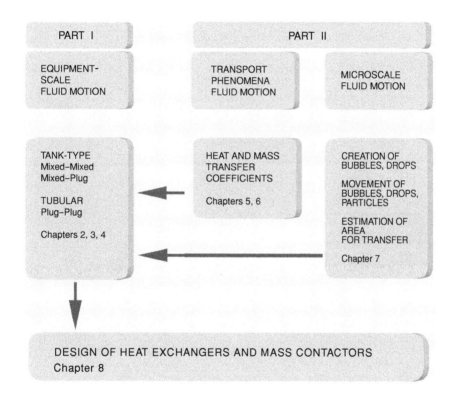

PART I	PART II	
EQUIPMENT-SCALE FLUID MOTION	TRANSPORT PHENOMENA FLUID MOTION	MICROSCALE FLUID MOTION
TANK-TYPE Mixed–Mixed Mixed–Plug TUBULAR Plug–Plug Chapters 2, 3, 4	HEAT AND MASS TRANSFER COEFFICIENTS Chapters 5, 6	CREATION OF BUBBLES, DROPS MOVEMENT OF BUBBLES, DROPS, PARTICLES ESTIMATION OF AREA FOR TRANSFER Chapter 7

DESIGN OF HEAT EXCHANGERS AND MASS CONTACTORS
Chapter 8

In Part I of this text we developed the model equations for analyzing experiments and for the technically feasible design of laboratory-, pilot-, and commercial-scale processing equipment including reactors, heat exchangers, and mass contactors. Our organization in terms of the macroscale fluid motions in such equipment (Table 1.1) has broader applicability because many systems of interest in living organisms and in the natural environment can also be similarly analyzed.

The constitutive equations used in the model equations in Part I are summarized in Table 1.5. The overall heat transfer coefficient U and the mass transfer coefficient K_m are engineering parameters defined by these constitutive equations. These transport coefficients depend on both the materials involved and the microscale and macroscale fluid motions of these materials, as well as their thermodynamic state (i.e., temperature and pressure). Our need to determine these parameters by experiment reflects our lack of understanding of the fluid mechanics affecting the transport of energy in a turbulent or laminar fluid to a solid surface, for example, or the transfer of a species at the interface between two phases with complex fluid motions. These *boundary layers* are critical regions at the fluid–fluid and fluid–solid interfaces where the dominant resistances to heat and mass transfer are located in flowing fluids. Transport coefficients deduced from analysis of existing equipment are accurate only if the model equations correctly describe the fluid motions in the experiment. The figure on the Part II opening page attempts to capture the logic of the preceding discussion.

In this part of the book we examine in detail the constitutive equations commonly referred to as *Fourier's and Fick's "Laws"* (Table 1.4). This requires us to carry out a Level III analysis on control volumes in which we only have conductive heat transfer or molecular diffusion. In this manner we will be able define the coefficients of heat and mass transfer in terms of the *material properties* of thermal conductivity and diffusivity. Material properties, such as thermal conductivities, are properties that are intrinsic to the material. For example, the thermal conductivity of dry air at standard temperature and pressure is a fixed value whether that air is the wind on a desert summer day or the air inside a sealed vessel in a laboratory. This is in contrast to the heat and mass transfer coefficients, which are engineering quantities that depend on the geometry of the system under analysis and the operating conditions, in addition to the material properties. For example, the heat transfer coefficient for the air in the wind will depend on the air's thermal conductivity as well as the wind speed and the geometry of the surface in thermal contact. Hence it will be necessarily different from the heat transfer coefficient for the quiescent air in the sealed laboratory vessel despite the fact both have the same thermal conductivity.

We begin in Chapter 5 by considering conduction and diffusion in homogeneous solids or completely stagnant liquids. Analysis of one-dimensional molecular transport establishes the relations between the rate of energy transport, thermal conductivity k, and temperature gradients, and the complementary set of rate of molecular diffusion, diffusivity D_{AB}, and gradients in species concentration. Analyses of relatively simple geometries relate the engineering transport coefficients to the material properties. We also consider convection, induced by mass transport, transport through membranes, and transient conduction and diffusion problems. The latter sets a basis for penetration theory, which is used to model engineering transport coefficients in flowing systems in Chapter 6.

In Chapter 6 we analyze the processes of energy and mass transfer in the presence of convective fluid motion. This development explores the nature of boundary layers at interfaces and their role in defining resistances to heat, mass and momentum transfer to and from flowing fluids. Methods introduced in Chapter 5 are extended to derive the relationships between the engineering transport coefficients, material properties, and fluid motion. These take the form of *transport correlations* relating critical dimensionless groups such as the Nusselt, Sherwood, Reynolds, Prandtl, and Schmidt numbers. The analysis enables proper design of experiments to gather useful engineering data, and the resultant correlations enable scale-up and the design of process equipment. The analyses also lead to the concept of transport analogies, which provide another route to estimating transport coefficients for technically feasible design.

The correlations developed in Chapter 6 are limited to systems with well-defined boundaries, such as those found in heat exchangers. However, two-phase mass contactors pose additional challenges as the area for mass transfer and the mass transfer coefficient are both determined by the mixing in the process equipment. Chapter 7 presents methods for estimating the interfacial area a in two-phase mass contactors. The analysis of mixing in two-phase mass contactors also leads to methods for independently estimating K_m, which leads to more effective methods for the analysis and technically feasible design. The developments are based on considerations of turbulent flow and energy dissipation in flowing systems.

Chapter 8 presents methods for the technically feasible design of selected heat and mass transfer unit operations. The analyses presented in Part I are combined with the ability to calculate transport coefficients, which enables predicting heat and mass transfer equipment performance and the rational design of such equipment to meet engineering performance requirements.

5 Conduction and Diffusion

5.1 Rate of Thermal Conduction

In Chapter 3 we presented model equations for heat exchangers with our mixed–mixed, mixed–plug, and plug–plug classifications. All these fluid motions generally require some degree of turbulence, and all heat exchangers, except for those for which there is direct contact between phases, require a solid surface dividing the two control volumes of the exchanger. To predict the overall heat transfer coefficient, denoted as U in the analyses in Part I, we must be able to determine how U is affected by the turbulent eddies in the fluids and the physical properties of the fluids and how the rate of heat transfer depends on the conduction of heat through the solid surface of the exchanger.

We begin our study of conductive transport by considering the transfer of heat in a uniform solid such as that employed as the boundary between the two control volumes of any exchanger. This requires a Level III analysis and verification of a constitutive equation for conduction. This is followed by a complementary analysis of molecular diffusion through solids and stagnant fluids.

5.1.1 Experimental Determination of Thermal Conductivity k and Verification of Fourier's Constitutive Equation

Consider an experiment whereby the heat flow through the wall between the tank and the jacket in Figure 3.7 is measured. For the purposes of this analysis, we consider the heat transfer to be essentially one dimensional in the y direction, with the barrier essentially infinite in the z–x plane. Consequently, we analyze a slice of the barrier with no heat flow (Q) in the x or z direction, depicted in Figure 5.1. Further, for this analysis we assume that the surface temperature at either end is equal to that of the well-mixed tank. This assumption will be relaxed shortly. As the wall of the tank is thin relative to its height and width, we expect heat transfer in one dimension only, namely in the y direction. The wall is essentially infinite in the x and z directions in the view shown in Figure 5.1 such that there is no heat flow in those directions. An equivalent system is a bar of wall material that is well insulated on the faces in the x and z directions. Such a bar of material will also have heat flows in only the y direction. Consequently, we consider the equivalent system of a "bar" of wall

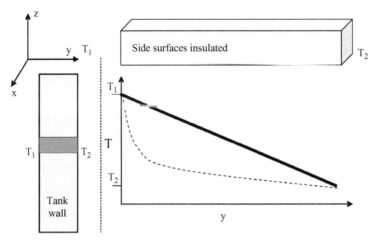

Figure 5.1. Definition of thermal conductivity. [The left-hand figure illustrates a section of the tank wall from Figure 3.7, along with the coordinate frame for analysis. The right-hand side shows the equivalent representation as an insulated bar with temperatures maintained at each end. The inset shows the temperature profile at some intermediate time (dashed curve) and at steady state (solid line).]

material initially at equilibrium temperature T_2 that is heated uniformly on one side to a temperature of T_1, as shown in Figure 5.1.

Initially, the wall (or bar) is at T_2, the temperature of the process fluid. The jacket or utility fluid is introduced on the other side with temperature T_1. At early times, the temperature profile through the bar rises (Figure 5.1) from left to right. At long times, the temperature profile in the bar approaches steady state and is observed to become linear, with one side of the bar at temperature T_1 and the other at T_2.

Analysis of this one-dimensional conduction problem proceeds by writing an energy balance, in which, for solids at normal pressures, the enthalpy is given by $H = \rho V \hat{C}_p T$ as discussed in Chapter 3. Then, taking a slice or *shell* in the direction orthogonal to heat transfer (y) with area a (see Figure 5.2), we can apply the statement of conservation of energy in a Level III analysis as follows. Note that, as the sides are insulated, we can assume that there are no heat flows Q_i in the other directions (x, z). Further, we assume that there is no thermal expansion such that the dimensions of the solid are constant. Applying the statement of the conservation of energy (the word statement illustrated in Figure 1.6) to the shell indicated in Figure 5.2 yields

$$(a\Delta y)\,(\rho\hat{C}_p T|_{t+\Delta t} - \rho\hat{C}_p T|_t) = \Delta t(Q_y|_y - Q_y|_{y+\Delta y}), \qquad (5.1.1)$$

taking the limit $\Delta y \to 0$, $\Delta t \to 0$,

$$\therefore \; \frac{\partial(\rho\hat{C}_p T)}{\partial t} = -\frac{1}{a}\frac{\partial Q_y}{\partial y}$$

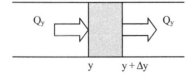

Figure 5.2. Control volume for shell balance.

The preceding equation can be further simplified as the area does not change with position through the material. Hence the heat flux ($q_y = Q_y/a$) can be substituted into the preceding equation to yield

$$\frac{\partial}{\partial t}(\rho \hat{C}_p T) = -\frac{\partial q_y}{\partial y}. \tag{5.1.2}$$

This Level III balance at steady state reduces to the statement that

$$\frac{dq_y}{dy} = 0, \qquad \therefore q_y = \text{constant}.$$

In words, at steady state the heat flux is constant for one-dimensional conduction in this geometry.

At equilibrium there will be no net heat fluxes in the system; thus $q_y = 0$. As is shown next with the aid of a constitutive equation, this corresponds to a uniform temperature profile throughout the solid.

To proceed with a Level III balance, a *constitutive equation* is required relating heat flux q_y to the temperature difference. Returning to the experiment depicted in Figure 5.1, we can measure the total rate of heat flow through the material, Q_y, and, knowing the area for heat transfer, calculate the heat flux. Through experimentation, we find that the heat flow through the barrier is, for small enough temperature differences across the barrier, linearly proportional to the temperature difference. Further experiments with barriers of varying thickness but similar material demonstrate that the heat flow varies inversely with the thickness L. Finally, for a given geometry and temperature difference, the heat flow will still depend on the material. These observations are captured in a simple mathematical relationship among the heat flow Q, temperature difference $\Delta T = T_1 - T_2$, material thickness L, and cross-sectional area a. The proportionality constant k is a function of the material itself and is denoted the *thermal conductivity*. This material property is defined by the following equation (which is the finite-difference form of Fourier's constitutive equation introduced in Table 1.4):

$$Q_y = -ka\frac{\Delta T}{L}. \tag{5.1.3}$$

In Eq. (5.1.3), Q_y is the flow of heat in the y direction in SI units of watts (W) (British thermal units per hour) (the subscript indicates the direction for the heat flow as heat flux is a vector quantity), q is the magnitude of the heat flux in watts per square meter (British thermal units per hour per square feet), a is the cross-sectional area for heat transfer in square meters (square feet), L is the thickness of the metal slab in meters (feet), and $\Delta T = T_1 - T_2$ is the temperature difference in the y direction in degrees Kelvin (degrees Farenheit). From Eq. (5.1.3) it is evident that the units of k are watts per meter times degrees Kelvin (British thermal units per hour per feet per degrees Farenheit). If measurements are made carefully, the material is homogeneous, and the temperature difference across the bar is linear and the gradient small, the thermal conductivity is a material property independent of the thickness of the material. Typical values for thermal conductivity are shown in Table 5.1.

Table 5.1. Typical thermal properties of various substances. (Note that a good source of thermophysical property data can be found at the National Institute of Science and Technology Web site: www.nist.gov.)

Substance (near 25 °C at 1 bar)	Thermal conductivity k (W/m K)	Density ρ (kg/m³)	Heat capacity $\hat{C}_p$ (J/kg K)	Thermal diffusivity $\alpha = k/\rho\hat{C}_p$ (m²/s) × 10⁶
Steel, stainless	16	7820	460.8	4.4
Window glass	0.78	2720	840	0.34
Brick (masonry)	0.66	1670	838	0.47
Wood	0.21	820	2390	0.11
Fiberglass insulation	0.03	400	1000	0.075
Mercury	8.2	13 500	140	4.34
Water	0.60	998	4182	0.14
Oil	0.14	880	1675	0.09
Air	0.026	1.18	1006	22.0
Cabon dioxide	0.017	1.76	846	11.4

Examine the values for the thermal conductivity presented in Table 5.1. Thermal conductivity is a measure of the ability of a sample to transmit heat. The molecular processes by which heat is transmitted are direct convection, molecular collisions, electron motion, and radiation. Thermal conductivity is a measure of the molecular transfer of kinetic energy by collision and electronic motion (for conductors). Therefore it depends on both the chemical composition and the physical state of a substrate. As seen, the thermal conductivity of solids is typically comparable with or greater than that of liquids, which are in turn greater than that for gases. The thermal conductivity depends on the mean free path between atoms, which is much shorter in solids and liquids relative to gases. Crystallinity also increases the rate of transfer as lattice vibrations contribute to conductive heat transfer. In metals, the valence electrons are free to move within the lattice and, consequently, thermal conductivity is proportional to electrical conductivity (the Wiedemann–Franz–Lorenz equation). Liquid metals are often used for special applications in heat transfer.

Jean-Baptiste-Joseph Fourier (1768–1830), a mathematical physicist, is credited with understanding that the flow of heat was proportional to a linear driving force given by the difference in temperature across a region of interest and inversely proportional to the resistance to the transfer of heat. Although others, including Biot (1804), reported earlier experiments exploring heat conduction and tests of Newton's "Law" of cooling, which was discussed in Example 5.1.2, Fourier developed the mathematical description of heat conduction now used. The simple *constitutive* relationship he proposed was already introduced in Chapter 1 (Table 1.4) as Fourier's "Law." Note that although there is a solid molecular theoretical basis for this linear relationship between heat flux and temperature gradient, it strictly applies to homogeneous materials with no convection and small thermal gradients. Hence it is not a Law of Nature, such as conservation of energy, and consequently we refer to it more properly as a constitutive equation.

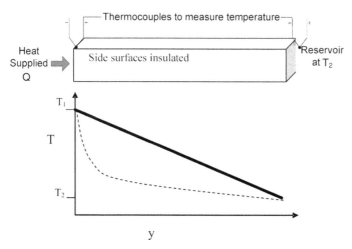

Figure 5.3. Measurement of thermal conductivity. (The graph shows the measurements and the assumed linear temperature profile used for the determination of the conductivity.)

The differential form of Fourier's constitutive equation for heat conduction is given in Table 1.4 and written here in one-dimensional form for conduction in the y direction:

$$q_y = \frac{Q_y}{a} = -k\frac{dT}{dy}. \tag{5.1.4}$$

This relationship has been verified through over 100 years of experimental testing. Many types of apparatus were designed to enable precise determination of k in both metals and nonmetal solids. Equation (5.1.3) shows that one must, at steady state, measure the heat flow, temperature difference and length to determine the material property of thermal conductivity. The concept of such a devise is sketched in Figure 5.3, where a sample of thickness L is placed between an electric heater of known heat generation and a heat reservoir that accepts the heat without significant change in temperature. A fixed quantity of heat energy is generated by the electric heater for a given time period. The sides of the sample and the heater are insulated so that, in principle, all of the energy is transferred across the y dimension of the sample. Temperature probes (thermocouples) on both sides of the sample enable measurement of the temperature difference ΔT. Observations are recorded only when the temperatures on both sides of the sample are constant (but unequal). Then k can be calculated from Eq. (5.1.3). It is much more difficult to determine conductivity of liquids and gases because the thermal gradients can drive convection, resulting in an overestimate of k. To reduce the possibility of fluid motions, devices have been constructed to study thin-liquid films and small temperature differences. Early adaptations used transient heat flow measurements (Ångström's method) to avoid having to measure the heat flux. Modern American Society for Testing and Materials (ASTM) standards exist for accurate measurements of thermal conductivity.

Returning to the energy balance given in Eq. (5.1.1), we find that inserting the one-dimensional form of Fourier's constitutive equation leads to a closed equation for the unknown temperature profile:

$$\frac{\partial(\rho\hat{C}_p T)}{\partial t} = \frac{\partial\left(k\dfrac{\partial T}{\partial y}\right)}{\partial y}.$$

For constant physical properties $(\rho, \hat{C}_p, k)$,

$$\frac{\partial T}{\partial t} = \frac{k}{\rho\hat{C}_p}\frac{\partial^2 T}{\partial y^2} = \alpha\frac{\partial^2 T}{\partial y^2}. \qquad (5.1.5)$$

A convenient grouping of terms defined in Eq. (5.1.5) is the *thermal diffusivity* $\alpha = k/\rho\hat{C}_p$. The final equation, (5.1.5), is often denoted as the one-dimensional version of Fourier's "Second Law" and is the simplified Level III balance for this problem.

There are several important simplifications inherent in the final result, Eq. (5.1.5): one-dimensional conductive heat transfer and constant thermal diffusivity independent of position and temperature. Applying this equation to the physical situation of *steady-state* transport in the barrier yields

$$0 = \alpha\frac{d^2 T}{dy^2}, \qquad \therefore \frac{d^2 T}{dy^2} = 0.$$

Integrating once shows the temperature gradient is a constant:

$$\frac{dT(y)}{dy} = A.$$

Integrating again yields

$$T(y) = Ay + B. \qquad (5.1.6)$$

This result confirms the intuitive result that the temperature profile in the bar should be linear at steady state. The constants of integration (A, B) are obtained from applying two boundary conditions. Notice that at thermal equilibrium the temperature is constant throughout the material, $A = 0$, and there is no net heat flux. Next, we consider various boundary conditions and the resulting steady-state temperature profiles and heat fluxes.

Constant-Temperature Boundary Conditions

First consider the example shown in Figure 5.1, with constant temperatures at the barrier surfaces:

$$y = 0, \quad T = T_1,$$
$$y = L, \quad T = T_2.$$

Using these boundary conditions, we determine the constants of integration A, B to yield the linear temperature profile along the barrier as

$$T(y) = \left(\frac{T_2 - T_1}{L}\right)y + T_1.$$

From this solution, we can determine the total heat flow through the barrier at steady-state by using Fourier's constitutive equation:

$$Q_y = q_y a = -ak\frac{dT}{dy},$$

$$\therefore Q_y = ak\frac{T_1 - T_2}{L}. \tag{5.1.7}$$

Thus the heat flow through the barrier is proportional to the thermal conductivity, cross-sectional area, and temperature difference, but inversely proportional to the material thickness.

Flux Boundary Condition

Now consider varying the boundary conditions, such as those in the experiment shown in Figure 5.3, where the heat flux into the bar is specified at $y = 0$. Such a condition can be realized, for example, if an electric heater is attached with good thermal contact to the bar and insulated in the other directions. Constant-heat-flux boundary conditions are often of interest in problems involving heat sinks, in which the heat from an electrical component, such as a computer chip, must be removed to avoid overheating. The boundary condition at the other end of the bar is in contact with a well-mixed fluid so as to fix the temperature at $y = L$ to be the temperature of the well-mixed fluid in the tank T_2. Then our new boundary conditions become

$$y = 0, \quad Q_y = -ak\frac{dT}{dy}\bigg|_{y=0},$$

$$y = L, \quad T = T_2.$$

These boundary conditions define the constants of integration in the solution, Eq. (5.1.6), as follows. The first constant of integration A is equal to the gradient in the temperature profile, and this gradient is defined by the heat flow into the bar, namely,

$$\frac{dT}{dy} = A, \quad A = -\frac{Q_y}{ak}.$$

The second boundary condition defines the integration constant B as

$$T(y = L) = T_2 = -\frac{Q_y}{ak}L + B,$$

$$\therefore B = T_2 + \frac{Q_y}{ak}L,$$

$$\therefore T(y) = \frac{Q_y}{ak}(L - y) + T_2.$$

We can calculate the unknown value of the temperature at the $y = 0$ face of the material as

$$T(0) = \frac{Q_y}{ak}L + T_2. \tag{5.1.8}$$

This temperature is directly proportional to the applied heat load and inversely proportional to the material's conductivity. For example, the temperature of an electronic component attached to a heat sink could be estimated with this result.

A Mixed Boundary Condition

An even more realistic condition for the experiment depicted in Figure 5.3 would be to specify the temperature of the mixed tank (T_2), but to allow the bar's temperature at $y = L$ to be determined by the heat flux from the end of the bar to the well-mixed fluid in the tank. The heat flux from the bar to the reservoir is characterized in terms of a constitutive equation for the rate of heat transfer defined in Table 1.5. This relation is often denoted as *Newton's "Law" of cooling*, although it was probably first considered by Fourier. Thus the boundary condition for this problem is specified as a heat flow determined by the temperature difference between the mixed tank and the surface of the bar as $Q = ha[T(L) - T_2]$. Notice here that, as we are concerned with the rate of heat transfer from a well-mixed tank to a surface of a solid wall, the heat transfer coefficient is a local heat transfer coefficient denoted as h. This is in contrast to the overall heat transfer coefficient denoted by U, which we used to express the rate of heat transfer between two well-mixed tanks and across multiple interfaces. As noted briefly in Chapter 3, the overall heat transfer coefficient depends on the local heat transfer coefficients at each interface as well as conduction across the materials in between. This will be derived in detail for specific geometries shortly.

Using this rate expression, we find that the boundary conditions of interest are

$$y = 0, \qquad Q_y = -ak\frac{dT}{dy}\bigg|_{y=0},$$

$$y = L, \qquad Q_y = ha[T(y = L) - T_2].$$

The constants of integration are found as in the preceding case. However, we should recognize that, for this problem, Q_y is specified by the first boundary condition and at steady state, conservation of energy requires this heat flow be constant everywhere along the bar. Thus Q_y is constant at steady state, and the second boundary condition can be rearranged to yield

$$y = L, \qquad T(y = L) = T_2 + \frac{Q_y}{ha}.$$

Then we find

$$T(y) = Ay + B,$$

$$T(y) = -\frac{Q_y}{ak}y + B,$$

$$T(y = L) = T_2 + \frac{Q_y}{ha} = -\frac{Q_y}{ak}L + B,$$

$$B = T_2 + \frac{Q_y}{ha} + \frac{Q_y}{ak}L,$$

$$T(y) = \frac{Q_y}{ak}(L - y) + T_2 + \frac{Q_y}{ha}.$$

The unknown temperature at the hot end is determined to be

$$T(0) = T_2 + \frac{Q_y}{a}\left(\frac{L}{k} + \frac{1}{h}\right). \qquad (5.1.9)$$

Notice that this temperature now depends on both the material's conductivity and the transfer of heat to the reservoir, as well as the reservoir temperature. The preceding result can be rewritten to express the heat flux as

$$Q_y = \frac{a}{\left(\dfrac{L}{k} + \dfrac{1}{h}\right)} [T(0) - T_2] = Ua \, [T(0) - T_2]. \qquad (5.1.10)$$

This solution defines an *overall heat transfer coefficient* U in terms of the resistances to heat transfer of the bar and the thermal connection to the reservoir:

$$\frac{1}{U} = \frac{L}{k} + \frac{1}{h}.$$

The concept of the addition of resistances to heat transfer will be elaborated on later in the chapter.

Mathematical Considerations

It is important that the student of heat and mass transfer recognize that the simplified equations for conductive heat transfer and convection are classic examples of partial differential equations, whose methods of solution are discussed in applied mathematics classes. Fourier's "Second Law" is an example of a *parabolic* partial differential equation. These are often amenable to solution by the method of *separation of variables*. When the boundary conditions are linear and homogeneous, the resulting eigenvalue problem is known as a *Sturm–Liouville Problem*, the solutions to which have certain, known properties. Students are encouraged to reexamine their applied mathematics textbooks for methods of solving parabolic partial differential equations, including separation of variables, integral transforms, and eigenfunction expansions before proceeding further to solutions of time-dependent problems.

For steady-state conditions, both of the preceding problems are homogeneous elliptical ordinary differential equations of second order (i.e., the *Laplace* equation).

We also discussed three types of boundary conditions, namely,

- fixed temperature,
- fixed flux,
- external reservoir temperature specified with Newton's "Law" of cooling.

These fall under the following conventional mathematical names:

- boundary conditions of the first kind—*Dirichlet*,
- boundary conditions of the second kind—*Neumann*,
- boundary conditions of the third kind—*Robin*.

A student with a solid basis in applied mathematics will recognize the value in categorizing the final, mathematical form of the model equations in order to choose the method of solution as well as estimation of model behavior even without solution.

5.1.2 Definition of the Biot Number for Heat Transfer

The previous example demonstrates that the relative magnitude of the external heat transfer at the boundary of the bar (as characterized by the heat transfer coefficient

h) to the thermal resistance inside the bar (as characterized by L/k) is important in determining the temperature profile. Equation (5.1.9) can be rearranged to yield

$$T(0) = T_2 + \frac{Q_y}{a}\frac{L}{k}\left(1 + \frac{1}{Bi}\right).$$ (5.1.11)

The dimensionless group that characterizes the relative rate of heat transfer from the object to the surroundings relative to internal heat conduction is given a special name in honor of Biot, a French physicist who contributed to our understanding of heat transfer:

$$Bi = \frac{hL}{k}\frac{[(W/m^2\,K)m]}{(W/m\,K)}.$$ (5.1.12)

Within the context of this problem, the temperature at the hot surface T(0) critically depends on the Biot number. Large values of the Biot number (i.e., large h) lead to the solution obtained in Eq. (5.1.8), where the surface temperature of the barrier was that of the fluid in the mixed tank. Small values can lead to much higher temperatures [note that $Bi \geq 0$, and, as $Bi \to 0$, $T(0) \to \infty$]. Note that, in the definition of the Biot number, the heat transfer coefficient is external to the material, which is in direct contrast to a related number, the *Nusselt* number, defined in the next section.

EXAMPLE 5.1 "LUMPED" ANALYSIS: COOLING BY A FIN. Consider the consequences of modifying the insulated bar shown in Figure 5.3 by removing the insulation and exposing the entire bar to the atmosphere. For simplicity, consider a rod, as shown in Figure E5.1; a highly thermally conductive rod is attached to a hot surface. This is typically done to increase the rate of heat transfer from a warm body, such as on air-cooled gasoline engines. Air flows along the exposed sides of the "fin." An example of this idealized geometry might be the cooling fins on a motorcycle engine. The goal of this analysis is to calculate the extra heat transfer achieved.

SOLUTION. Consider the case in which the rate of heat conduction along the rod will be fast relative to the rate of convection from the free surface (that is, $Bi \ll 1$). A balance on the shell indicated in Figure E5.1 yields (where we assume a

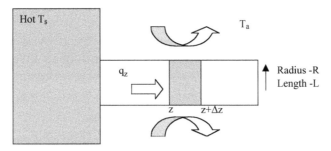

Figure E5.1. Cylindrical fin.

constant heat transfer coefficient from all surfaces and apply Fourier's constitutive equation),

$$0 = q_z \pi R^2|_z - q_z \pi R^2|_{z+\Delta z} - h(2\pi R\Delta z)(T - T_a),$$

$$\frac{dq_z}{dz} = -\frac{2h}{R}(T - T_a), \tag{E5.1}$$

$$\frac{d^2T}{dz^2} = \frac{2h}{Rk}(T - T_a).$$

The relevant boundary conditions are that, at $z = 0$, $T = T_s$, and for simplicity, we assume that the end of the fin is insignificant enough such that at $z = L$, $q_z = 0$.

This model equation is readily integrated. However, it is beneficial to express all quantities in dimensionless form. Nondimensional equations are easier to program for computational solution as well as more compact and more easily translated to other related problems. Hence we choose a natural length scale as the rod length L and the natural temperature scale as the difference between the hot-body temperature and the air temperature. Defining the following dimensionless variables,

$$\hat{T} = (T - T_a)/(T_s - T_a), \quad \hat{z} = z/L, \quad \hat{R} = R/L, \quad Bi = 2Rh/k,$$

enables writing the model equation in dimensionless form as

$$\frac{d^2\hat{T}}{d\hat{z}^2} = \frac{Bi}{\hat{R}^2}(\hat{T}), \tag{E5.2}$$

where

$$\hat{z} = 0, \quad \hat{T} = 1,$$
$$\hat{z} = 1, \quad \frac{d\hat{T}}{dz} = 0.$$

This equation can be readily integrated to yield the following solution. Defining $\Gamma^2 = Bi/\hat{R}^2$ yields

$$\hat{T}(\hat{z}) = \cosh(\Gamma\hat{z}) - \tanh(\Gamma)\sinh(\Gamma\hat{z}) \tag{E5.3}$$

This result is shown graphically for a parameter value of $\Gamma = 3$ in Figure E5.2. Notice that the temperature at the end of the fin does not approach the ambient temperature,

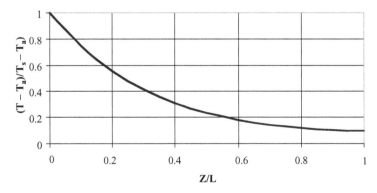

Figure E5.2. Temperature profile in fin for $\Gamma = 3$.

which is a consequence of our no-flux boundary condition. Further, note that the temperature varies rapidly along the fin, such that far from the hot surface the driving force for heat transfer is low. This implies that there is a law of diminishing returns with regards to the choice of fin length. Increasing the fin length does not commensurately increase the rate of heat transfer. This is calculated next.

The rate of heat conduction from the hot body is most easily calculated by determining the heat flow at the surface in contact to the hot body, i.e., Q_z at $z = 0$. From our dimensionless solution, this is determined as

$$Q_z = -\pi R^2 k \frac{dT}{dz}\bigg|_{z=0} = -\pi R^2 k \left(\frac{T_s - T_a}{L}\right) \frac{d\hat{T}}{d\hat{z}}\bigg|_{z=0}$$

$$= \pi R^2 k \left(\frac{T_s - T_a}{L}\right) \Gamma \tanh(\Gamma). \tag{E5.4}$$

Fin efficiency is defined as the heat transferred from the fin relative to that transferred if the entire fin surface were at the temperature of the hot body. This "ideal fin" has the heat transfer rate of

$$Q^{\text{ideal-fin}} = h 2\pi RL (T_s - T_s).$$

Consequently, the fin efficiency η is calculated as

$$\eta = \frac{Q_z}{Q^{\text{ideal-fin}}} = \frac{\pi R^2 k \left(\dfrac{T_s - T_a}{L}\right) \Gamma \tanh(\Gamma)}{h 2\pi RL (T_s - T_s)} = \frac{\tanh(\Gamma)}{\Gamma}. \tag{E5.5}$$

This function is plotted in Figure E5.3, showing that increasing the fin length (i.e., increasing gamma) leads to lower and lower fin efficiency. Thus, given economic and other design constraints, there will be an optimum fin design to achieve a given amount of heat transfer.

Finally, note that the entire analysis is based on the assumption that the temperature profile inside the fin is uniform in the radial direction. For this to occur, conduction within the fin must be much greater than convection from the fin surface. This requirement is captured in the Biot number, Bi, which must be small for the lumped-parameter analysis to be valid. Consequently, Γ can be large only if the fin has a high aspect ratio.

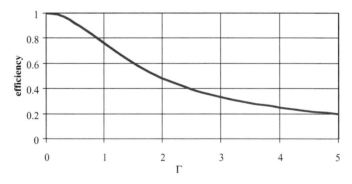

Figure E5.3. Fin efficiency calculated from Eq. (E5.5).

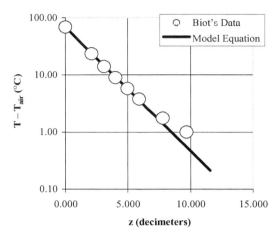

Figure E5.4. Comparison of Biot's data and fin model.

EXAMPLE 5.1 CONTINUED: COMPARISON WITH EXPERIMENT. Biot reported data (Table E5.1) in his investigations of heat conduction and Newton's "Law" of cooling by placing thermometers along a bar, with one end heated to 82 °C and the bar exposed to air at 13 °C. In Biot's experiment, the bar was much longer (27.342 dm) such that the free end was found to be at ambient temperature. Biot successfully analyzed his data by assuming an exponential decay; show if this follows from the prior analysis.

SOLUTION. We again assume the Biot number is sufficiently small such that a lumped analysis applies. The derivation leading to Eqs. (E5.2) is modified because the bar is essentially infinitely long for this experiment such that the heat never propagates to the far end. We can obtain a solution either by assuming that the parameter $\Gamma \rightarrow \infty$ or by replacing the second boundary condition such that

$$\hat{z} \rightarrow \infty, \quad \frac{d\hat{T}}{d\hat{z}} = 0.$$

The solution becomes simply $\hat{T}(\hat{z}) = e^{-\Gamma \hat{z}}$, which was the functional form assumed by Biot. Figure E5.4 shows a best fit of this expression to the data, with the parameter value of

$$\Gamma = \frac{1}{2},$$

confirming Biot's calculations (see Table E5.1).

5.1.3 Definition of the Nusselt Number

So far, we have employed two different constitutive equations to model the rate of heat transfer. Compare Fourier's constitutive equation written in finite-difference form, Eq. (5.1.3), with our earlier rate law for heat transfer, $Q = ha\Delta T$, where the temperature difference is written such that heat flows from hot to cold, i.e., along the temperature gradient. We can use our solution of the temperature profile and

Table E5.1. Biot's experimental data (Biot JB. Mémoire
sur la propagation de la chaleur. Bibliotheque
Britannique 1804; 27: 310–329)

Therm. #	Position (dm)	$T - T_{air}$ (°C)
0	0.000	69.00
1	2.115	23.50
2	3.115	14.00
3	4.009	9.00
4	4.970	5.75
5	5.902	3.75
6	7.777	1.75
7	9.671	1.00
8	11.556	*

Fourier's constitutive equation to compute a heat flow through the wall. Then we can describe this heat flow in terms of an effective heat transfer coefficient for the wall h_{wall} by equating the heat flow to the engineering constitutive equation. For one-dimensional conduction through a perfect solid we have

Fourier's constitutive equation:

$$\frac{Q_y}{a} = q_y = -k\frac{dT}{dy} = k\frac{(T_1 - T_2)}{L};$$

Rate of heat transfer:

$$\frac{Q_y}{a} = h_{wall}(T_1 - T_2),$$

$$\therefore h_{wall} = k/L.$$

Note that, in the preceding equation, the heat transfer coefficient is defined for the solid itself (as opposed to the external heat transfer coefficient, such as appearing in the Biot number previously defined). Thus the empirical heat transfer coefficient h_{wall} is seen to be a combination of the material property of thermal conductivity and the geometry of the experiment.

In general, a dimensionless group can be defined for any heat transfer coefficient, termed the *Nusselt* number, Nu. It is defined in terms of the material's thermal conductivity and a characteristic length L for heat transfer:

$$Nu = \frac{hL}{k} \frac{[(W/m^2\ K)\,m]}{(W/m\ K)}. \qquad (5.1.13)$$

So, for one-dimensional heat transfer through a perfect solid,

$$Nu_{wall} = \frac{h_{wall}L}{k} = 1 \quad \text{for a perfect solid.} \qquad (5.1.14)$$

Although trivial, this result is the first example of a *heat transfer correlation*, of which we will have much to say in Chapter 6. Further, although the thermal conductivity, length, and effective heat transfer coefficient will vary from material to material and geometry to geometry, all such solids can be represented by the simple relationship that Nu = 1. This is an example of how engineering analysis can be used to show the similarities between experiments on different materials and geometries. For liquids and gases in which the fluid is moving, convection leads to a more complex functionality relating the Nusselt number to dimensionless groups associated with the fluid mechanics and ratios of transport coefficients. However, note that the methodology for calculating the overall heat transfer coefficient involves equating the heat flux derived for a specific geometry to the engineering rate constitutive equation. We will employ this method extensively in the next chapter. However, before continuing with our study of heat conduction, we turn to the analogous process of molecular diffusion in solids and quiescent fluids.

5.2 Rate of Molecular Diffusion

In Chapter 4 the model equations for mass contactors were developed for the mixed–mixed, mixed–plug, and plug–plug types of mass contactors. When developing the analogous model equations for heat exchangers in Chapter 3 we had to estimate the heat transfer coefficient U. However, with mass contactors we need to estimate both the mass transfer coefficient K_m and the interfacial area to complete analysis and design problems. In the following subsection, we examine the fundamentals of mass transfer in idealized solids and quiescent fluids as precursors to considerations of mass transfer in flowing fluids, which will be discussed in Chapter 6. The interfacial area prediction is more complex, and its prediction will be covered in Chapter 7.

Following the development of heat conduction in Section 5.1, here we begin by examining an experiment to determine binary diffusivities. We proceed by analyzing the experimental situation and propose a constitutive equation, which will be verified by comparison with experiment.

5.2.1 Experimental Determination of Binary Diffusivities D_{AB} and Verification of Fick's Constitutive Equation

Adolf Fick (1829–1901), a medical researcher, was concerned with oxygen transport in tissues and blood. He was aware of Fourier's work and hypothesized that diffusive mass transfer should have a constitutive relation that relates the flux of a species linearly to its concentration gradient, in direct analogy to Fourier's constitutive equation that relates the heat flux to the temperature gradient. Fick tested the hypothesis by setting up two containers of water with soluble crystals in the bottom. The solute (A) diffused through the water in a narrow channel up to a large reservoir that was maintained to be at nearly zero concentration in A. The crystals dissolved, setting the concentration at the bottom of the channel to be that in equilibrium with the solid crystals, and, with time, a steady concentration profile in the channel region was obtained. The concentration profile was deduced by sampling of the fluid at various points in the water channel between the crystals and the top reservoir and measuring the specific gravity. The data are shown in Figure 5.4, where two types of channels were tested, a straight cylinder and a funnel shape. For the straight cylinder, the expected linear profile was observed, whereas, for the funnel shape, a nonlinear density gradient was measured.

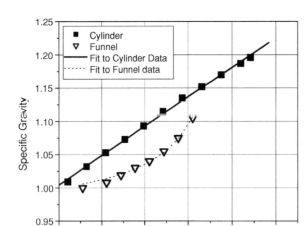

Figure 5.4. Fick's experiment (Fick, 1855).

First consider Fick's experiment with the cylindrical channel. The model equations are derived for a control volume depicted in Figure 5.5. This cylinder is filled with a liquid (denoted as species B) through which species A diffuses. We assume that the concentration of species A is very low. There is no pressure-driven flow or any net fluid motion. We idealize the experiment as a channel connecting two well-mixed reservoirs of nearly infinite extent. If $C_{A,0} > C_{A,L}$, it would be reasonable to expect that species A will be transported in direction z and that the flux of species A, denoted as a molar flux $J_{A,z}$, will be proportional to a driving force $\Delta C_A = C_{A,0} - C_{A,L}$. Note that, given that the tube is capable of flow, there could be transport in the tube by convection, as given by the volumetric flow rate q. A steady-state concentration profile (and transport rate) is expected to develop after an initial transient. At this point we wish to note that we will, in keeping with the general literature, use a *molar concentration* for C_A (in units of moles per cubic meter) such that $J_{A,z}$ is a *molar flux of* A (in moles per square meter per seconds). Lower case c_A and $j_{A,z}$ are often used to denote mass concentration and mass flux, respectively. These are related, but, for bookkeeping purposes, keep in mind that molar quantities are used throughout this chapter.

Application of the statement of conservation of mass introduced in Chapter 1 by use of the nomenclature just introduced for molar quantities leads to the following

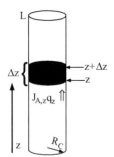

Figure 5.5. Control volume for analysis of Fick's experiment with a cylinder of radius R_C and length L.

Level III balance for the moles of species A:

$$\pi R_C^2 \left(C_A|_{t+\Delta t} - C_A|_t\right) \Delta z = \pi R_C^2 \Delta t \left(J_{A,z}|_z - J_{A,z}|_{z+\Delta z}\right)$$

$$+ \left(q_z C_A|_z - q_z C_A|_{z+\Delta z}\right) \Delta t,$$

$$\therefore \lim_{\Delta z, \Delta t \to 0} \frac{\partial C_A}{\partial t} = -\frac{\partial (N_{A,z})}{\partial z} \tag{5.2.1}$$

where

$$N_{A,z} = J_{A,z} + v_z C_A, \qquad v_z = \frac{q_z}{\pi R_C^2}.$$

Here, we introduce the standard nomenclature of the total flux of A, which is the sum of the diffusive flux plus the convective flux:

$$N_{A,z} = J_{A,z} + v_z C_A, \tag{5.2.2}$$

{total flux of A} = {diffusive flux} + {convective flux}.

The velocity in the z direction is written in terms of the volumetric flow rate and the cross-sectional area as

$$v_z = \frac{q_z}{a}.$$

We have written the balance equation for a general case in which diffusion and convection may occur, but simplifications are possible for Fick's experiment, as discussed next.

In Fick's experiment he was careful to prevent any mixing in the tube. Further, as the concentrations are dilute, there will be no significant flow of mass in the tube so we can set $q_z = 0$ (we revisit this assumption later in the chapter). Then the steady-state solution shows that, for this geometry, the flux of species A is constant:

$$\frac{d J_{A,z}}{dz} = 0, \qquad \therefore J_{A,z} = \text{constant}. \tag{5.2.3}$$

The Level II balance for this problem shows that the species mass flux is zero, which, as will be shown shortly, demonstrates that the concentration of A is uniform in the tube.

To proceed with a Level III analysis, a constitutive equation is required relating the molar flux $J_{A,z}$ to the concentration gradient. Fick proposed, in direct analogy to Fourier's constitutive equation, a relation of the form introduced in Chapter 1, Table 1.4. It states

$$J_{A,z} = -D_{AB} \frac{dC_A}{dz}. \tag{5.2.4}$$

This relationship is denoted as Fick's constitutive equation of diffusion, or popularly as Fick's "Law" of diffusion. In the preceding equation, D_{AB} is the coefficient of diffusion, or *diffusivity*. This is a binary diffusivity as species A is diffusing through a material of species B. In analogy to the thermal conductivity k, the diffusivity is expected to be determined experimentally, from empirical correlations, or, in limited cases, from molecular properties. D_{AB} has SI units of square meters per second

Table 5.2. *Typical diffusion coefficients at 1 atm (from Bird et al., 2002, Table 17-1-1,2,3 and other sources)*

Gas Pair A–B	Temperature (K)		D_{AB} ($\times 10^4$) (m²/s)
CO_2–N_2	273.2		0.144
N_2–O_2	273.2		0.181
H_2 N_2	273.2		0.674
H_2O–N_2	308		0.259
C_3H_8–nC_4H_{10}	378.2		0.0768
O_2–Water	298		1.80×10^{-5}
Ethanol–water (liquid)	298	$X_A = 0.026$	1.076×10^{-5}
"	"	$X_A = 0.408$	0.405×10^{-5}
"	"	$X_A = 0.944$	1.181×10^{-5}
He–SiO_2 (solid)	293.3		2.4–5.5×10^{-10}
H_2–Ni (solid)	348		1.16×10^{-8}
O_2–Polystyrene (solid)	298		0.11×10^{-6}
O_2–Polybutadiene (solid)	298		1.5×10^{-6}

(square feet per second). However, unlike heat conductivities, the diffusivity is a property of both diffusing species A and bulk material B. Note that $D_{AB} = D_{BA}$.

Typical values of binary diffusivities are given in Table 5.2. Diffusivities in gases are of the order of 10^{-5} m²/s, liquids drop to $\sim 10^{-9}$ m²/s, and gases in polymers to $\sim 10^{-10}$ m²/s. Note that the diffusivities will, in general, depend on relative concentration. Molecular theories for estimating binary diffusivities can be found in the references given at the end of this chapter.

Analysis of the one-dimensional diffusion problem defined by Fick's experiment proceeds in direct analogy to the solution of general problems in one-dimensional heat transfer. When Fick's constitutive equation is used the species mole balance becomes

$$\frac{\partial C_A}{\partial t} = D_{AB} \frac{\partial^2 C_A}{\partial z^2}. \tag{5.2.5}$$

The striking similarity to Fourier's "Second Law," i.e., Eq. (5.1.5), is apparent where the thermal diffusivity α plays the role corresponding to the mass diffusivity D_{AB}. Just as for problems of heat conduction considered in Subsection 5.1.1, we envision three types of boundary conditions, constant temperature, constant flux, or coupling to an external mass transfer coefficient. As the mathematics will be essentially the same, we restrict our attention to the case of constant-concentration boundary conditions, as this corresponds most closely to Fick's experiment.

There are several important simplifications inherent in Eq. (5.2.5): one-dimensional mass transfer, no bulk flow (i.e., dilute in A), and a constant diffusivity independent of position. Applying this equation to the idealized model of Fick's experiment, i.e., steady-state transport in the cylindrical channel, yields

$$0 = D_{AB} \frac{d^2 C_A}{dz^2}. \tag{5.2.6}$$

Notice the mathematical similarity with Fourier's analysis of one-dimensional heat conduction, Eq. (5.1.6). Following the same analysis, the equation can be integrated twice to yield $C_A(z) = Az + B$. The constants of integration are obtained from applying the two boundary conditions:

$$z = 0, \quad C_A = C_{A,0},$$
$$z = L, \quad C_A = C_{A,L},$$

yielding a linear concentration profile in the channel:

$$C_A(z) = \left(\frac{C_{A,L} - C_{A,0}}{L} \right) z + C_{A,0}. \tag{5.2.7}$$

As shown in Figure 5.4, the solution, Eq. (5.2.6), corresponds to the experimental measurement of a linear concentration profile. Given a calibration curve of specific gravity versus concentration, the preceding solution of the model can be directly compared with experiment. Note that, in Fick's experiment, the concentration at the surface of the crystals at the bottom of the device is taken to be the saturated solubility at equilibrium at the experimental temperature $C_{A,sat}(T)$, whereas the concentration at the top of the tube is taken to be zero as the large reservoir at the top maintains a low concentration.

EXAMPLE 5.2. Consider how one uses the solution previously derived to experimentally determine the diffusivity from measurements of the weight loss of the crystals.

SOLUTION. From the linear profile, Eq. (5.2.7), the molar flux can be calculated from Fick's constitutive equation if the binary diffusivity were known. A mole balance on the crystals in the bottom of the cylindrical channel enables connecting the change in mass of the crystals to the flux. Given the flux is proportional to the binary diffusivity, this enables estimating it from the mass change of the crystals. The procedure of analysis is analogous to the problem of dissolution of the salt tablets in the stirred tank discussed in Chapter 4, except here the flux is given solely by diffusion and not convection. The Level III mass balance for the crystals is simply given by

$$\frac{d\rho^{II} V^{II}}{dt} = -J_{A,z} a MW^{II},$$

where a is the area of the crystals of volume V, density ρ, and molecular weight MW^{II}. The flux is given by Fick's constitutive equation as

$$J_{A,z} = -D_{AB} \frac{dC_A}{dz} = -D_{AB} \left(\frac{0 - C_{A,sat}}{L} \right) = D_{AB} \frac{C_{A,sat}}{L}.$$

Integration of the mole balance requires specifying the relationship between area and volume, i.e., specifying the shape of the crystals as

$$a = N\alpha (V^{II}/N)^{2/3} = N^{1/3} \alpha (V^{II})^{2/3},$$

as was done in Chapter 4, where N is the number of crystals with shape factor α. The balance equation for the volume of the crystals becomes

$$\frac{d\left(V^{II}\right)^{1/3}}{dt} = -D_{AB}\frac{C_{A,sat}}{L}\alpha N^{1/3}\frac{MW^{II}}{\rho^{II}}.$$

Integration yields the following equation:

$$V^{II}(t) = \left\{[V^{II}(0)]^{1/3} - \left(\alpha D_{AB}N^{1/3}\frac{C_{A,sat}}{3L}\frac{MW^{II}}{\rho^{II}}\right)t\right\}^3.$$

This can be directly compared with the solution for the dissolution of the salt tablets, Eq. (4.2.28). Comparison shows that defining an effective mass transfer coefficient as

$$K_m \Leftrightarrow \frac{D_{AB}}{L}$$

makes the two expressions equivalent. We will have more to say about this shortly. Often, the size of the crystals is more readily measured. To compare with experimental data for the size of the crystals in the bottom of the container with time, the equation can be rearranged to yield

$$[V^{II}(0)]^{1/3} - [V^{II}(t)]^{1/3} = \left(\alpha D_{AB}N^{1/3}\frac{C_{A,sat}}{3L}\frac{MW^{II}}{\rho^{II}}\right)t.$$

Note that, as the cube root of the volume of the crystals is a measure of the crystal radius, this result shows that the crystal radius decreases linearly with time. This result, of linear dissolution, has an analog in crystal growth, whereby crystals often grow by diffusion of new material to the surface with a steady linear rate of growth.

5.2.2 Definition of the Biot Number for Mass Transfer

The preceding analysis assumes that the rate of mass transfer from the crystals to the surrounding fluid is sufficiently fast such that the local concentration of A in the liquid phase will be in equilibrium with the solid. In direct analogy to the discussion of the various boundary conditions for heat transfer, in Section 5.1, there are also three types of boundary conditions for mass transfer problems, namely specification of the concentration, flux, or the use of a constitutive equation linking the concentration to the surrounding concentration. For mass transfer, the constitutive equation relating the rate of mass transfer was introduced in Table 1.5 and employed extensively in Chapter 4.

The rate of mass transfer between the crystal surface and the surrounding fluid is defined by a linear relationship involving the mass transfer coefficient: $r_A = k_m\left(c_A^I - Mc_A^{II}\right)$, where r_A is the mass flux of species A, M is the distribution coefficient, and $c_A = MW_A C_A$ is the mass concentration in keeping with the notation in this chapter. Further, the lowercase k_m denotes the mass transfer coefficient across one interface, as opposed to the overall mass transfer coefficient K_m, which, as discussed in Chapter 4, is related to a sum over the individual local mass transfer coefficients.

For this specific problem, the rate of moles of A dissolving from the crystal must diffuse through the cylinder for the system to be at steady state. The rate of mass transfer at the boundary $z = L$, taken to be the bottom of the cylinder where the crystals are located, must therefore be equal to the mass flux of A through the cylinder, or $r_A = M_{w,A} J_{A,z}$, where sign is correct for the fact that r_A has been written with "I" denoting the liquid and "II" the solid, and that the solute A moves from $z = 0$ (the bottom of the cylinder) toward the clear fluid at $z = L$.

In analogy to the development for heat transfer, if we include resistance to mass transfer from the crystals to the quiescent fluid in the cylinder, it will take the form of a flux boundary condition. This new boundary condition will lead to a modification of the equation for the concentration profile, Eq. (5.2.7). The Level III balance, Eq. (5.2.5), applies and the steady-state concentration profile will still be linear. This modification follows from identifying the boundary conditions for the problem as

$$z = 0, \quad C_A = C_{A,0},$$

$$z = L, \quad J_{A,z} = -D_{AB}\frac{dC_A}{dz} = r_A = k_m[C_A(L) - C_{A,sat}].$$

Notice that, in the preceding equations, we have dropped the superscript $(^1)$ as it is understood we are considering the concentration in the liquid phase (which is now given in molar units), as well as recognizing that the driving force is proportional to the difference in concentration with respect to the saturation concentration of the crystal material. The constants of integration are found from the new boundary conditions, as was done in Section 5.1 for the corresponding heat transfer problem [Eq. (5.1.11)]. A linear concentration profile again results, with the following form:

$$C_A(z) = \left(\frac{C_{A,sat} - C_{A,0}}{1 + Bi_m}\right)\frac{Bi_m}{L}z + C_{A,0}. \tag{5.2.8}$$

As with the analogous example in heat transfer, the solution can be rewritten in a form that combines mass transfer and diffusion terms into a dimensionless group that is known as the Biot number for mass transfer, in direct analogy to the Biot number for heat transfer Equation (5.1.12).

$$Bi_m = \frac{k_m L}{D_{AB}}\frac{(m/s)m}{m^2/s}. \tag{5.2.9}$$

Notice how this dimensionless group characterizes the importance of mass transfer resistance at a boundary (i.e., at the crystal surface) to the diffusive resistance through the control volume (which, in this case, is the liquid column). For the case in which the crystal is indeed at equilibrium with the surrounding fluid at the bottom of the column,

$$k_m \gg \frac{D_{AB}}{L} \quad \text{and then} \quad Bi_m \gg 0.$$

In this limit, Eq. (5.2.8) reduces to the simpler expression derived for no mass transfer limitations, Eq. (5.2.7), where $C_{A,L} = C_{A,sat}$.

EXAMPLE 5.3. Estimate the importance of mass transfer limitations in a typical crystal dissolution experiment such as that of Fick.

SOLUTION. Here it is sufficient to consider typical orders of magnitude of the various terms in Bi_m. For a 1-cm (0.01-m) length L, typical mass transfer coefficients in the liquid phase of 10^{-5} m/s and binary diffusivities of the order of 10^{-9} m²/s (see Table 5.2),

$$Bi_m \simeq \frac{10^{-2}10^{-5}}{10^{-9}} \simeq 100$$

Clearly, for such situations, the Biot number for mass transfer is sufficiently large such that resistance to dissolution at the surface of the crystals is negligible and the rate of dissolution is controlled by diffusion in the stagnant liquid column. Note that the situation can be different for gases, for which the diffusivities are much higher.

5.2.3 Definition of the Sherwood Number

Again by analogy to one-dimensional steady-state heat transfer in a homogeneous solid, for the idealized problem considered by Fick of diffusion through the quiescent liquid column, we can readily identify the mass transfer coefficient in terms of the material transport property D_{AB} and the geometry. The overall flux through the column can be written in analogy to the rate of heat conduction through a solid (Section 5.1.3):

$$J_{A,z} = k_m(C_{A,0} - C_{A,L}). \tag{5.2.10}$$

This form of the rate expression for mass transfer written for the specific experiment at hand can now be compared with Fick's constitutive equation for mass transfer to relate the local mass transfer coefficient k_m to the binary diffusivity. Equating this flux with Fick's constitutive equation yields

Fick's constitutive equation:

$$J_{A,z} = -D_{AB}\frac{dC_A}{dz} = \frac{D_{AB}}{L}(C_{A,0} - C_{A,L});$$

Rate of mass transfer:

$$J_{A,z} = \frac{r_A}{M_{w,A}} = k_{m,Fick}(C_{A,0} - C_{A,L}),$$

$$\therefore k_{m,Fick} = \frac{D_{AB}}{L}.$$

Just as with heat transfer, we can expect that the mass transfer coefficient will be determined by regression of experimental data, correlations based on data, or, in limited cases, a more fundamental analysis of the mass transport. However, just as for the heat transfer coefficient, the mass transfer coefficient k_m is expected to be a function of the geometry of the experiment as well (which in this case is the length of the channel). Finally, the SI units of $k_m = $ m/s are those of velocity.

Often it will be useful to define a dimensionless mass transfer coefficient, termed the *Sherwood* number Sh, in analogy to the Nusselt number Nu, as

$$\text{Sh} = \frac{k_m \, L}{D_{AB}} \frac{(\text{m/s}) \, \text{m}}{\text{m}^2/\text{s}}. \tag{5.2.11}$$

The Sherwood number is the rate of mass transfer to the effective rate for diffusion, and it depends on material properties in the phase for which the mass transfer coefficient is defined.

For Fick's experiments, we see that we can define an effective mass transfer coefficient and Sherwood number as

$$\text{Sh}_{\text{Fick}} = \frac{k_{m,\text{Fick}} L}{D_{AB}}$$

$$= 1 \quad \text{for diffusive mass transfer.} \tag{5.2.12}$$

Thus, for the idealized, one-dimensional diffusive mass transfer problem considered by Fick, Sh $= 1$, in analogy to the simple, one-dimensional heat transfer conduction in a homogeneous solid, for which Nu $= 1$. However, more generally, when fluid motion occurs, the Sherwood number will depend on the relative rate of convective motion as well as on material properties, as will be seen in Chapter 6.

5.3 Geometric Effects on Steady Heat Conduction and Diffusion in Solids and Quiescent Fluids

In this section, we briefly consider the effects of geometry on the steady-state conduction of heat and molecular diffusion in perfect solids and quiescent fluids. The geometry of the material through which heat is conducted or through which diffusion occurs will affect the rate of transfer as well as the temperature or concentration profile. An example of this is apparent in Fick's experiment, in which the concentration profile is not linear in the conical geometry, as opposed to the linear profile observed for the cylindrical geometry (see Figure 5.4).

5.3.1 One-Dimensional Heat Conduction in Nonplanar Geometries

Consider steady, uniform heat conduction through a spherical shell. Heat is supplied in the interior and is conducted away by the surrounding fluid. Figure 5.6 illustrates the physical situation as well as the control volume for analysis. The outer surface temperature will be taken to be that of the the the bath T_{bath}, whereas there is a fixed heat generation in the interior Q_0. A Level I analysis of the sphere yields

$$H_{t+\Delta t} = H_t + Q_0 \Delta t - Q \Delta t,$$

$$\frac{dH}{dt} = Q_0 - Q. \tag{5.3.1}$$

This relates the heat flows to the change in enthalpy of the sphere. At steady state, the inner and outer heat flows must balance.

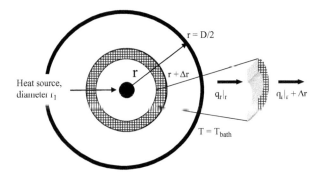

Figure 5.6. Sketch of the control volume for heat conduction in a spherical object.

To proceed, a Level III balance inside the object is required for determining how the temperature varies with position. Using the word statement of the conservation of energy applied to the spherical shell of thickness Δr illustrated in Figure 5.6 yields

$$(\rho \pi r^2 \Delta r \, \hat{H})_{t+\Delta t} = (\rho \pi r^2 \Delta r \, \hat{H})_t + (q_r \pi r^2)_r \Delta t - (q_r \pi r^2)_{r+\Delta r} \Delta t.$$

Dividing by $\pi r^2 \Delta r \Delta t$ and taking the limits as Δr and $\Delta t \to 0$, we obtain

$$\frac{\partial(\rho \, \hat{H})}{\partial t} = -\frac{1}{r^2}\frac{\partial \left(r^2 q_r\right)}{\partial r}.$$

Again, the enthalpy can be written in terms of the temperature, and for situations in which the material properties are approximately constant, we find

$$\rho \hat{C}_p \frac{\partial T}{\partial t} = -\frac{1}{r^2}\frac{\partial(r^2 q_r)}{\partial r}.$$

Fourier's constitutive equation can be applied if the material is homogeneous and there is no internal motion, leading to

$$\rho \hat{C}_p \frac{\partial T}{\partial t} = \frac{k}{r^2}\frac{\partial \left(r^2 \dfrac{\partial T}{\partial r}\right)}{\partial r},$$

$$\frac{\partial T}{\partial t} = \frac{\alpha}{r^2}\frac{\partial \left(r^2 \dfrac{\partial T}{\partial r}\right)}{\partial r}. \tag{5.3.2}$$

This constitutes a Level III model equation for the temperature profile as a function of radial distance in the sphere. The partial differential equation is first order in time and second order in spatial coordinate r, requiring one initial condition and two boundary conditions to be well specified. We will assume that the surface of the sphere is in good thermal contact with a well-mixed bath such that the Biot number is very large, and so the surface temperature will be that of the bath. If the system is at steady state and heat is being generated in the interior (such as by electrical heating) at the rate Q_0, the equation and boundary conditions simplify to

$$\frac{d\left(r^2 \dfrac{dT}{dr}\right)}{dr} = 0,$$

$$r = r_1, \quad -4\pi r_1^2 k \frac{dT}{dr} = Q_0,$$

$$r = D/2, \qquad T = T_{bath}. \tag{5.3.3}$$

Integration of the Eq. (5.3.3) yields

$$r^2 \frac{d}{dr} T(r) = A,$$

$$T(r) = \frac{-A}{r} + B.$$

Application of the boundary conditions enables solving for the constants as

$$T(r) = \frac{Q_0}{4\pi k}\left(\frac{1}{r} - \frac{2}{D}\right) + T_{bath}. \tag{5.3.4}$$

The solution illustrates that the profile is curved in form with the highest temperature in the interior surface, as expected. The nonlinear profile results from the changing area for heat transfer with increasing radius. This is illustrated by calculation of the heat flux in the interior as

$$q_r = -k\frac{dT}{dr} = \frac{Q_0}{4\pi r^2}.$$

As seen, the heat flux decreases with increasing radial distance. Of course, the total heat flow, which is the product of the heat flux and the area across which the flux traverses $(4\pi r^2)$, yields a constant heat flow, Q_0. This is a consequence of energy conservation. In general, steady-state conduction in nonplanar geometries leads to nonlinear temperature profiles and spatial variations in the heat flux, such that the total heat flow through any region of the material is at steady state.

5.3.2 One-Dimensional Diffusion in a Conical Geometry

Returning to Fick's experiment, analysis of the second experiment is complicated by the varying area along the conical diffusive region. We can readily handle this effect by deriving a mass balance appropriate for this geometry, as subsequently shown. Again, the physical effect of a varying cross section along the diffusive path is qualitatively similar as to that treated in the preceding heat transfer example. Denoting a(z) as the cross-sectional area of a differential slice of the conical region, or "funnel," leads to the following mathematical statement of conservation of mass:

$$a(z)\,(C_A|_{t+\Delta t} - C_A|_t)\,\Delta z = \Delta t[a(z)\,(v_z C_A + J_{A,z})\,|_z - a(z)\,(v_z C_A + J_{A,z})\,|_{z+\Delta z}],$$

dividing by $\Delta t\,\Delta z$ and taking the limits, $\lim_{\Delta t\to 0}$, $\lim_{\Delta z\to 0}$ yields,

$$\therefore\ \frac{\partial C_A}{\partial t} = -\frac{1}{a}\frac{\partial a\,(N_{A,z})}{\partial z}.$$

Again, for this experiment there will be no convection as the fluid in the diffusion region is quiescent and the concentration is very dilute. Using Fick's constitutive equation, we find that the resulting balance equation for the molar concentration of species A along the conical diffusion pathway is

$$\frac{\partial C_A}{\partial t} = D_{AB}\left(\frac{\partial^2 C_A}{\partial z^2} + \frac{1}{a}\frac{\partial a}{\partial z}\frac{\partial C_A}{\partial z}\right). \tag{5.3.5}$$

This equation can be integrated for steady-state diffusion to yield

$$C_A(z) = A\left(\frac{1}{L_{max} - z} - \frac{1}{L_{max}}\right). \tag{5.3.6}$$

The constants of integration A and L_{max} relate to specific parameters of the experiment. A fit of Eq. (5.3.6) to Fick's data for the funnel-shaped tube is also shown in Figure 5.4, which validates the model equations and therefore provides additional evidence supporting Fick's constitutive equation for diffusion. In the original report of his findings, Fick (1855) compares his calculated solution with the measurements to support the validity of his hypothesis.

5.4 Conduction and Diffusion Through Composite Layered Materials in Series

5.4.1 Overall Heat Transfer Coefficient for Composite Walls: Resistance Formulation

Consider heat conduction through a composite layered material, such as the insulated wall shown in Figure 5.7. With reference to our example of two well-stirred chambers in Section 3.4, this could be a composite wall, or a wall with a coating of paint or fouling on one side. Our objective is to determine the heat flux for a given overall temperature difference. Here again, we initially assume that the surfaces of the barrier are at the same temperature of the stirred tank (Biot numbers are large for both sides of the wall). This assumption will be subsequently relaxed. This problem becomes a paradigm for understanding the overall heat transfer coefficient that was employed in the heat exchanger design equations. Hence we use this analysis to define an overall heat transfer coefficient for a composite material arranged in series.

The model equation for each material (I or II) is identical to that derived in Eq. (5.1.5), which has as a general steady-state solution a linear temperature profile, Eq. (5.1.6). Proceeding as before, the constants of integration are to be determined by application of the boundary conditions. However, concerning the control volume consisting of material I, the temperature at T_1 is specified to be that of the bath on the left, but T_2 is unknown. A similar problem arises for the control volume consisting of material II, in which T_3 is specified to be that of the bath on the right, but T_2 is unknown. Finally, if we take the composite wall as the control volume for which the temperatures are known, the solution [Eq. (5.1.7)] is not valid as the thermal conductivity k was assumed constant, whereas for this control volume it varies with position.

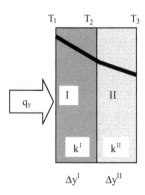

Figure 5.7. Geometry for resistances in series calculation.

To proceed, an additional equation must be derived that relates the internal temperature T_2 to the known parameters of the problem. Turning again to the statement of conservation of energy, we consider as a system the interface between Regions I and II of area a, which will be of differential thickness δ. Writing the law of energy conservation for this control volume with the heat fluxes previously defined yields the Level III balance:

$$\frac{d\hat{H}}{dt} = q_y^I a - q_y^{II} a. \tag{5.4.1}$$

For steady-state heat conduction (i.e., no time-dependent quantities), Eq. (5.4.1) shows that there can be no accumulation of energy at the interface. Further, as the area is constant for this one-dimensional geometry, $q_y^I = q_y^{II}$. This important result should be intuitive—the heat flow into the interface must be equal to the heat flow out at steady state as there are no sources or sinks. Consequently we can express the heat fluxes in terms of the temperatures from our earlier solution to Fourier's experiment in Eq. (5.1.7) as

$$q_y^I = q_y^{II} \Rightarrow -k^I(T_2 - T_1)/\Delta y^I = -k^{II}(T_3 - T_2)/\Delta y^{II}.$$

We can obtain the overall flux through the composite by solving for T_2 and substituting into either expression for the heat flux yields. Another method to obtain this expression is to rearrange the flux expressions to solve for the temperature differences, and then add them to obtain the overall temperature difference, as follows:

$$(a) \qquad T_3 - T_2 = -q_y \frac{\Delta y^{II}}{k^{II}},$$

$$(b) \qquad T_2 - T_1 = -q_y \frac{\Delta y^I}{k^I},$$

$$(a + b) \quad T_3 - T_1 = -q_y \left(\frac{\Delta y^I}{k^I} + \frac{\Delta y^{II}}{k^{II}} \right).$$

Thus the overall heat flux through the composite expressed in terms of known temperatures is

$$q_y = -\left(\frac{1}{\dfrac{\Delta y^I}{k^I} + \dfrac{\Delta y^{II}}{k^{II}}} \right) (T_3 - T_1). \tag{5.4.2}$$

The heat flux can also be used to define an overall heat transfer coefficient as

$$q_y = \frac{Q_y}{a} = U(T_1 - T_3).$$

The overall heat transfer coefficient for the composite becomes

$$U = \left(\frac{1}{\dfrac{\Delta y^I}{k^I} + \dfrac{\Delta y^{II}}{k^{II}}} \right)$$

We obtain a useful extension of this result by rearranging the finite-difference form of Fourier's constitutive equation:

$$Q_y = ka\frac{\Delta T}{\Delta Y} = \frac{\Delta T}{(\Delta y/ka)}.$$

This leads to the identification of the "thermal resistance" R_T:

$$R_T = \frac{\Delta T}{Q_y} = (\Delta y/ka). \tag{5.4.3}$$

Similar to electrical circuits, materials layered in series yield a total resistance to heat transfer that is the sum of the individual resistances of each layer of material,

$$R_{total} = R_1 + R_2 + R_3 + \cdots = \sum_{i=1}^{N} R_i.$$

Hence the total thermal resistance of the composite material can be written as the sum of resistances of the individual layers as

$$R_t = \left(\frac{1}{Ua}\right) = \left(\frac{\Delta y^I}{k^I a} + \frac{\Delta y^{II}}{k^{II} a}\right).$$

This formula can be generalized to any number of layers of varying thickness $L^{(i)}$ and thermal conductivity $k^{(i)}$ as

$$R_t = \frac{1}{Ua} = \sum_i \frac{L^{(i)}}{k^{(i)}a}. \tag{5.4.4}$$

EXAMPLE 5.4 INSULATION OF A FURNACE. As an illustration of application of Eq. (5.4.4), consider the reduction in heat loss in a steel furnace by adding 10 cm of fiberglass insulation. The furnace is constructed of 1-cm-thick steel and operates at 800 °C (see Figure E5.5). For the purpose of this calculation, assume that the outside of the furnace (or the insulation) will be at 25 °C. Thermal properties are found in Table 5.1.

SOLUTION. The model equations that describe the temperature profile in each section have been previously derived, and they predict a linear temperature profile in each section. One could solve sequentially for the temperature profiles by using the additional relationship of conservation of energy at the interface between the furnace wall and the fiberglass insulation. This approach yielded general equations (5.4.3) and (5.4.4), which enable direct calculation of the heat flux through the composite wall.

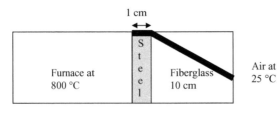

Figure E5.5. Temperature distribution in furnace wall.

The heat loss per square meter of furnace is, under the preceding assumptions,

$$q = \frac{(16 \text{ W/m K})}{0.01 \text{ m}}(800 - 25 \text{ K}) = 1240 \frac{\text{kW}}{\text{m}^2}.$$

The same calculation can be used to determine the heat flux when fiberglass insulation is added. Now the overall resistance is calculated as

$$R_T a = \frac{0.01 \text{ m}}{16 \text{ W/m K}} + \frac{0.1 \text{ m}}{0.03 \text{ W/mK}} = 3.33 \frac{\text{m}^2 \text{ K}}{\text{W}}.$$

Thus,

$$q = \frac{Q}{a} = \frac{\Delta T}{R_T a} = (800 - 25)/3.33 = 0.232 \text{ kW/m}^2.$$

This is a reduction in the next heat flux of $(1240 - 0.232)/1240)100\% = 99.98\%$! Clearly, a little insulation goes a long way. Note that we have neglected some important effects in this simple problem, such as radiative heat transfer and, more important, the fact that the outer and inner furnace wall temperatures will not be specified, but will rather be given through calculations involving heat transfer coefficients from the wall to the inside and outside gases. We will visit the latter issue shortly.

Once the heat flux is known, the internal temperature can be determined, as previously illustrated. From the integrated form of Fourier's constitutive equation, $q = -k(T_2 - T_1)/0.01$, we can solve for the temperature at the interface between the insulation and the steel as $T_2 = T_1 + q(L^{(i)}/k^{(i)}) = 800 \,^{\circ}\text{C}$. That is, nearly the entire temperature drop occurs in the fiberglass. The steady-state temperature distribution is sketched in Figure E5.5. Note that as before the profile in each separate material is linear. However, the temperature gradient varies inversely with the material's conductivity so as to maintain a constant heat flow at steady state. Deciding on the optimum amount of insulation depends on, among other factors, the economics of the process.

In analyzing the previous example, we should also reconsider the problem statement as given. Our experience is that the outside of a steel furnace will be dangerously hot and not at room temperature. In fact, considering the conservation of energy, it is obvious that the temperature of the furnace must be warmer than that of the surrounding air in order to convect the heat away from the furnace into the surrounding atmosphere. In the next section, we consider overall heat transfer coefficients for which the surface temperatures are not directly measured.

One-Dimensional Heat Conduction with Convection

Returning to the previous example, consider a more realistic problem statement in which the furnace temperature and the outside ambient air temperature are known. Consequently the inside and outside temperatures of the steel wall of the furnace are unknown, but can be determined from the energy balance [Eq. (5.1.1)], Fourier's constitutive equation [Eq. (5.1.4)], and appropriate boundary conditions that will involve a local rate of heat transfer constitutive equation. The system under consideration now includes convective heat transfer at the surfaces, as shown in Figure 5.8.

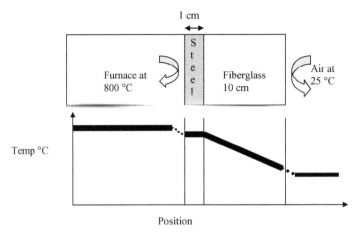

Figure 5.8. Temperature profile for furnace with internal and external resistance to heat transfer.

Again, we can analyze the steady-state heat transfer in the manner as before, in which the thermal resistance between each well-mixed gas and the internal and external surfaces of the insulated wall is given by the rate of heat transfer constitutive equation (Table 1.5). The rate of heat transfer expresses a resistance to heat transfer in terms of a local heat transfer coefficient h as

$$Q_y = ha\Delta T = \frac{(\Delta T)}{1/ha}, \quad \therefore R = 1/ha. \tag{5.4.5}$$

Consequently we find that the total resistance of our insulated wall defined by the temperature difference between the internal and external temperatures is

$$R_{total} = \frac{1}{h_{in}a} + \frac{L^{steel}}{k^{steel}a} + \frac{L^{insulation}}{k^{insulation}a} + \frac{1}{h_{out}a}. \tag{5.4.6}$$

Thus, given values for the internal and external heat transfer coefficients (h_{in} and h_{out}), we can readily determine the total resistance to heat transfer by Eq. (5.4.6). The flux is calculated from Eq. (5.4.3), and, once the flux has been calculated, the internal temperature distribution can be determined stepwise by application of the procedure outlined in the previous example. The results are sketched in Figure 5.8, where we note that the temperature profiles in the gas near the solid surfaces have been sketched as curves connecting the surface and bulk temperatures, but we do not actually know the shapes of these curves. Rather, all that can be specified is the temperature difference between the well-mixed gas and the surface of the wall. The calculation of the heat transfer coefficients and the details of that temperature profile in the boundary layer of the fluid near the walls of the furnace are the subject of the next chapter.

This analysis allows us to define the *overall heat transfer coefficient* U that has been employed in heat exchanger design. Consider a simple wall, rather than a furnace, with one material, but with both internal and external convection as an example of the wall of a tank-type heat exchanger, such as those studied in Chapter 3. Then the total heat flux can be written as

$$Q = Ua\Delta T = \frac{\Delta T}{R_t} = \frac{\Delta T}{\dfrac{1}{h_{in}a} + \dfrac{L^{steel}}{k^{steel}a} + \dfrac{L^{insulation}}{k^{insulation}a} + \dfrac{1}{h_{out}a}},$$

$$\therefore U = \frac{1}{a\sum_i R_i} = \frac{1}{\dfrac{1}{h_{in}} + \dfrac{L^{steel}}{k^{steel}} + \dfrac{L^{insulation}}{k^{insulation}} + \dfrac{1}{h_{out}}}. \tag{5.4.7}$$

Notice that we have been careful to keep the interfacial area a explicit in all of the preceding equations. This is important when we consider geometries other than the simple, planar arrangement of the furnace, such as the cylindrical geometry found in a tubular heat exchanger.

5.4.2 Overall Heat Transfer Coefficient for a Tubular Exchanger

When a tubular heat exchanger is considered, calculation of the overall heat transfer coefficient is complicated by the geometric effects of varying area for heat transfer with radial distance, as discussed in Section 5.3. To reinforce the general paradigm for solving problems in one-dimensional heat conduction with emphasis on materials with multiple resistances, we rederive the overall heat transfer coefficient appropriate for radial transport in a pipe, such as shown in Figure 5.9. Note that there is convective heat transfer on both the inside and outside boundaries of the tube. This could, for example, be a tube in a tube or shell-and-tube heat exchanger.

To start, we analyze the steady-state one-dimensional heat transfer in the radial direction in a pipe. In Section 5.1 we derived the equation governing heat conduction (with internal heat generation) by using the statement of the conservation of energy and Fourier's constitutive equation. The key difference to the planar geometry was recognizing that the area for heat transfer increases linearly with radial distance from the center of the wire; hence the area must be kept inside the differential with position. Using this result and Newton's constitutive equation for cooling, we will derive an expression for the overall heat transfer coefficient.

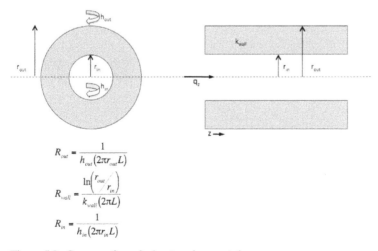

$$R_{out} = \frac{1}{h_{out}(2\pi r_{out}L)}$$

$$R_{wall} = \frac{\ln\left(\dfrac{r_{out}}{r_{in}}\right)}{k_{wall}(2\pi L)}$$

$$R_{in} = \frac{1}{h_{in}(2\pi r_{in}L)}$$

Figure 5.9. Cross section of a heat exchanger tube.

Consider the heat flow through the wall of the pipe only. Using the statement of energy conservation and Fourier's constitutive equation, we derived a Level III energy balance equation appropriate for this geometry. Assuming steady state and no internal heat generation leads to

$$0 = \frac{u}{r} \frac{d\left(r \frac{dT}{dr}\right)}{dr}. \tag{5.4.8}$$

Integrating twice yields

$$T(r) = A \ln(r) + B.$$

For now, we assume the following boundary conditions for which the unknown wall temperatures will be labeled as T_w. This enables defining the heat flux in the tube wall, which, as seen before, must be coupled to the heat flows in and out of the wall into the surrounding fluids:

$$r = r_{out}, \quad T = T_{w,out}$$

$$r = r_{in}, \quad T = T_{w,in}.$$

Solving for the constants leads to the temperature profile,

$$T(r) = \frac{(T_{w,out} - T_{w,in})}{\ln(r_{out}/r_{in})} \ln\left(\frac{r}{r_{in}}\right) + T_{w,in}, \tag{5.4.9}$$

and the total heat flux for a section of pipe of length L is given by

$$Q_r = -k_{wall} a_{out} \frac{dT}{dr}\bigg|_{r=r_{out}} = -k_{wall} 2\pi r_{out} L \left[\frac{(T_{w,out} - T_{w,in})}{r_{out} \ln(r_{out}/r_{in})}\right],$$

$$\therefore Q_r = \frac{k_{wall} 2\pi L}{\ln(r_{out}/r_{in})} (T_{w,in} - T_{w,out}) = \frac{1}{R_{wall}} (T_{w,in} - T_{w,out}). \tag{5.4.10}$$

This last line in Eq. (5.4.10) defines the thermal resistance that is due to the wall, R_{wall}.

Conservation of energy will necessitate that the total heat flow across any cylindrical shell be constant at steady state (note that, because of the changing area, the heat *flux* is not equal at each material boundary). Therefore we can write

$$Q_r = h_{out} a_{out} (T_{w,out} - T_{out})$$

$$= \frac{k_{wall} 2\pi L}{\ln(r_{out}/r_{in})} (T_{w,in} - T_{w,out})$$

$$= h_{in} a_{in} (T_{in} - T_{w,in}) = U_{out} a_{out} (T_{in} - T_{out}), \tag{5.4.11}$$

where the overall heat transfer coefficient U_{out} is defined in terms of the outer wall area. Note that we can define the overall heat transfer coefficient in terms of any area, as long as we keep track of the choice and use it consistently. Then we can solve for U_{out} as follows:

(a) $\qquad (T_{w,out} - T_{out}) = Q_r / h_{out} a_{out},$

(b) $\qquad (T_{w,in} - T_{w,out}) = Q_r \ln (r_{out}/r_{in}) / 2\pi L k_{wall},$

(c) $\qquad (T_{in} - T_{w,in}) = Q_r / h_{in} a_{in},$

$(a + b + c) \Rightarrow (T_{in} - T_{out}) = Q_r / U_{out} a_{out},$

$\therefore 1/U_{out} a_{out} = 1/h_{out} a_{out} + r_{out} \ln(r_{out}/r_{in})/k_{wall} a_{out} + 1/h_{in} a_{in}.$

The final result is an expression for the overall heat transfer coefficient in terms of the inner and outer areas:

$$\therefore U_{out} = \frac{1}{\dfrac{1}{h_{out}} + \dfrac{r_{out} \ln(r_{out}/r_{in})}{k_{wall}} + \dfrac{r_{out}}{h_{in} r_{in}}} \qquad (5.4.12)$$

This is the working equation for calculating the overall heat transfer coefficient for heat exchangers. Additional corrections for effects such as *fouling* can be added in as resistances. Fouling refers to the accumulation of minerals or other deposits on the heat exchanger tube. Cleaning of heat exchangers is important to keep the resistance to heat transfer as low as possible.

EXAMPLE 5.5 OVERALL HEAT TRANSFER COEFFICIENT FOR A TUBULAR HEAT EXCHANGER. Process oil is to be heated by saturated steam at 1 bar in a shell-and-tube heat exchanger consisting of steel pipes of 2-cm ID (internal pipe diameter) and 3-cm OD (external pipe diameter). The oil flows inside the pipe. Estimate the heat transfer rate per meter of pipe if the oil is delivered at 20 °C.

SOLUTION. The rate of heat transfer is given by Eq. (5.4.11), with the overall heat transfer coefficient defined by Eq. (5.4.12). The diameters are known, as well as the inside and outside temperatures (at least at the inlets), and, given the material of construction, we can get the resistance in the heat exchanger wall from the conductivity. However, the heat transfer coefficients in the saturated steam and the oil are not known. Estimates can be found, however, in the tables presented in Chapter 3 as well as those found in handbooks and reference textbooks. Table E5.2 (adapted from Bird et al., 2002, Table 14.1–1) provides some order-of-magnitude estimates. Notice that there is a large uncertainty in these estimates. Taking intermediate values for the steam $h \approx 10$ kW/m^2 K and the oil $h \approx 0.25$ kW/m^2 K, the overall heat transfer coefficient can be estimated as

$$U_{out} = \frac{1}{\dfrac{1}{10} + \dfrac{0.015 \ln(0.015/0.01)}{0.016} + \dfrac{0.015}{0.01 \times 0.25}} = 0.15 \text{ kW/m}^2 \text{ K}.$$

Analysis of this result shows it to be dominated by the resistance on the oil side. As a quick check of this result, we can also turn to tables of the overall heat transfer coefficients for various systems, such as those found in the *Chemical Engineers' Handbook* (1997, Table 10-10). The estimates for overall heat heat transfer coefficients in tubular heat exchangers for steam–fuel oil range from 15 to 90 Btu/(ft^{2}h °F),

Table E5.2. Estimates of heat transfer coefficients
(from Gröber H, Erk S, Grigull U.
Wärmeübertragung, 3rd ed. Berlin: Springer, 1995)

System	U (kW/m^2 K)
Free convection	
Gases	0.003–0.02
Liquids	0.1–0.6
Boiling water	1–20
Forced convection	
Gases	0.01–0.1
Liquids	0.05–0.5
Water	0.5–10
Condensing vapors	1–100

which is in the range of 0.1–0.5 kW/m^2 K, giving us some confidence that our estimate is of the right order of magnitude.

The heat flux per length of pipe (L) can now be estimated from Eq. (5.4.11), where the sign is negative to indicate that heat flows into the pipe from the shell side:

$$\frac{Q_r}{L} = 0.15 \times \pi(0.03)(1)(20 - 100) = -1.1 \text{ kW/m pipe.}$$

A more extensive discussion of heat exchanger design follows in later chapters.

5.4.3 Overall Mass Transfer Coefficient for Diffusion Through a Composite Wall

The previous examples should give you confidence that, at least for some problems, there is a one-to-one correspondence between heat conduction in homogeneous solids and diffusive mass transfer. As a further example of the utility of this analogy, and its limitations, consider one-dimensional diffusion through a composite wall (Figure 5.10), by analogy to the simple heat transfer problem considered in Subsection 5.4.1. A practical example of this might be diffusion of a drug through a layered gel capsule (see Figure 5.10). It will be left as an exercise to the reader to prove the following resistance formulations for simple diffusive mass transport:
Mass transfer coefficient:

$$a J_A = a K_m \Delta C_A = \frac{\Delta C_A}{R_m},$$
$$R_M = 1/a K_m;$$

Diffusive mass transfer:

$$a J_A = a D_{AB} \frac{\Delta C_A}{L} = \frac{\Delta C_A}{R_m},$$
$$R_M = L/a D_{AB}.$$

A complication arises because of the thermodynamic partitioning of solute A between Phases I and II. Consider diffusion through the wall where materials I

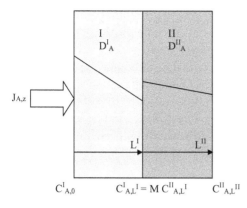

Figure 5.10. Diffusion through a composite medium. (The profile shown is for M < 1 and $D_A^{II} > D_A^I$.)

and II are indicated. The equilibrium condition at the interface between materials I and II is defined by the equilibrium relationship $C_A^I = MC_A^{II}$. A proper formulation requires careful bookkeeping. Applying the conservation of mass to species A in both phases and employing Fick's constitutive equation will lead to linear concentration profiles in each phase. However, at the interface, the flux must be equal and the concentrations across the interface are assumed to be related by the distribution coefficient M. Mathematically, these conditions are

$$J_{A,z}^I = D_A^I \frac{C_{A,0}^I - C_{A,L^I}}{L^I} = J_{A,z}^{II} = D_A^{II} \frac{C_{A,L^I}^{II} - C_{A,L^{II}}^{II}}{L^{II}},$$

$$C_{A,L}^I = MC_{A,L}^{II}.$$

Now we have to make a choice of concentrations to define the overall mass transfer coefficient K_m^{eff}. Here, the defining equation will become

$$J_{A,z} = K_m^{eff}\left(C_{A,0}^I - MC_{A,L^{II}}^{II}\right). \tag{5.4.13}$$

Again, using the same strategy employed for the analysis of the analogous heat conduction problem, we find that the effective or overall mass transfer coefficient becomes

$$C_{A,0}^I - C_{A,L}^I = J_{A,z}^I \frac{L^I}{D_A^I},$$

$$MC_{A,L^I}^{II} - MC_{A,L^{II}}^{II} = J_{A,z}^I \frac{ML^{II}}{D_A^{II}},$$

$$\therefore C_{A,0}^I - MC_{A,L^{II}}^{II} = J_{A,z}^I \left(\frac{L^I}{D_A^I} + \frac{ML^{II}}{D_A^{II}}\right),$$

$$\therefore J_{A,z}^I = \frac{C_{A,0}^I - MC_{A,L^{II}}^{II}}{\left(\dfrac{L^I}{D_A^I} + \dfrac{ML^{II}}{D_A^{II}}\right)} = K_m^{eff}\left(C_{A,0}^I - MC_{A,L^{II}}^{II}\right),$$

$$\therefore K_m^{eff} = \frac{1}{\left(\dfrac{L^I}{D_A^I} + \dfrac{ML^{II}}{D_A^{II}}\right)}. \tag{5.4.14}$$

Equation (5.4.14) shows that the thermodynamic distribution coefficient enters into the formula for the effective mass transfer coefficient for the series composite wall. As with the heat conduction formulation, a more general formula can be derived for multiple "resistances" in series and in parallel, although the development becomes algebraically tedious because of the need to multiply each subsequent layer by the distribution coefficient that converts the concentration to that of the reference layer (which was layer I in our example). Additional resistances, such as external mass transfer, can be included under the same logic. However, it must be noted that each interface will have a specific distribution coefficient M that will have to be specified to achieve a solution.

5.5   Molecular Conduction and Diffusion with Generation

5.5.1 Radial Heat Conduction with Generation

Consider the following problem of heat conduction in an insulated wire, shown in Figure 5.11. Heat is generated internally in the wire because of resistive heating, and the heat is conducted in the radial direction to the surface of the wire, where it is convected into the surrounding air. Further, the wire is assumed to be uniform in the z direction. The approach to solving this problem is the same as that of the previous examples and starts with writing a conservation equation for energy as per the word statement given in Figure 1.6. The system will be a coaxial cylindrical shell of thickness Δr. The resistive heating is given in terms of the volumetric heating rate $\dot{q}_v$. Further, the wire is considered to be infinitely long in the z direction, but we can solve the equations for a unit length of wire, L. Written for the differential shell, the energy balance becomes

$$2\pi r\,\Delta r\,L\left(\rho\hat{C}_p T|_{t+\Delta t}-\rho\hat{C}_p T|_t\right)=2\pi L[(r\,q_r)|_r-(r\,q_r)|_{r+\Delta r}]\Delta t$$
$$+\,\dot{q}_v\Delta t 2\pi r\,\Delta r\,L.$$

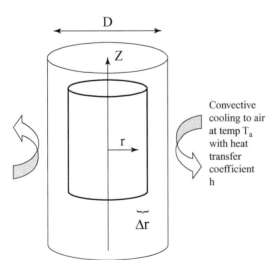

Convective cooling to air at temp T_a with heat transfer coefficient h

Figure 5.11. Geometry of wire with internal heat generation and outside cooling by convection. (The internal structure shows the geometry for the shell balance, which extends in the z direction.)

In the balance we are careful to recognize that the shell surface area and volume increase linearly with radius by grouping the radius with the heat flux on the right-hand side. We convert this to a partial differential equation by dividing by the shell volume and the time increment and taking both the shell thickness and time increment to be differential:

$$\frac{\partial(\rho \hat{C}_p T)}{\partial t} = -\frac{1}{r}\frac{\partial(r q_r)}{\partial r} + \dot{q}_v. \tag{5.5.1}$$

Further, the flux is assumed to be given by Fourier's constitutive equation, written in the appropriate coordinates, as $q_r = -k(dT/dr)$. Then, for constant physical properties, the energy equation becomes

$$\frac{\partial T}{\partial t} = \alpha \frac{1}{r}\frac{\partial}{\partial r}\left(r\frac{\partial T}{\partial r}\right) + \frac{\dot{q}_v}{\rho \hat{C}_p}. \tag{5.5.2}$$

This equation is to be solved for the steady-state temperature profile in the wire given the following boundary conditions:

$$r = 0, \quad \frac{dT}{dr} = 0, \quad r = \frac{D}{2}, \quad q_r = -k\frac{dT}{dr} = h[T(r) - T_a].$$

The first boundary condition is just an expression of the symmetry of the problem, i.e., there should not be a heat flow through the center of the rod. Equation (5.5.2) can be solved at steady state as follows:

$$\frac{1}{r}\frac{d\left(r\dfrac{dT}{dr}\right)}{dr} = -\frac{\dot{q}_v}{k},$$

$$d\left(r\frac{dT}{dr}\right) = -r\frac{\dot{q}_v}{k}dr,$$

$$r\frac{dT}{dr} = -\frac{\dot{q}_v}{k}\frac{r^2}{2} + A,$$

$$dT = \left(-\frac{\dot{q}_v}{k}\frac{r}{2} + \frac{A}{r}\right)dr,$$

yielding the radial temperature profile as

$$T(r) = -\frac{\dot{q}_v}{k}\frac{r^2}{4} + A\ln(r) + B. \tag{5.5.3}$$

Application of the first boundary condition requires that A be zero [as $\ln(0)$ is singular]. Application of the second boundary condition gives the undetermined integration constant B indirectly as follows. From the heat flux at the surface of the wire, we find

$$r = \frac{D}{2}, \quad q_r = -k\frac{dT}{dr} = h[T(r) - T_a] = -k\left(-\frac{\dot{q}_v}{k}\frac{r}{2}\right),$$

$$T\left(r = \frac{D}{2}\right) = T_a + \frac{\dot{q}_v}{h}\frac{D}{4},$$

$$T\left(r = \frac{D}{2}\right) = T_a + \frac{1}{Bi}\frac{\dot{q}_v}{k}\frac{D^2}{4},$$

where

$$Bi = \frac{hD}{k},$$

but, from the solution

$$T(r) = -\frac{\dot{q}_v}{k}\frac{r^2}{4} + A\ln(r) + B,$$

and because $A = 0$, equating these at the boundary yields

$$-\frac{\dot{q}_v}{k}\frac{D^2}{16} + B = T_a + \frac{1}{Bi}\frac{\dot{q}_v}{k}\frac{D^2}{4},$$

$$\therefore B = T_a + \frac{\dot{q}_v}{k}\frac{D^2}{16}\left(1 + \frac{4}{Bi}\right),$$

$$\therefore T(r) = T_a + \frac{\dot{q}_v}{16k}\left[D^2\left(1 + \frac{4}{Bi}\right) - 4r^2\right]. \qquad (5.5.4)$$

One important result of the preceding analysis is that the centerline temperature of the wire is given by the constant B, which is seen to depend inversely on the Biot number Bi. That is, the higher the external rate of heat transfer, the cooler the interior temperature of the wire, which is an intuitive result. Further, the temperature profile inside the wire is parabolic, not linear, which is partly a consequence of the varying area for heat transfer with increasing radial position. Note that we can also determine the temperature at the wire surface directly from a Level I analysis by taking the wire as the system, without solving for the interior temperature. More complex geometries can be envisioned, which may require considering multiple spatial dimensions.

5.5.2 Diffusion with Chemical Reaction

A problem analogous to that of the heated wire arises in diffusive mass transfer, as follows. Consider a capsule (I) within which a compound reacts slowly as $B \rightarrow A$ with a rate constant that is, for the duration of the experiment, found to be independent of the local concentration of either A or B. The A that is produced diffuses out of the capsule, which we can consider to a first approximation as an infinite cylinder, into the surrounding solution (II). Application of the conservation of mass leads to the following expression (where we assume that A is dilute, that we can neglect any convection, and that the volumetric rate of reaction to produce A is given by r_{A+}):

$$2\pi r\,\Delta r\,L\left(C_A^I|_{t+\Delta t} - C_A^I|_t\right) = 2\pi L[(r\,J_{A,r})|_r - (r\,J_{A,r})|_{r+\Delta r}]\Delta t$$
$$+ r_{A+}\Delta t 2\pi r\,\Delta r\,L.$$

We convert this finite-difference equation to a partial differential equation by dividing by the shell volume and the time increment and taking both the shell thickness and time increment to be differential. Further, the flux is assumed to be given by Fick's constitutive equation, written in the appropriate coordinates, as $J_{A,r} = -D_{AB}(dC_A/dr)$. Then, for constant physical properties, the species balance

Table 5.3. *Correspondence between heat transfer with generation and mass with reaction*

Heat transfer with generation, Eq. (5.5.2)	Diffusion with reaction, Eq. (5.5.5)
T	C_A^I
T_a	MC_A^{II}
α	D_{AB}
h	k_m
$\dot{q}_v / \rho \hat{C}_p$	r_{A+}

equation for A becomes

$$\frac{\partial C_A^I}{\partial t} = D_{AB} \frac{1}{r} \frac{\partial}{\partial r} \left(r \frac{\partial C_A^I}{\partial r} \right) + r_{A+}. \tag{5.5.5}$$

This equation is to be solved for the quasi-steady-state concentration profile in the cylinder given the following boundary conditions:

$$r = 0, \quad \frac{dC_A^I}{dr} = 0,$$

$$r = \frac{D}{2}, \quad J_{A,r} = -D_{AB} \frac{dC_A^I}{dr} = k_m \left[C_A^I(r) - MC_A^{II} \right],$$

where we have defined the mass transfer coefficient relative to the concentrations in the solid phase, i.e., $M = C_A^I / C_A^{II}$ in order to make the analogy to Eq. (5.5.2) more transparent. Close examination of model equation (5.5.5) and the boundary conditions just given show an exact correspondence, with the equivalent variables given in Table 5.3. Therefore, because the model equation and the boundary conditions are identical on change of variables, the solutions must necessarily be identical upon change of variables. Table 5.3 is an extension of the equivalence table presented in Appendix 4.B.

Transforming the solution of the heat transfer problem, Eq. (5.5.4), into mass transfer variables leads to the solution

$$C_A^I(r) = MC_A^{II} + \frac{r_{A+}}{16 D_{AB}} \left[D^2 \left(1 + \frac{4}{Bi_m} \right) - 4r^2 \right], \tag{5.5.6}$$

where $Bi_m = k_m D / D_{AB}$ is a Biot number for mass transfer.

5.6 Diffusion-Induced Convection: The Arnold Cell

Now, given the parallel development, one might be tempted to extend the analogy between heat conduction in solids and diffusive mass transfer so as to expect that mathematically equivalent solutions will always exist for both heat and mass transfer. For many problems involving *dilute* mass transfer of gases or liquids in other gases or liquids this analogy will hold. Two complications arise, however, that when present require modification to the simple mass transfer balances given in the preceding example. The first is that isothermal mass transfer may result in *convection*, which, except for very dilute systems, cannot generally be neglected as was done previously.

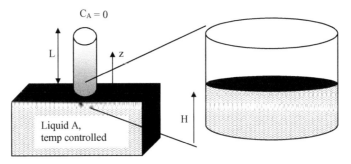

Figure 5.12. Arnold cell for measuring diffusivities. (The expanded view on the right-hand side shows the liquid level in the tube.)

Second, mass transfer through solids, such as in membranes, involves partitioning of the solute between the gas of the liquid phase and the resistive phase, such as the membrane. The former is handled by proper accounting in the balance equation, whereas the latter is handled by proper specification of the boundary conditions. Both effects are illustrated in the next example.

Binary diffusivities in gases can be measured in an Arnold cell, as depicted in Figure 5.12. A pure liquid A is placed in a container and a tube inserted. A is volatile and will evaporate (the temperature is controlled by a thermal bath) in the long tube, leaving at the top ($z = L$) into a large surrounding air space, such that the concentration at the top of the tube is effectively maintained at zero. For the discussion to follow, z is taken from the gas–liquid interface in the tube.

If the vapor pressure of A is small, the analysis in Section 5.2 could be employed, with the modification that the concentration of A in the gas phase in direct contact with the pure liquid A (i.e., $z = 0$) is calculated by assuming thermodynamic equilibrium at the interface. For an ideal-gas mixture in the gas phase, $C_A = p_A/RT$, with p_A the partial pressure of A at the temperature of the experiment. If the carrier gas (air) were soluble in liquid A then a multicomponent liquid–vapor equilibrium calculation would be required for determining the concentrations in the gas phase. The inherent assumption underlying thermodynamic equilibrium is that the rate of transport does not disturb the establishment of thermodynamic equilibrium.

However, we know from experience that as A evaporates the liquid level will slowly drop, and, consequently, there must be a net transfer of moles or mass of A from the liquid to the upper reservoir. Therefore, there is a finite velocity of gas in the tube from the liquid surface to the upper reservoir. Thus the net flux of A moving through the stationary control volume ($N_{A,z}$) consists of two parts, a diffusive flux described by Fick's constitutive equation plus a convective flux, given by the velocity v_z times the local molar concentration of A, as given in Eq. (5.2.2). Notice that in this problem there is no forced convection; rather, the convection arises naturally as species A is diffusing up the tube. Thus the velocity in the z direction is given by the fluxes of all species, which in this case reduces to

$$v_z = \frac{1}{C}(N_{A,z} + N_{B,z}), \tag{5.6.1}$$

where $C = C_A + C_B$. Equation (5.2.2) is the general statement of the molar flux in one dimension and, in the limit of negligible velocities, reduces to the Fickian expression. A way to visualize this result is to consider mass transfer through the capillary shown in Figure 5.12. If there were also a plug flow with velocity v_z moving along the capillary, then, in a coordinate frame *moving* with the flow (i.e., flowing along with the fluid), the flux of A would be due only to molecular diffusion. However, in the laboratory coordinate frame, where we do our shell balance, the net flux of A consists of the velocity of the flow times the local concentration, $v_z C_A$ plus the diffusive flux J_A. In the Arnold cell, this flow is due to the volatility of A. As B is not soluble in A, there is no comparable return of B into the container. In the experiment performed by Fick, this velocity was negligible because a crystal was slowly dissolving in water, as compared with the relatively fast evaporation of A into air.

As a further simplification of the problem, let's assume that the rate of drop of the liquid level is insignificant on the timescale of diffusion. This is not a bad assumption as liquids are typically $\sim 10^3$ times as dense as vapors near STP (standard temperature, pressure), and hence a substantial amount of gas diffusion is required for liquid motion. Then we can assume that the gas-phase concentrations in the tube are at *quasi steady state* and solve for the concentration profiles and then the flux of A as follows.

The one-dimensional balance equation for a slice orthogonal to the z direction (Figure 5.4) becomes

$$\pi R_C^2 \left(C_A|_{t+\Delta t} - C_A|_t \right) \Delta z = \pi R_C^2 \Delta t (N_A|_z - N_A|_{z+\Delta z})$$

$$\lim_{\Delta t \to 0}, \lim_{\Delta z \to 0}$$

$$\therefore \frac{\partial C_A}{\partial t} = -\frac{\partial N_{A,z}}{\partial z}.$$

Using Fick's constitutive equation and the definition of the net flux yields

$$\frac{\partial C_A}{\partial t} = D_{AB} \frac{\partial^2 C_A}{\partial z^2} - v_z \frac{\partial C_A}{\partial z}. \tag{5.6.2}$$

This result assumes that the diffusivity and the velocity are constant in the tube. To proceed, we require an expression for the molar velocity, which we can obtain by considering the behavior of species B in the capillary tube.

A similar balance equation is performed on species B, the inert gas in the capillary, leading to the same equation (as $D_{AB} = D_{BA}$) with just a change in subscript,

$$\frac{\partial C_B}{\partial t} = -\frac{\partial N_{B,z}}{\partial z} = D_{AB} \frac{\partial^2 C_B}{\partial z^2} - v_z \frac{\partial C_B}{\partial z}, \tag{5.6.3}$$

the solution of which plus the boundary conditions gives the profile of B in the tube. The boundary conditions are

$$z = 0, \quad C_A = C_{A,sat}, \quad N_{B,z} = 0,$$

$$z = L, \quad C_A = 0, \quad C_B = C_{B,0}. \tag{5.6.4}$$

Notice that the flux of B is known exactly at the liquid surface, for if B is insoluble in liquid A, the total flux of B must be zero. Then the net flux of B is given by

$N_{B,z} = J_{B,z} + v_z C_B$, which requires that the molar velocity balance the diffusive flux of B moving to replace the evaporating species A. Therefore, as $N_{B,z} = 0$ at $z = 0$, we find

$$v_z = \frac{D_{AB}}{C_B}\frac{dC_B}{dz} = D_{AB}\frac{d\ln C_B}{dz}.$$

At steady state, Eq. (5.6.3) can be integrated to show that $N_{B,z}$ = constant. Note that, as the total flux of species B ($N_{B,z}$) must be zero at the surface, it is zero everywhere in the tube at quasi steady state. Note that this does not imply that B does not diffuse, for indeed, it does in order to replace the moles of A that leave (remember that the capillary is assumed to be at constant pressure, which requires a uniform molar concentration). Rather, this analysis shows that the molecular diffusion of B down the tube balances the convective flux that is due to the evaporating A. In fact, the motion of A out of the tube "drags" B along, and, as a consequence, B will have a concentration gradient that drives a diffusive flux to replenish B in the tube. This balance of fluxes is required for the system to operate at steady state.

Given this solution, i.e., $N_{B,z} = 0$, the velocity out of the tube can be determined in terms of the flux of A alone, as $v_z = N_{A,z}/(C_A + C_B) = N_{A,z}/C$, where C the total molar concentration is a constant. With this, our balance equation for species A, Eq. (5.6.2), can be rewritten and solved as follows. We also choose to express the concentrations in terms of mole fractions (y_i) for both convenience and to enable comparison of model predictions for various partial pressures of species A:

$$N_{A,z} = J_{A,z} + \frac{C_A}{C}N_{A,z} = J_{A,z}/(1-y_A),$$

$$y_A = C_A/C, \quad J_{A,z} = -CD_{AB}\frac{dy_A}{dz},$$

$$\therefore N_{A,z} = -\frac{CD_{AB}}{1-y_A}\frac{dy_A}{dz}$$

$$\therefore \frac{\partial y_A}{\partial t} = D_{AB}\left[\frac{1}{1-y_A}\frac{\partial^2 y_A}{\partial z^2} - \frac{1}{(1-y_A)^2}\left(\frac{\partial y_A}{\partial z}\right)^2\right]. \tag{5.6.5}$$

In arriving at final expression (5.6.5), both the diffusivity (D_{AB}) and total molar concentration are assumed to be constant. The latter is sensible if the pressure in the tube is constant and ideal-gas mixture properties are assumed.

A steady-state solution is derived by integration of the flux expression directly, with the following boundary conditions:

$$z = 0, \quad y_A = y_{A,0},$$
$$z = L, \quad y_A = y_{A,L}.$$

Integration yields (recognizing that, for steady state, $N_{A,z}$ = constant)

$$N_{A,z}\int_0^z dz' = -CD_{AB}\int_{y_{A,0}}^{y_A}\frac{1}{1-y_A}dy_A',$$

$$\therefore N_{A,z} = \frac{-CD_{AB}}{z}\ln\left(\frac{1-y_A}{1-y_{A,0}}\right),$$

$$\text{so,} \quad N_{A,z} = \frac{-CD_{AB}}{L} \ln\left(\frac{1-y_{A,l}}{1-y_{A,0}}\right),$$

$$\therefore \left(\frac{1-y_A(z)}{1-y_{A,0}}\right) = \left(\frac{1-y_{A,L}}{1-y_{A,0}}\right)^{z/L}. \tag{5.6.6}$$

Thus the mole fraction (or concentration) profile is not simply linear. Note that the mole fraction of component B can be determined from a Level I analysis, yielding $y_B = 1 - y_A$.

As with the Fick experiment, the flux is all we need now to use this device to measure diffusivity. Given thermodynamic knowledge of the equilibrium vapor pressure and the overall pressure, this quasi-steady-state flux expression can be used to derive a Level I balance equation for the change in the liquid level [H(t)] with time as

$$\rho A \frac{dH}{dt} = -AN_{A,z},$$

where the only complication is that L, the diffusion path length, changes with H. The flux becomes

$$N_{A,z} = \frac{-CD_{AB}}{L} \ln\left(\frac{1-y_{A,L}}{1-y_{A,0}}\right). \tag{5.6.7}$$

In the limit of $y_{A,0} \to 0$, the flux reduces to the expected expression:

$$N_{A,z} = -cD_{AB}\frac{y_{A,L} - y_{A,0}}{L}.$$

Clearly, for volatile solvents, significant errors would be incurred by assuming a linear concentration profile in relating the liquid height change to the molecular diffusivity.

For purposes of understanding the original issue, that of the role of diffusion-induced convection in the total mass flux, solutions for the concentration profile are graphed in Figure 5.13 as a function of the volatility of the liquid (expressed as

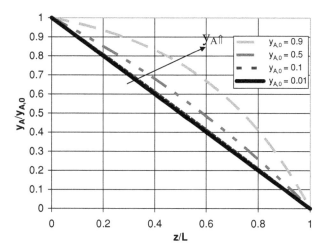

Figure 5.13. Mole fraction profiles for the Arnold cell for varying equilibrium partial pressures. (Note that, for relatively moderate volatilities, the solution closely approximates the linear profile that would be obtained by neglecting diffusion-induced convection.)

$y_{A,0}$). Note that, as expected, for very low volatilities, the solution tends toward the simple linear form we would have expected by application (incorrectly) of Fick's constitutive equation without accounting for the net mass flow. This can be derived mathematically from the solution, for $\ln(1-x)$ with x small is approximately equal to $-x$. This yields

$$y_A(z) \approx y_{A,0}\left(1 - \frac{z}{L}\right).$$

Clearly, this would lead to incorrect fluxes and an error in the diffusion coefficient for values of $y_{A,0} \approx 1$.

5.7   Basics of Membrane Diffusion: The Sorption–Diffusion Model

Consider the simple gas membrane shown in Figure 5.14. The driving force is the concentration difference between the two phases, denoted by superscripts [I] and [II]. The flux expression is written as the linear product of the concentration times the mass transfer coefficient. If we write the driving force for the concentration profile *inside* the membrane, the result is, in the limit of low concentration of A (i.e., no convection),

$$J_{A,z} = \frac{D}{L}[C_A(0) - C_A(L)].$$

This concentration profile is linear, and the diffusivity D is that of species A in the membrane. However, the concentration inside the membrane is generally not measurable, but rather, some specific separation is required, and hence the concentrations in the two reservoirs or baths are specified and measured. The concentrations at the surfaces of the membrane are assumed to be related by a thermodynamic partition coefficient, similar to the partition coefficient M employed in the mass contactor design in Chapter 4. Here, in keeping with the literature on membranes, we use the nomenclature H for the partition coefficient, which is defined as

$$H = \frac{C_A(0)}{C_A^I} = \frac{C_A(L)}{C_A^{II}}.$$

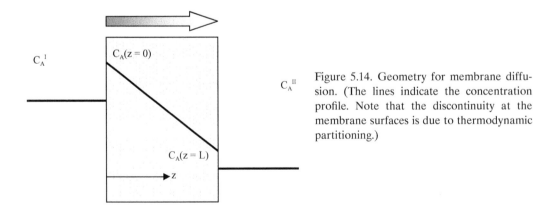

Figure 5.14. Geometry for membrane diffusion. (The lines indicate the concentration profile. Note that the discontinuity at the membrane surfaces is due to thermodynamic partitioning.)

Table 5.4. Permeability values for some typical systems (Cussler, 1997) (SI units)

Film	$P_{CO_2} \times 10^{11}$	$P_{N_2} \times 10^{11}$	Selectivity
Saran	0.29	0.009	31
Mylar	1.5	0.05	31
Nylon	1.6	0.010	16
Neoprene	250	12	21
Polyethylene	350	19	19
Natural rubber	1310	80	16

The assumption made here is that the partition coefficient is constant and equal for both fluids I and II. For glassy polymer membranes that do not swell in the gases of interest, the solubility of A in the membrane is often very low and hence these are good assumptions. Using this, we find that the final result for the flux across the membrane is

$$J_{A,z} = \frac{DH}{L}\left[C_A^I(0) - C_A^{II}(L)\right]. \tag{5.7.1}$$

If the membrane is for gas separation, the ideal-gas law can be used to rewrite the concentrations in terms of partial pressures (assuming ideal-gas mixtures) as $p_A = C_A RT$, leading to the result

$$J_A = \frac{P}{L}\left(p_A^I - p_A^{II}\right). \tag{5.7.2}$$

Here, $P = DH/RT$ is the permeability [SI units (mol m/m^2 s Pa)]. The quantity (P/L) is denoted the *permeance*, which is often the operational quantity if the membrane thickness is unknown. Some typical permeabilities are given in Table 5.4.

The values are in mixed units that are often expressed as barrers, where

$$1 \text{ barrer} = \frac{10^{-10} \text{ cm}^3 \text{ gas(STP)(cm thickness)}}{(\text{cm}^2 \text{ membrane area})(\text{cm Hg pressure}) \text{ s}}.$$

Of course, the model equations for membranes that swell and extensions of these ideas to reverse osmosis and pervaporation require additional considerations. The interested student is referred to the book by Matsuura (1994) for additional information on membrane processes.

5.8 Transient Conduction and Diffusion

Consider a wall (or insulated bar) that is initially at a temperature T_0 when the surface of the wall at $y = 0$ is suddenly raised to temperature T_1. Heat will flow into the material, raising the temperature inside. The evolution of the temperature is given by the solution of the Level III balance given by Eq. (5.1.5):

$$\frac{\partial T}{\partial t} = \alpha \frac{\partial^2 T}{\partial y^2}.$$

It is useful to rewrite this equation into a dimensionless form so that general solutions can be given that are valid for all one-dimensional heat transfer problems of this type. We cast the equation in dimensionless form by recognizing the following scaled variables:

$$\theta\,(y, t) = \frac{T\,(y, t) - T_0}{T_1 - T_0} \quad \zeta = y/L, \quad \tau = t\alpha/L^2.$$

The logic behind this scaling is important to state as we will make significant use of this procedure in the next chapter. The goal is to develop dimensionless variables so that variations in properties and parameters can be expressed in a convenient scale. Here, for example, the temperature in the bar is taken relative to the reservoir temperature T_0, and scales relative to the maximum difference expected, $T_1 - T_0$. Hence we expect dimensionless temperatures to vary between 0 and 1, and similarly for the length, which is scaled on the wall thickness. Finally, we require a dimensionless time to judge what is "long" or "short" and, consequently, when steady state is achieved. We find the natural time scale by using the thermal diffusivity α, which has SI units of square meters per seconds, and the length scale we already used, L. The value of this time scale will become apparent shortly. Note that the temperature, length, and time scales are defined on the "natural" variables of the problem.

Often, the dimensionless time τ is denoted as the *Fourier* number (Fo). However, unlike the Biot number, which is solely a function of material properties, the Fourier number depends on an independent parameter, time.

With these definitions, the Level III balance becomes

$$\frac{\partial\theta}{\partial\tau} = \frac{\partial^2\theta}{\partial\zeta^2}, \tag{5.8.1}$$

with the boundary conditions

$$\tau = 0, \quad \theta = 0,$$
$$\zeta = 0, \quad \theta = 1,$$
$$\zeta = 1, \quad \theta = 0. \tag{5.8.2}$$

Equation (5.8.1) can be solved numerically or in terms of a series solution derived by the method of separation of variables, which leads to a series solution (see, for example, Carslaw and Jaeger, 1959). It will have a solution of the form $\theta\,(\zeta, \tau)$.

Some approximate analytical results are possible. For example, for a sphere with constant surface temperature, the average temperature in the sphere is found to be (for Fo > 1)

$$\bar{\theta} = 1 - 0.81\exp(-9.87\,\text{Fo}).$$

Numerical integration is difficult for this particular problem because of the singular nature of the boundary conditions; however, finite-difference methods can be applied. Alternatively, charts of solutions exist of the insulated bar (i.e., "slab" geometry) as well as cylinders and spheres in Carslaw and Jaeger (1959), which are easy to use as well for calculating temperatures at specific locations and times. Solutions have been plotted for other boundary conditions, in which the temperature profile will also be a function of the Biot number [i.e., $\theta(\zeta, \text{Bi}, \tau)$]. As noted throughout

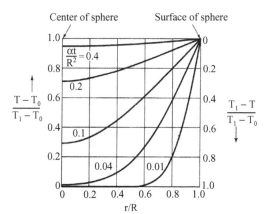

Figure 5.15. Transient heat conduction in a sphere. [The initial temperature of the sphere is T_0, and T_1 is the temperature imposed at the sphere surface for time $t > 0$. (Carslaw and Jaeger, 1959. Reproduced with permission.)]

this chapter, solutions for diffusive mass transfer can be found by analogy with the corresponding heat transfer problem. A typical example of such a solution in the form of a chart is shown in Figure 5.15.

The chart shows the solution for Eq. (5.3.2) for the heating of a uniform sphere in which the surface temperature is suddenly raised. Note that, at short times (i.e., small Fo), the temperature differs from the initial temperature only in the vicinity of the surface of the sphere, and the profiles advance into the sphere with a marked reduction in slope near the surface. Recognizing that the rate of heat transfer to the sphere is given by the slope

$$\left(Q = -ak \frac{\partial T}{\partial r} \bigg|_{r=D/2} \right),$$

it is apparent that the rate of heat transfer continually decreases.

5.8.1 Short-Time Penetration Solution

A problem of particular interest is to estimate the rate of heat or mass transfer that occurs at short times when a boundary is suddenly changed, such as by contact with a hot or cold reservoir (or solution with concentration of species A). As seen in the prior section, the heat (or mass of A) does not penetrate into the material very far for small Fo, and the problem resembles heat (or mass) penetration into an effectively infinite material. To account for this, the second boundary condition in Eqs. (5.8.2) is replaced such that $T \to T_0$ as $y \to \infty$. This problem has an analytical solution in terms of the error function as

$$\frac{T - T_0}{T_1 - T_0} = 1 - \mathrm{erf}\left(\frac{y}{2\sqrt{\alpha t}} \right) \tag{5.8.3}$$

A plot of this solution is shown in Figure 5.16.

The *thermal penetration depth* δ_T is defined as when the temperature drops 99% of the difference to the bath temperature, which is approximately

$$y = \delta_T = \sqrt{4\alpha t}. \tag{5.8.4}$$

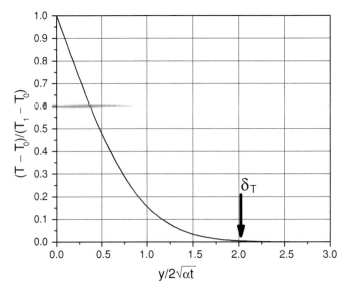

Figure 5.16. Penetration of heat into an infinite slab [Eq. (5.8.3)].

This is indicated in Figure 5.16. Interestingly, the preceding approximation shows that the depth of heat penetration grows as $t^{1/2}$, such that doubling the depth of heat or mass penetration into an object requires not two, but four, times as long. This result is of particular value for understanding how temperature profiles and concentration profiles penetrate into quiescent fluids and solids, and will be revisited as the *penetration model* in Chapter 6 when models for mass and heat transfer coefficients are discussed.

The flux of heat into the bar can be calculated from the temperature profile and Fourier's law. The result is

$$q_y|_{y=0} = -k\frac{\partial T}{\partial y}\Big|_{y=0} = \frac{k}{\sqrt{\pi \alpha t}}\,(T_1 - T_0). \qquad (5.8.5)$$

This equation shows that heat flow into the bar decreases as $t^{-1/2}$. Notice that the heat flow is approximately given by

$$q_y|_{y=0} \approx k\frac{(T_1 - T_0)}{\delta_T}.$$

In fact, substituting $\delta_T = L$ yields the exact result for the heat flux through the bar at steady state [Eq. (5.1.7)].

EXAMPLE 5.6 TRANSIENT HEAT FLOW. A hot (80 °C), spherical rock 10 cm in diameter is dropped into a very large, well-stirred tank of ice water. What is the total amount of heat given off by the rock after 1 min?

SOLUTION. We recognize that the rock has a relatively low thermal conductivity (from Table 5.1, $k \sim 0.66$ W/m K) such that an estimate of the Biot number for the process,

$$\mathrm{Bi} = \frac{hD}{k} \approx \frac{10^3 \times 0.1}{0.66} = 150,$$

suggests approximating the surface of the rock to have the temperature of the well-stirred tank of ice water, $0\,^\circ\mathrm{C}$, thus simplifying the problem. Further, we can estimate the depth of penetration into the sphere from approximation (5.8.4) as

$$\delta_T = 4\sqrt{0.47 \times 10^{-6} \times 60} = 0.021 \text{ m}$$

or $\delta_T/D = 0.2$. This is small enough such that the solution we obtain from the penetration approximation will be a reasonable first estimate, but not so small such that curvature effects can be entirely ignored. Alternatively, the Fourier number can be evaluated for the center of the sphere, as:

$$\mathrm{Fo} = \frac{4t\alpha}{D^2} = \frac{4 \times 60 \times 0.47 \times 10^{-6}}{0.1^2} = 1.1 \times 10^{-6}.$$

This very small value indicates that the center of the sphere will be unaffected by the cooling at the surface. Consequently, as we desire an estimate only of the heat transferred from the rock, we use the penetration solution and the heat flux given by Eq. (5.8.5). The total heat flow for 1 min is obtained by integration:

$$Q_{\text{total}} = \int_0^{60} q_y|_{y=0}\, dt = \int_0^{60} \frac{k}{\sqrt{\pi\alpha t}}\,(T_1 - T_0)(\pi d^2)\, dt = \frac{-k}{2\sqrt{\pi\alpha}}\,(T_1 - T_0)\, t^{\frac{1}{2}}\big|_0^{60}(\pi d^2),$$

$$Q_{\text{total}} = \frac{-0.66}{2\sqrt{\pi 0.47 \times 10^{-6}}}\,(80 - 0)\, 60^{\frac{1}{2}}(\pi 0.1^2) = -5.3 \text{ kJ}.$$

Thus we estimate that the rock can provide a significant amount of heat. Indeed, a Level I balance on the rock shows that the total amount of heat that can be transferred to the ice water is given by

$$Q_{\text{load}} = m\hat{C}_p\,(T_1 - T_0) = 1670 \times \pi\frac{0.1^3}{6} \times 838 \times 80 = 586 \text{ kJ}.$$

This demonstrates that the rock has lost only $\sim 1\%$ of its available heat to the water bath in the first 60 s. Heating water by hot rocks was a common method of cooking by Native Americans without metal cooking pots, is used in saunas, and is the basis of geothermal energy sources.

5.8.2 Small Biot Numbers—Lumped Analysis

If the Biot number is small, such that the rate of heat transfer to the object is limiting, then the internal temperature inside the object under consideration will be nearly uniform. An example would be the heating or cooling of a metal sphere in which poor thermal contact is made with the surroundings. In the limit of $\mathrm{Bi} \to 0$, a *lumped analysis* can be performed in which the temperature inside the object is assumed to be uniform. An example of this analysis is illustrated below.

EXAMPLE 5.7 LUMPED ANALYSIS. Reconsider the previous example if the sphere is made of stainless steel and is placed in a room ($T_a = 20\,^\circ\text{C}$) in where the air is only gently mixed. The heat transfer coefficient between the sphere and the surroundings is found to be $h = 50\ \text{W/m}^2\ \text{K}$. What is the rate at which the sphere's center temperature changes?

SOLUTION. From Table 5.1, we obtain the necessary physical properties for stainless steel and calculate the Biot number as

$$\text{Bi} = \frac{hD}{k} = \frac{50 \times 0.1}{16} = 0.3.$$

This is small enough to use a lumped analysis. When the sphere is treated as the system, an energy balance yields

$$\frac{dH}{dt} = -Q = -ha\,(T - T_a).$$

Notice that this balance is equivalent to the Level III analysis of a batch well-mixed heat exchanger, Eq. (3.2.1). The assumptions of being well mixed and of uniform temperature are mathematically equivalent. Using $H \approx \rho V \hat{C}_p T$ and simplifying yields

$$\frac{dT}{dt} = -\frac{ha}{\rho V \hat{C}_p}\,(T - T_a),$$

$$\therefore\ T - T_a = (T_i - T_a)\,e^{-\beta t},$$

$$\beta = \frac{6h}{\rho D \hat{C}_p}.$$

Plugging in the parameters shows $\beta = 8.3 \times 10^{-4}\ \text{s}^{-1}$. For 1 min, the temperature in the steel sphere has dropped to only 77 $^\circ$C. This small temperature change is a consequence of the low rate of convective heat transfer from the surface of the sphere. The uniformity of the temperature inside the sphere is a consequence of the relatively fast conduction of heat within the sphere.

Nomenclature

A	Area (m^2)
Bi	Biot number (hL/k, k_mL/D$_{AB}$)
C	Concentration (mol/m^3)
$\hat{C}_p$	Heat capacity (kJ/kg K)
D	Diffusivity (m^2/s)
Fo	Fourier number
h	Local heat transfer coefficient (J/s m^2 K, W/m^2 K)
H	Enthalpy (kJ)
H	Partition coefficient
J	Molar flux (mol/m^2 s)
k	Thermal conductivity (W/m K)
k_m	Local mass transfer coefficients (m/s)
K_m	Mass transfer coefficient (m/s)

Le	Lewis number
L	Length (m)
M	Distribution coefficient
M_w	Molecular weight (kg/mol)
N	Number of particles
$N_{A,z}$	Total flux $J_{Az} + v_z C_A$ (mol/m^2 s)
Nu	Nusselt number
P = DH/RT	Permeability (mol m/m^2 s Pa)
p_A	Partial pressure of component A
Pr	Prandtl number
$\dot{q}_v$	Volumetric heating rate
q	Flow rate (m^3/s)
q_i	Heat flux i direction (kJ/s m^2)
Q	Heat flow (kJ/s)
r_A	Rate of reaction (mol/m^3 s)
r	Radius (m)
R_m	Mass transfer resistance
R_T	Thermal resistance
S	Schmidt number
Sh	Sherwood number
t	Time (s)
T	Temperature (K)
U	Overall heat transfer coefficient (kJ/s m^2 K, kW/m^2 K)
v	Velocity (m/s)
V	Volume (m^3)
x	Distance in x direction (m)
y	Distance in y direction (m)
z	Distance in z direction (m)

Greek

Δ	Difference
ρ	Density (kg/m^3)
α	Thermal diffusivity $K/\rho C_p$ (m^2/s)
α	Shape factor
δ	Penetration depth (m)

Subscripts

A	Species A
B	Species B
O	Initial
sat	Saturation

Superscripts

I	Phase I

II	Phase II
eff	Effective

Important Dimensionless Groups

The following dimensionless groups were introduced in this chapter.
- *Biot:*

$$Bi = \frac{hL}{k} \qquad (5.1.12)$$

expresses the ratio of external rate of heat transfer (h) to internal rate of heat transfer (k/L).
- *Nusselt:*

$$Nu = \frac{hL}{k} \qquad (5.1.13)$$

is the dimensionless heat transfer coefficient (note, h, L, and k are all in the same material).
- *Biot for mass transfer:*

$$Bi_m = \frac{k_m L}{D_{AB}} \qquad (5.2.9)$$

expresses the ratio of external rate of mass transfer to internal rate of mass transfer.
- *Sherwood:*

$$Sh = \frac{k_m L}{D_{AB}} \qquad (5.2.11)$$

is the dimensionless mass transfer coefficient (k_m and D_{AB} refer to the same species A being transported in material B with path length L).

Nondimensionalization of the conservation of momentum equation, as developed in Chapter 6, leads to the identification of the ratio of inertial forces to viscous forces as the *Reynolds* number:

$$Re = \frac{\rho VD}{\mu}.$$

Boundary-layer theory, introduced in Chapter 6, suggests that the transport coefficients depend on molecular parameters. These appear in correlations in the form of dimensionless groups. Although viscosity and thermal conductivity have different units, both have units of square meters per second when expressed as "diffusivities" for momentum and heat, i.e., as the kinematic viscosity and thermal diffusivity. This ratio is a dimensionless group that is a characteristic of the material for a given thermodynamic state. Further, one might expect this ratio to be relatively constant for a broad range of chemicals in a similar thermodynamic state (i.e., by analogy to the idea of corresponding states). The *Prandtl* [Ludwig Prandtl (1875–1953)] number relates kinematic viscosity to thermal diffusivity:

$$Pr = \frac{\nu}{\alpha} = \frac{\hat{C}_p \mu}{k},$$

$$Pr = \frac{\hat{C}_p}{\hat{C}_V} = \frac{5}{3} \quad \text{(simple kinetic theory for a monatomic gas).}$$

In the last line, we used the relationship $\hat{C}_p = \hat{C}_V + R$ and the fact that $\hat{C}_V = 3/2R$ for an ideal, monatomic gas.

Similarly, we can derive two additional ratios, the ratio of kinematic viscosity to diffusivity, the *Schmidt* [Ernst Schmidt (1892–1975)] number,

$$Sc = \frac{\nu}{D_{AB}} = \frac{\mu}{\rho D_{AB}},$$

$$Sc = 1 \quad \text{(simple kinetic theory for a monatomic gas),}$$

and, finally, the *Lewis* [Warren K. Lewis (1882–1975)] number as the ratio of thermal diffusivity to species diffusivity:

$$Le = \frac{\alpha}{D_{AB}} = \frac{k}{\rho \hat{C}_p D_{AB}},$$

$$Le = \frac{Sc}{Pr} = \frac{3}{2} \quad \text{(monatomic gas).}$$

In discussing transport analogies, the following dimensionless groups will be introduced:

• *Stanton number:*

$$St = \frac{Nu}{Pr\,Re}. \tag{6.2.14}$$

• *Colburn j factors:*

$$j_H = \frac{Nu}{Re(Pr)^{1/3}},$$

$$j_D = \frac{Sh}{Re(Sc)^{1/3}}. \tag{6.2.15}$$

• *Coefficient of skin friction:*

$$f = \frac{\mu \frac{\partial v_x}{\partial y}}{\frac{1}{2}\rho V_\infty^2} \tag{6.1.7}$$

The latter is defined as the ratio of the viscous surface drag per area on a body divided by the specific kinetic energy of the bulk flow.

REFERENCES

Bird RB, Stewart WE, Lightfoot EN. *Transport Phenomena* (2nd ed.). New York: Wiley, 2002.

Carslaw HS, Jaeger JC. *Conduction of Heat in Solids* (2nd ed). Oxford: Oxford University Press, 1959.

Chemical Engineers' Handbook (7th ed.). New York: McGraw-Hill Professional, 1997.

Cussler EL. *Diffusion: Mass Transfer in Fluid Systems* (2nd ed.). Cambridge: Cambridge University Press, 1997.

Fick A. Ueber diffusion. Ann. Phys. 1855; 94: 59–86.

Matsuura T. *Synthetic Membranes and Membrane Separation Processes*. Boca Raton, FL: CRC Press, 1994.

National Institute of Science and Technology Web site: www.nist.gov

PROBLEMS

5.1 (Adapted from William J. Thomson, *Introduction to Transport Phenomena*, Prentice-Hall, 2000, Problem 3.18). A pressurized spherical storage tank contains liquid oxygen and oxygen vapor. As the oxygen vaporizes (with $\Delta \hat{H}_v$), it is vented off, compressed, refrigerated, and returned to the tank as liquid. Assume that the ambient temperature is T_a and that heat is conducted into the tank from all directions equally.

  a. Derive an expression for the steady-state temperature distribution within the tank wall as a function of radius r and the boil-off rate $\dot{m}$.

  b. Determine the boil-off rate if the tank is constructed of stainless steel and if the temperature of the inner shell is constant at –60 °C. The following data apply (see Figure P5.1):

$$T_a = 25\,°C,\ R_0 = 50\,ft.,\ R_i = 49.5\,ft.,\ \hat{H}_v = 2.14 \times 10^5\,J/kg.$$

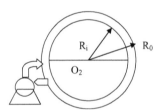

Figure P5.1.

5.2 (Adapted from William J. Thomson, *Introduction to Transport Phenomena*, Prentice-Hall, 2000, Problem 3.17). A lubrication oil with thermal conductivity k fills the annulus between a long, cylindrical rotating shaft and an outer sleeve, as shown in Figure P5.2.

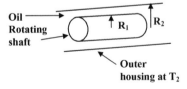

Figure P5.2.

The friction that is due to the rotation generates a uniform energy source within the oil,

$$\dot{Q}_v = \mu \left(\frac{\Omega R_1}{R_2 - R_1} \right)^2,$$

where Ω is the shaft rpm and μ is the viscosity of the oil. The sleeve is cooled and the outer temperature maintained at T_2. You may assume the shaft is sufficiently long such that all of the heat is conducted in the radial direction.

    a. Derive the model equation for the steady-state temperature distribution in the annulus. For this part you may assume that the shaft is a perfect insulator. Solve and sketch the temperature distribution in terms of the variables in the problem.

    b. Solve for the energy flux at the outer surface and show that this is consistent with an overall energy balance.

5.3 Reconsider Problem 5.2. Relax the assumption of the shaft being an insulator.

    a. Derive the model equations for the steady-state temperature distribution in the shaft and the oil now assuming that there is a thin film between the oil and both surfaces, with a heat transfer coefficient of $h = 100$ W/m^2 K (independent of rotation speed for the range of interest).

    b. Will the temperature vary in the shaft? Will the temperature of the shaft depend on the material the shaft is made of? Clearly explain why or why not to both questions using the model equations.

    c. Plot out the maximum temperature in the shaft as a function of rotation rate if the shaft is made of stainless steel, the oil has the properties listed in Table 3.3, a viscosity of 0.1 Pa s, and $T_2 = 25\,^\circ$C. The rotation rate ranges from 10 to 10 000 rpm and $R_1 = 1$ cm, $R_2 = 1.2$ cm.

    d. Determine the overall heat transfer coefficient defined in terms of the temperature difference from the surface of the shaft to the outer housing.

5.4 Explain how the heat tiles on the underside of the space shuttle function. Using publicly available data perform a calculation to justify the thickness of the tiles.

5.5 You are a detective investigating a murder, and need to know when the murder occurred. There are two suspects: Howard, who was seen leaving town the previous day, and Mickey, who could have committed the crime only in the last 30 hours. The victim's body *(which is of cylindrical geometry)* is very long, and has a radius of 15 cm. The average thermal conductivity, density, and heat capacity of a human body are 1.5×10^{-3} cal/cm sec K, 1.1 g/cc, and 0.83 cal/g K, respectively, and the outside heat transfer coefficient is approximated to be 8 W/m^2 K. The room temperature has been constant at $20\,^\circ$C, and the temperature at the center of the body is measured to be $29\,^\circ$C. How long ago was the crime committed, and who is the culprit?

5.6 A thermal pane window consists of two 8-mm window glass panes ($k = 0.78$ W/m K) separated by a gas layer. Determine the spacing of the two layers if an $R = 4$ value is required. The gas used is argon and the outside temperature is $0\,^\circ$C. Use the Chapman–Enskog theory for calculating the argon properties required for solving the problem.

5.7 a. Derive the solution to Fick's experiment in the funnel geometry, Eq. (5.3.6). Use R_0 and R_L to denote the radii at the bottom ($z = 0$) and top ($z = L$) of the funnel. Clearly state all assumptions.

b. Plot the concentration profile as compared with the straight cylinder of radius R_0 and compare with Figure 5.4. Comment on Fick's conclusions.

c. Determine how the steady-state rate of crystal dissolution compares with the rate obtained for a straight cylinder of equal height and constant radius R_0.

Specifically, plot the rate of dissolution in the funicular geometry as compared with the straight geometry as a function of the ratio R_L/R_0. If one used this funnel form device to determine the diffusion coefficient, what would be the error in assuming a linear concentration profile?

5.8 Crystals of Agent X are on the bottom of a 20-cm-long sealed tube filled with stagnant water containing bacteria (see Figure P5.3). The saturation concentration is reported to be 1 M, and the diffusivity of Agent X is 3.4×10^{-6} cm^2/s. The bacteria are evenly dispersed in solution and consume Agent X at a zeroth-order volumetric rate of 5.3×10^{-9} M/s. Given this information, do the following:

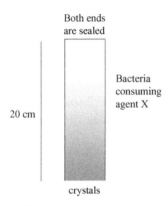

Both ends
are sealed

Bacteria
consuming
agent X

20 cm

crystals

Figure P5.3.

a. Derive a one-dimensional, steady-state shell balance for the concentration in the tube. Clearly state all assumptions and boundary conditions.

b. Solve the differential equation for the concentration profile in the tube.

c. Determine the steady-state concentration of Agent X at the top of the tube.

d. Suggest a dimensionless number that would describe the uniformity of the concentration profile.

5.9 To achieve steady-state (zeroth-order) drug delivery, "reservoir devices" are used, in which a mixture of drug crystals and saturated aqueous drug solution is encapsulated by a thin membrane with finite drug permeability (see Figure P5.4). Consider a spherical capsule 5 mm in radius and a membrane thickness of 0.2 mm in your TA's stomach *(which you can assume to be infinitely large in respect to the pill)*. You have been told that the effective diffusivity of the drug through the membrane is 6.7×10^{-8} cm^2/s and the partition coefficient between the membrane and the aqueous phase is 0.14 M (in membrane)/M(aqueous). You also know that the saturation concentration is 0.2 M, the

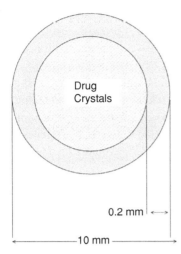

Figure P5.4.

capsule initially contains 1.0×10^{-3} mol of drug, and drug consumption in the stomach is very fast (i.e., drug concentration in the stomach is negligible).

a. Derive a one-dimensional steady-state shell balance equation for the drug diffusing through the membrane.
b. Solve for the steady-state concentration profile in the membrane.
c. Calculate the flux through the membrane.
d. Determine how often the drug needs to be administered.
e. What is a potential disadvantage of this drug delivery system?

5.10 After graduating from the University of Delaware, Joe Blue Hen installs an indoor swimming pool in his house. The length and the width of this swimming pool are 30 m and 10 m, respectively. The temperature is held constant at 20 °C, and the humidity of the air 1 m above the water is set to 50%. To determine the rate at which the pool must be replenished with water, Joe conducts an experiment using an Arnold diffusion cell. In this experimental setup, bone-dry air flows over the top of the cell, while liquid water resides on the bottom of the vessel. The initial distance between the bone-dry air and the water is 10 cm, and the pressure in the cell is 1 bar. After conducting an experiment, Joe finds that the liquid–vapor interface falls at a rate of 20 μm/hr. Help Joe with the following:

a. Derive a model equation that relates the rate of drop in liquid level in the Arnold cell to the diffusivity.
b. Find the diffusivity of water vapor in air.
c. Compare the value that you calculated with the literature value of 0.260 cm²/s.
d. Determine the rate at which Joe should be adding water to his swimming pool.
e. Suggest a source of uncertainty for this estimate. Is it most likely high or low?

5.11 In Ångström's method to determine thermal conductivity, a periodic heat flow is applied to a bar (insulated in the x, z directions and thermistors located at two y positions along the bar a length L apart). At steady state, the thermistors read the local temperature as a function of time, which will be sinusoidal. Analyze this experiment and show how the thermal conductivity can be calculated from the temperature measurements.

5.12 Y. Geng and D. E. Discher (Visualization of degradable worm micelle breakdown in relation to drug release. Polymer. 2006; 47: 2519–25) proposed using wormlike micelles as a novel delivery vehicle for the controlled release of the cancer drug taxol (see Figure P5.5). The wormlike micelles can be thought of essentially as infinitely long cylinders with two compartments: a core with a nominal diameter of 11 nm, and a shell through which the taxol must diffuse in order to be released to the surrounding environment. The following table gives experimental data taken for the total percent of taxol released versus time at 4 °C and pH 5 for a 1 wt.% solution of wormlike micelles initially loaded with 4.5 wt.% taxol.

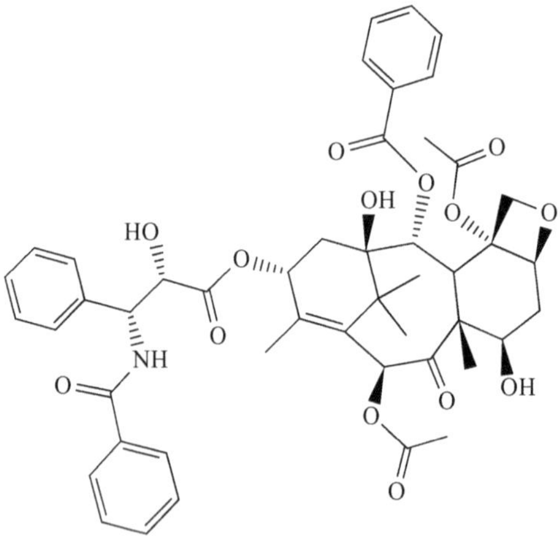

Taxol

Figure P5.5.

Time (days)	% taxol release
0	0
1	17
2	32
3	37
6	51
11	59

a. Draw and label a diagram that can be used in modeling the release of taxol from the wormlike micelles, assuming that the resistance to mass transfer is through the shell (which you can assume is approximately 10 nm in thickness). The surrounding water phase is well mixed.

b. Assume that the Biot number for mass transfer for taxol between the core and the shell is small such that a lumped-sum analysis can be used. Derive a model equation for the total percent release of taxol from the micelles as a function of time in terms of an overall mass transfer coefficient K_m between the micelle core and the water.

c. Using the data in the preceding table, determine if the assumption of a lumped analysis is valid. (The partition coefficient of taxol between the water and micelle phases is 0.005.) Does the model provide a good fit to the data? Does the value of K_m you obtained make physical sense?

d. If there were an external resistance to mass transfer in the stirred water surrounding the micelle, would this improve the comparison between model equation and data?

e. Relax the assumption of a low Biot number for mass transfer and derive a model equation that describes the rate of diffusion of taxol from the micelle if there is diffusion in the core, through the shell, and an external resistance to mass transfer in the water phase. Do not solve. However, clearly indicate what information you would require to compare the new model against experiment.

5.13 The murder investigation, subject of Problem 5.5, is being reconsidered by the CSI team, and additional analyses of the data to determine time of death are called for.

a. Repeat the analysis using the short-time penetration method discussed in Section 5.8.

b. Repeat the problem using the lumped-analysis method discussed in Subsection 5.8.2.

c. Will you accuse the same person in each case? Which method (penetration or lumped analysis) is more accurate, and why?

6 Convective Heat and Mass Transfer

The coefficients of heat and mass transfer rate expressions depend on any fluid flows in the system. Our personal experience with "wind-chill" factors on chilly winter days and in dissolving sugar or instant coffee in hot liquids by stirring suggests that the rate of heat and mass transfer can be greatly increased with increasing wind speed or mixing rates. The technically feasible design of heat and mass transfer equipment requires calculating the transport coefficients and their variation with the fluid flows in the device, which depend intimately on the design of the device. For example, the area for heat transfer calculated for a tubular–tubular heat exchanger can be achieved by an infinite combination of pipe diameters, lengths, and for shell-and-tube exchanges, the number of tubes. However, selecting a pipe diameter for a given volumetric flow rate sets the fluid velocity in the pipe and the type of flow (i.e., laminar versus turbulent), which sets the overall heat transfer coefficient. This is why the design of heat and mass transfer equipment is often an iterative process. This chapter presents methods for estimating transport coefficients in systems with fluid motion.

The *central hypothesis* for flowing systems is that the friction, resistance to heat transfer, and resistance to mass transfer are predominately located in a thin *boundary layer* at the interface between the bulk flowing fluid and either another fluid (liquid or gas) or a solid surface. Consequently, understanding how to estimate heat and mass transfer coefficients starts with an analysis of the mass, momentum, and energy transport in an idealized, laminar boundary-layer. The following outline indicates the structure of this chapter.

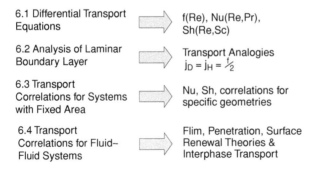

6.1 Differential Transport Equations ⟹ f(Re), Nu(Re,Pr), Sh(Re,Sc)

6.2 Analysis of Laminar Boundary Layer ⟹ Transport Analogies $j_D = j_H = \frac{1}{2}$

6.3 Transport Correlations for Systems with Fixed Area ⟹ Nu, Sh, correlations for specific geometries

6.4 Transport Correlations for Fluid–Fluid Systems ⟹ Flim, Penetration, Surface Renewal Theories & Interphase Transport

Section 6.5 concludes the chapter by summarizing the key results. The boundary-layer analysis requires *differential* balances for momentum, energy, and species mass flows. The derivation of the differential transport equations is presented in Appendix A, where we use the word statement in Figure 1.6 along with constitutive equations from Table 1.4. *Dimensional analysis* and *scaling* of the transport equations simplified for the laminar boundary layer provide the general dependencies of the dimensionless transport coefficients on flow rates, system geometry, and fluid properties through the Reynolds, Prandtl, and Schmidt dimensionless groups. Further analysis of the boundary-layer transport equations in Section 6.2 yields *transport analogies* that are shown to have broader validity beyond the simplified boundary layer problem. The development of analogies between transport processes, determining the range of their validity, and their use in analysis of experiment and design of equipment is one central theme of this chapter.

These results obtained from analyzing a boundary layer are applicable to systems with defined areas for transport, such as generally found in heat exchangers. A survey of the experimental evidence for transport analogies and empirical correlations for varying geometries and flow regimes are presented in Section 6.3.

Transport coefficients for situations involving direct fluid–fluid contacting, as is typical of mass transfer, are more difficult to estimate as the area and relative fluid velocities are often poorly known. Semiempirical models for estimating and correlating transport coefficients for these systems are developed from the boundary-layer analysis in Section 6.4. Further methods for handling systems in which the area for transport depends on the fluid motions will be presented in Chapter 7.

6.1 The Differential Transport Equations for Fluids with Constant Physical Properties in a Laminar Boundary Layer

The derivations of the differential transport equations for momentum, heat, and mass transfer follow a common paradigm, which starts with a differential control volume (Figure 6.A1 in Appendix A) and the word statement of conservation, Figure 1.6. A complete derivation of the differential transport equations is provided in Appendix A and the interested student is referred to that derivation plus the references cited in this chapter for more information. In the first part of this section we refer to material covered in an undergraduate fluid mechanics class which, as such, is a review for most students. Nonetheless, full derivations of the continuity and Navier–Stokes equations are provided in Appendix A. In this section, we simplify the differential transport equations for the analysis of the laminar boundary layer considered in this chapter and perform a dimensional analysis to study the effects of flow on the rate of mass and heat transfer.

The laminar boundary layer is shown in Figure 6.1, where the fluid flowing uniformly in the x direction impinges on a thin plate oriented in the y–z plane. There is a no-slip or "stick" boundary condition at the surface of the plate, such that the fluid velocity in all three directions is zero at the surface of the plate (note that only the upper half is shown in the figure for clarity). As a consequence, a fluid velocity gradient will develop along the plate, as well as a velocity component normal to the plate (in the y direction). The symmetry of the problem determines that $v_z = 0$ everywhere. Furthermore, there are no body forces, such as pressure or gravity, acting on the fluid.

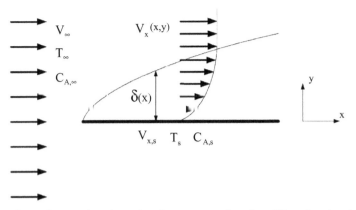

Figure 6.1. Laminar boundary layer over a flat plate. [The plate is taken to be infinite in the z direction and infinitely thin. There is a corresponding boundary layer on the bottom of the plate not shown here, as well as a velocity component $v_y(x,y)$ not shown for clarity.]

Finally, we consider steady-state flow, so the equations will not be dependent on time. These specifications enable greatly simplifying the differential transport equations that are derived for a general, three-dimensional flow in Appendix A. These simplified transport equations are presented next for conservation of mass, x momentum, energy, and species A.

6.1.1 Mass Conservation—Continuity Equation

For the problems that follow, we assume that the density of the fluid is constant. The continuity equation, Equation (6.A3), for a constant-density fluid simplifies to the following for the laminar boundary layer with no z velocity:

$$\frac{\partial v_x}{\partial x} + \frac{\partial v_y}{\partial y} = 0. \tag{6.1.1}$$

This equation expresses in mathematical form that changes in the x-velocity component that are due to drag at the surface of the plate will require a corresponding y-velocity component in the boundary layer. It is *scaled* to put it into a useful form for further analysis. Scaling refers to reducing the equation to a nondimensional form by carefully choosing length, time, temperature, concentration, and other scales that correspond to *characteristic scales* of the problem at hand. Equation (6.1.1) contains velocities and distances. The characteristic velocity for this problem is given in Figure 6.1 as V_∞, the upstream, uniform fluid velocity. This makes for a good velocity scale as the dimensionless x velocity, denoted by the tilde above the symbol, as $\tilde{v}_x$ would be expected to range from zero at the surface of the plate to one far from the plate. Hence we can immediately see the value of scaling as, regardless of the initial velocity, the relative magnitude of the velocity in the boundary layer, in dimensionless form, will be $0 \le \tilde{v}_x \le 1$. The choice of length scale is less obvious for this problem as the plate is assumed to be infinite in breadth and length, and infinitely thin. However, as shown in Figure 6.1, there will be a "boundary-layer thickness," defined as δ, which grows along the length of the plate. This thickness will be defined as where the x velocity reaches 99% of its initial value, or, in other words, where $\tilde{v}_x = 0.99$. The velocities and other properties we consider, such as temperature and concentration, are expected to vary significantly over the boundary layer and hence over length scales of the order of δ. Thus the characteristic scales become

$$\tilde{x} = \frac{x}{\delta}, \quad \tilde{y} = \frac{y}{\delta}, \quad \tilde{v}_x = \frac{v_x}{V_\infty}, \quad \tilde{v}_y = \frac{v_y}{V_\infty}. \tag{6.1.2}$$

Substituting these definitions into Equation (6.1.1) leads to the scaled or dimensionless form of the continuity equation:

$$\frac{\partial \tilde{v}_x}{\partial \tilde{x}} + \frac{\partial \tilde{v}_y}{\partial \tilde{y}} = 0. \tag{6.1.3}$$

6.1.2 Momentum Transport—Navier–Stokes Equation

Momentum is a vector quantity that is calculated as mass multiplied by velocity. Expressed in terms of the density, the momentum in the x direction for fluid in a volume V is $\rho v_x V$. As shown in Appendix A, conservation of momentum leads to the Navier–Stokes equation when Newton's constitutive equation is used to relate shear stress to velocity gradients. Simplification of Equation (6.A10) for steady flow of fluid with a constant density and shear viscosity μ in the laminar boundary layer leads to the following expression of x-momentum conservation:

$$v_x \frac{\partial v_x}{\partial x} + v_y \frac{\partial v_x}{\partial y} = \nu \frac{\partial^2 v_x}{\partial y^2}. \tag{6.1.4}$$

Remember that $\nu = \mu/\rho$ is the kinematic viscosity. In simplifying the Navier–Stokes equation, the second derivative of the x velocity along the x direction is assumed to be much smaller than the second derivative along the y direction, i.e.,

$$\frac{\partial^2 v_x}{\partial y^2} \gg \frac{\partial^2 v_x}{\partial x^2}.$$

This assumption can be checked once a solution for the velocity field for this problem is found, which will be shown in the next section. However, the rationale for this difference in magnitude is clear if one considers the x velocity and how it varies near the wall. Clearly, the x velocity decreases rapidly as the plate is approached in the y direction anywhere in the boundary layer. However, the x velocity varies only slowly along the x direction. Notice that in the inertial terms (the left-hand side of the equation) we keep both first derivatives. They are of similar magnitude because the smaller derivative $\partial v_x/\partial x$ is multiplied by the larger velocity v_x, whereas the larger derivative $\partial v_x/\partial y$ is multiplied by the smaller velocity v_y, Finally, by symmetry, there is no variation of the x velocity in the z direction.

This equation can also be scaled by use of the same characteristic scales of the problem defined in Eqs. (6.1.2). Substituting into Eq. (6.1.4) and rearranging yields

$$\tilde{v}_x \frac{\partial \tilde{v}_x}{\partial \tilde{x}} + \tilde{v}_y \frac{\partial \tilde{v}_x}{\partial \tilde{y}} = \mathrm{Re}^{-1} \frac{\partial^2 \tilde{v}_x}{\partial \tilde{y}^2}. \tag{6.1.5}$$

Here, scaling the equation leads to the identification of a new dimensionless group that is important for the problem, the Reynolds number as the ratio of inertial to viscous forces acting on the fluid (Reynolds, 1883):

$$\mathrm{Re} = \frac{\rho V_\infty \delta}{\mu} = \frac{\rho V_\infty^2}{\mu \left(\dfrac{V_\infty}{\delta}\right)} = \frac{\text{inertial forces}}{\text{viscous forces}}. \tag{6.1.6}$$

In the preceding equation the forces are written out for clarity. Solutions of the dimensionless velocity profile will, in general, depend on the Reynolds number.

The engineering transport coefficient associated with changes in x momentum that are due to flow over the plate is the *coefficient of skin friction*. The change in fluid velocity in the boundary layer will exert a drag force on the plate. Defining F_x as the force the plate exerts on the fluid at location x from the leading edge of the plate, the coefficient of skin friction f is defined as

$$f = \frac{F_x/(\text{area})}{\left(\frac{1}{2}\rho V_\infty^2\right)} = \frac{\mu \frac{\partial v_Y}{\partial y}}{\left(\frac{1}{2}\rho V_\infty^2\right)} = \frac{2\mu}{\rho V_x \delta} \frac{\partial \tilde{v}_x}{\partial \tilde{y}} = \frac{2}{\text{Re}} \frac{\partial v_x}{\partial \tilde{y}}. \tag{6.1.7}$$

The dimensionless velocity gradient at the surface of the plate (where the frictional force is evaluated) will depend on the Reynolds number of the flow. Thus we can ascertain that, in general,

$$f = F(\text{Re}). \tag{6.1.8}$$

This relation states that the friction coefficient will be a function (F) of the Reynolds number. Specification of that function requires a solution for the velocity profile. The same procedure just used to calculate the friction coefficient is now used to determine the rate of heat and mass transfer, in dimensionless forms of Nusselt and Sherwood numbers in the following. But before proceeding, we can explore a simple flow that yields results for the skin friction.

EXAMPLE 6.1 FRICTION COEFFICIENT FOR DRIVEN LAMINAR FLOW. Consider a laminar one-dimensional flow between parallel plates separated by a distance δ. The fluid flows between an upper parallel plate moving in the x direction relative to a stationary lower plate. Determine the coefficient of friction on the plate, which is the force required for driving the plate, if the motion is sufficiently slow such that the flow is laminar and the temperature and density are constant.

SOLUTION. For laminar flow, the continuity equation simplifies further to the statement $(\partial \tilde{v}_x/\partial \tilde{x}) = 0$ because there is no flow in the y or z direction. The Navier–Stokes equation simplifies to the statement that $0 = \text{Re}^{-1}(\partial^2 \tilde{v}_x/\partial \tilde{y}^2)$, where the continuity equation has been used along with the fact that there is flow in only the x direction. Two boundary conditions are required. There are no-slip, or "stick," boundary conditions such that $\tilde{y} = 0$, $\tilde{v}_x = 0$ at the lower plate, and $\tilde{y} = 1$, $\tilde{v}_x = 1$ at the upper plate. A solution that satisfies this equation is given by $\tilde{v}_x = \tilde{y}$, or, in dimensional form, $v_x = (V_\infty y/\delta)$. Given this solution, the coefficient of skin friction can be calculated from Eq. (6.1.7) as $f = 2/\text{Re}$. For a more complex flow, such as that over the boundary layer, a different dependence on the Reynolds number is to be expected as the velocity profile will not, in general, be linear.

6.1.3 Energy Conservation

Consider the boundary layer shown in Figure 6.1, where the plate and fluid are not at the same temperature, $T_S \neq T_\infty$. Heat transfer will occur between the plate and fluid, and the temperature profile and rate of heat transfer are of engineering interest. The energy content of a flowing fluid is written in terms of the enthalpy density, $\hat{H}$.

Enthalpy is convected in and out of the boundary layer by fluid motion $(\rho \hat{H} v_i)$ as well as by heat conduction (q_i), which can be described by Fourier's constitutive equation. As discussed in Chapter 3, the enthalpy can be written in terms of the heat capacity times the temperature, $\hat{H} \approx \hat{C}_p T$. Then the conservation equation for energy is derived in Appendix A of this chapter as Equation (6.A13). Simplification to the steady, laminar boundary layer for fluids of constant physical properties yields

$$v_x \frac{\partial T}{\partial x} + v_y \frac{\partial T}{\partial y} = \alpha \frac{\partial^2 T}{\partial y^2}. \tag{6.1.9}$$

The thermal diffusivity $(\alpha = k/\rho \hat{C}_p)$ naturally appears in the equation. Similar to the velocity profile, only the second derivative of the temperature variation in the y direction in the boundary layer is significant. At this point it is worthwhile to note the similarity of Eq. (6.1.9) to the simplified conservation equation for the x-momentum density, Eq. (6.1.4). This similarity will be exploited shortly to develop analogies between the transport processes.

As with the Navier–Stokes equation, the energy equation is scaled by identifying the characteristic scales of the problem. The velocity and length scales are defined by Eqs. (6.1.2). The issue at hand is heat transfer to (or from) the fluid to the plate. The heat transfer is assumed to take place in the boundary layer. Defining T_S as the temperature of the surface of the plate, which will be taken to be uniform, and T_∞ as the temperature of the fluid far from the plate, enables defining a dimensionless temperature as

$$\tilde{T} = \frac{T - T_S}{T_\infty - T_S}. \tag{6.1.10}$$

This definition of temperature ensures that the dimensionless temperatures will range between zero (at the surface of the plate) and one (far from the plate) anywhere in the fluid. With some algebraic manipulations, Eq. (6.1.9) can be written in the following dimensionless form:

$$\tilde{v}_x \frac{\partial \tilde{T}}{\partial \tilde{x}} + \tilde{v}_y \frac{\partial \tilde{T}}{\partial \tilde{y}} = \frac{1}{\mathrm{Re}\,\mathrm{Pr}} \frac{\partial^2 \tilde{T}}{\partial \tilde{y}^2}. \tag{6.1.11}$$

In addition to the Reynolds number, the Prandtl number (introduced in Chapter 5) appears in the scaled equation:

$$\mathrm{Pr} = \frac{\nu}{\alpha} = \frac{\mu \hat{C}_p}{k} = \frac{\text{momentum diffusivity}}{\text{thermal diffusivity}}.$$

In Section 5.1 we examined Fourier's experiment for heat conduction. In the absence of flow, equation (6.1.11) simplified to the relationship $0 = k(\partial^2 T/\partial y^2)$. This led to a linear temperature profile between the boundaries. Now, however, consider the boundary layer depicted in Figure 6.1, where there is steady, laminar flow in the x direction in addition to the temperature difference in the y direction. The fluid flow in the x direction will affect the transport of energy in the y direction as heat is both conducted along and convected away from the plate. With specification of appropriate boundary conditions for the temperature profile, Eq. (6.1.11) can be simplified to develop a solution for the temperature profile. Then we can determine the *local heat transfer coefficient* from the lower plate (h_x) by equating the heat flow leaving the plate, as given by Fourier's constitutive equation, to the constitutive equation for the rate of heat flow (Table 1.5):

$$q_y|_x = -k \frac{\partial T}{\partial y}\bigg|_{y=0,x} = h_x \Delta T,$$

$$h_n \equiv -\frac{k}{b} \frac{\partial \tilde{T}}{\partial y}\bigg|_{\tilde{y}=0,x}, \qquad (6.1.12)$$

Thus, given a solution for the temperature profile, the local heat transfer coefficient can be calculated. Even without an explicit solution, however, the scaled form of energy equation (6.1.11) shows us that the solution for the dimensionless temperature profile will be a function of the two dimensionless groups appearing in the equation, namely $\tilde{T}(\tilde{x}, \tilde{y}) = F(Re, Pr)$. As a consequence, we can determine that any solution will be expressed in the general functional form:

$$Nu_x = \frac{hx}{k} = -\frac{d\tilde{T}}{d\tilde{y}} = F(Re, Pr). \qquad (6.1.13)$$

The relationship states that the local Nusselt number at position x along the plate will be a function (F) of the two dimensionless groups in the scaled equation, or, in other words, the heat transfer coefficient will depend on the Reynolds number of the flow as well as on specific fluid properties through the Prandtl number. This is in contrast to the result obtained for conduction through a homogeneous solid, Nu = 1, Eq. (5.1.14). In the presence of flow we find that the solution will now depend on the flow rate through the Reynolds number and the material properties through the Prandtl number. However, a specific solution for the temperature gradient in the fluid in contact with the plate is required for specifying the actual function. This is derived in the next subsection.

6.1.4 Species Mass Conservation

Consider the situation again shown in Figure 6.1 in which now the concentration of species A is such that $C_{A\infty} \neq C_{As}$. Therefore, A is being swept away from the plate by the fluid, or, in the opposite case, A is deposited into the plate from the flowing liquid. The conservation of molar species A in the flowing fluid includes fluxes of species A in and out of the control volume that are due to diffusion, as discussed in the previous chapter, and by fluid motion, as discussed in the derivations for mass, momentum, and energy just performed. Fick's constitutive equation, discussed in Chapter 5, is applied to relate the diffusive flux to the concentration gradients. Further, assuming constant diffusivity and constant physical properties leads to the simplified form of the species conservation equation. Further, the net amount of mass transfer is assumed to be small enough that there is no significant velocity driven by the mass transfer, such as in the Arnold cell considered in Chapter 5, and not enough mass transfer to significantly change the geometry, such as the salt-tablet example considered in Chapter 4. Note that, for simplicity, we write D_A for the binary diffusivity of A in the fluid (B) flowing over the plate. Then the balance equation for the mass of species A, Equation (6.A17), simplified for this problem becomes

$$v_x \frac{\partial C_A}{\partial x} + v_y \frac{\partial C_A}{\partial y} = D_A \frac{\partial^2 C_A}{\partial y^2}. \qquad (6.1.14)$$

These concentrations are written in molar units, and multiplication by the molecular weight of A yields an equation for the mass concentration of A.

Similar to the case just considered for heat transfer, only the second derivative of concentration in the y direction is significant, and there are no variations in the z direction. This equation can also be scaled with the same dimensionless variables used for velocity and distance. A dimensionless concentration can be defined as the relative difference between the concentration in the fluid, anywhere, and that in the fluid at the surface of the plate, $C_{A,s}$. As we have done consistently throughout the text, we assume local equilibrium between the concentration of A in the plate with that in the fluid in direct contact with the plate (y = 0). Dividing by the maximum possible concentration difference, namely, that far from the plate, $C_{A,\infty}$ minus that at the surface of the plate defines

$$\tilde{C}_A = \frac{C_A - C_{As}}{C_{A\infty} - C_{As}} \tag{6.1.15}$$

The dimensionless equation for the conservation of the mass of A applicable to the laminar boundary layer becomes

$$\tilde{v}_x \frac{\partial \tilde{C}_A}{\partial \tilde{x}} + \tilde{v}_y \frac{\partial \tilde{C}_A}{\partial \tilde{y}} = \frac{1}{\text{Re}\,\text{Sc}} \frac{\partial^2 \tilde{C}_A}{\partial \tilde{y}^2}. \tag{6.1.16}$$

The additional dimensionless group is the Schmidt number Sc, which was defined in Chapter 5 as the ratio of kinematic viscosity to diffusivity:

$$\text{Sc} = \frac{\nu}{D_A} = \frac{\text{momentum diffusivity}}{\text{species A diffusivity}}$$

In Fick's experiment considered in Section 5.1, there is no flow and the steady state equation simplifies to $(\partial^2 \tilde{C}_A / \partial \tilde{y}^2)$. Integration for constant boundary conditions results in the linear concentration profile observed by Fick as derived in the previous chapter. For the boundary layer, the convection in the x direction will affect the rate of mass flux from or to the surface of the plate as Eq. (6.1.16) shows that the concentration profile depends on the fluid velocity. Using the same procedure as for heat transfer, we calculate the *local mass transfer coefficient* at position x along the plate by equating the flux at the surface of the plate to the engineering expression for mass transfer introduced in Table 1.5:

$$J_{Ay}|_x = -D_A \frac{\partial C_A}{\partial y}\bigg|_{y=0,x} = k_{m,x} \Delta C_A,$$

$$\therefore k_{m,x} = -\frac{D_A}{\delta} \frac{\partial \tilde{C}_A}{\partial \tilde{y}}\bigg|_{\tilde{y}=0,x}. \tag{6.1.17}$$

As with the previous example for heat transfer with convection, inspection of Eq. (6.1.16) shows that solutions will have the form $\tilde{C}_A(\tilde{x}, \tilde{y}) = F(\text{Re}, \text{Sc})$. Therefore the mass transfer coefficient will also depend on the fluid flow through the Reynolds number as well as the fluid properties through the Schmidt number:

$$\text{Sh}_x = \frac{k_{m,x} x}{D_A} = -\frac{\partial \tilde{C}_A}{\partial \tilde{y}}\bigg|_{\tilde{y}=0,x} = F(\text{Re}, \text{Sc}). \tag{6.1.18}$$

Compare this with the result of Chapter 5, where Sh = 1 was derived for diffusion through a uniform solid. Although general relations between dimensionless groups have been derived, specific quantitative results that are useful for solving transport problems require solving for the velocity, temperature, and concentration profiles. In the next section, we develop a solution for the laminar boundary layer and explore the consequences for mass and heat transfer.

6.2   Boundary Layer Analysis and Transport Analogies

From Section 6.1 it should be apparent that there is a fundamental similarity in the differential transport equations for momentum, heat, and mass. In this section we analyze a model problem as flow over a flat plate both to provide a calculation of heat and mass transfer coefficients for laminar flow over a flat plate, as well as to illustrate this similarity in the three basic transport mechanisms. This mathematical similarity of the equations governing the fields of velocity, temperature, and concentration in the steady boundary layer leads to the concept of a *transport analogy*, whereby results for one transport process, such as a friction coefficient, can be used to predict another, such as a heat or mass transfer coefficient.

Although rigorous results are generally available only for such simplified problems as flow over a flat plate, the analogies identified through this analysis are postulated to have much wider applicability. The basis for this postulate is that complex flows in heat exchangers and mass contactors generally have a *boundary layer* near the surface through which energy and mass transfer. It is often the situation such that the dominant resistance to heat and mass transfer is in this relatively thin boundary layer, hence the emphasis on analysis of transport in boundary layers. We start with the simplest, but perhaps least accurate, analogy, that of the Reynolds number for laminar flow over a flat-plate, and then we explore more robust analogies that may be applied over a broader range of fluid properties.

6.2.1 Laminar Boundary Layer

From Section 6.1, the dimensionless equations expressing conservation of total mass and x momentum that apply to this geometry shown in Figure 6.1 are

continuity:

$$\frac{\partial \tilde{v}_x}{\partial \tilde{x}} + \frac{\partial \tilde{v}_y}{\partial \tilde{y}} = 0; \tag{6.1.3}$$

x momentum:

$$\tilde{v}_x \frac{\partial \tilde{v}_x}{\partial \tilde{x}} + \tilde{v}_y \frac{\partial \tilde{v}_x}{\partial \tilde{y}} = \frac{1}{Re} \frac{\partial^2 \tilde{v}_x}{\partial \tilde{y}^2}. \tag{6.1.5}$$

These equations are to be solved with the following boundary conditions:

$$\tilde{y} = 0, \quad \tilde{v}_x = \tilde{v}_y = 0,$$
$$\tilde{y} \to \infty, \quad \tilde{v}_x = 1.$$

A method using a *similarity solution* is employed to yield the solution subsequently shown. The similarity transform is given by

$$\eta = \frac{y}{2}\sqrt{V_\infty/x\nu} = \frac{y}{2x}\sqrt{Re_x},$$

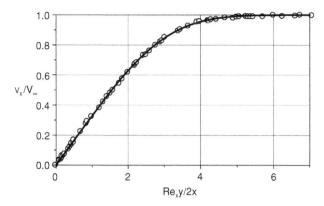

Figure 6.2. Blasius solution for the velocity profile in the laminar boundary layer compared with experimental data (Schlichting, 1979).

where

$$\mathrm{Re}_x = \frac{V_\infty x}{\nu}. \tag{6.2.1}$$

We find the solution by substituting in the similarity variable and converting the partial differential equation to an ordinary differential equation. The solution method can be found in texts on fluid mechanics, such as that by Schlichting (1979). The result is known as the *Blasius solution* and is plotted in Figure 6.2. The dimensionless velocity is plotted as a function of the similarity variable defined in Eq. (6.2.1). This variable contains both the x and y coordinates relative to the plate surface. The similarity solution indicates that the velocity profile is the same shape at each x location along the plate, but simply becomes "stretched" in the y direction as the boundary layer evolves with x distance. Figure 6.2 compares the Blasius solution with experimental measurements of the velocity profile for air flowing over a flat plate (Schlichting, 1979), where excellent agreement is observed.

The shear stress on the surface of the plate that is due to the fluid flow is evaluated from Newton's constitutive equation (Table 1.4) as

$$\tau_{y,x} = -\mu \frac{\partial v_x}{\partial y}\bigg|_{y=0}. \tag{6.2.2}$$

The derivative of the velocity field at the surface of the plate has been evaluated and is found to be

$$\frac{\partial v_x}{\partial y}\bigg|_{y=0} = 0.332 V_\infty \sqrt{\frac{V_\infty}{x\nu}} = 0.332 \frac{V_\infty}{x}\sqrt{\mathrm{Re}_x}. \tag{6.2.3}$$

Substituting into the definition of the coefficient of friction, Eq. (6.1.7), yields the following result:

$$f_x = \frac{-\tau_{yx}}{(1/2)\rho v_\infty^2} = \frac{0.664}{\sqrt{\mathrm{Re}_x}}. \tag{6.2.4}$$

This is the local friction coefficient, denoted as the "Fanning" friction factor, evaluated at the point x along the plate. Notice that this local friction coefficient *decreases* with increasing distance along the plate as $1/\sqrt{x}$. However, as the area

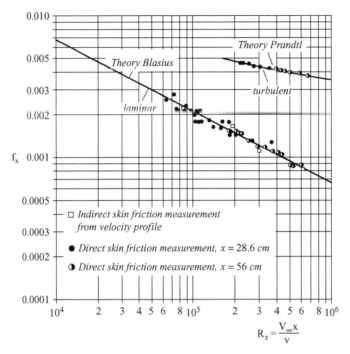

Figure 6.3. Local coefficient of friction f_x for incompressible flow over a flat plate. (Schlichting H. *Boundary Layer Theory*, 6th ed., 1979. Copyright McGraw-Hill Companies, Inc. Reproduced with permission.)

increases in proportion to distance along the plate, the total friction still increases as we consider longer distances along the plate.

For engineering calculation the average value of the friction coefficient is required, which, on integrating over the length of the plate L, becomes

$$f_L = \frac{1}{L} \int_0^L f_x dx = \frac{1.328}{\sqrt{Re_L}}, \qquad (6.2.5)$$

where the Reynolds number is defined by the total length of the plate:

$$Re_L = \frac{\rho V_\infty L}{\mu}. \qquad (6.2.6)$$

Figure 6.3 shows this result in comparison to experimental measurements. Excellent agreement is observed. The figure also contains data and theory for turbulent boundary layers, which exhibit a much higher coefficient of friction, a situation that will be discussed in the following subsection.

Another useful result from this analysis is the thickness of the boundary layer, which is evaluated by convention as where the velocity field reaches ~99% of the bulk value:

$$\delta(x) = \frac{5x}{\sqrt{Re_x}}. \qquad (6.2.7)$$

Table 6.1. *Relevant transport quantities for heat, mass, and momentum transport*

Transport	Field	Flux	Material property	Transport coefficient	Dimensionless group
x momentum	ρv_x	$\tau_{yx} = -\mu \dfrac{dv_x}{dy}$	$\nu = \mu/\rho$	$\dfrac{f}{2} = \dfrac{-\tau_{yx}}{\rho V_\infty^2}$	$\dfrac{f}{2}$
Heat	T	$q_y = -k\dfrac{dT}{dy}$	$\alpha = k/\rho\hat{C}_p$	$h = \dfrac{q_y}{\Delta T}$	$Nu = hL/k$
Mass	C_A	$J_{A,y} = -D_A\dfrac{dC_A}{dy}$	D_A	$k_m = \dfrac{J_{A,y}}{\Delta C_A}$	$Sh = k_m\,L/D_{AB}$

6.2.2 Reynolds Transport Analogy

The procedure for formulating the coefficient of friction from the solution of the continuity and Navier–Stokes equations in the boundary layer over a flat, thin plate is as follows:

• Derive and simplify the microscopic transport equations.
• Solve for the field of interest (velocity).
• Evaluate the flux at the surface (x momentum).
• Calculate the engineering transport coefficient (friction factor) by equating the flux to the engineering constitutive equation.

This same procedure is now applied to heat and mass transfer to obtain solutions for the temperature and concentration profiles in the boundary layer, and hence, for the rates of heat and mass transfer to or from the plate in the presence of the macroscopic flow field. Table 6.1 shows the key components of each transport mechanism.

For the problem of incompressible, laminar flow over the infinitely thin, infinitely wide plate, the following equations apply in the steady-state boundary layer (where the assumption of constant physical properties is made). The length scales of x and y are made nondimensional by dividing by the distance along the plate L, as

$$\tilde{x} = \frac{x}{L}, \quad \tilde{y} = \frac{y}{L}.$$

The simplified transport equations from Section 6.1 applied to the laminar boundary layer are shown in Table 6.2, in both dimensional and dimensionless forms. The associated boundary conditions applicable for this boundary layer problem are given in Table 6.3.

Reynolds noted that the equations and boundary conditions (Tables 6.2 and 6.3), when written in dimensionless form, are identical if the following holds: $Sc = 1, Pr = 1$. This condition restricts the physical properties of the fluids for which the analysis is valid. Under these conditions, the boundary layers for x-momentum, heat, and mass transfer coincide. Defining

$$\tilde{\theta} = \frac{\tilde{\rho}\tilde{v}_x - \tilde{\rho}\tilde{v}_{x,s}}{\tilde{\rho}\tilde{v}_{x,\infty} - \tilde{\rho}\tilde{v}_{x,s}} = \frac{T - T_s}{T_\infty - T_s} = \frac{C_A - C_{A,s}}{C_{A,\infty} - C_{A,s}}, \tag{6.2.8}$$

Table 6.2. *Transport equations for the laminar boundary layer*

	Simplified balance	Dimensionless field	Dimensionless form
Continuity	$\dfrac{\partial v_x}{\partial x} + \dfrac{\partial v_y}{\partial y} = 0$	$\tilde{v}_x = \dfrac{v_x}{V_\infty},\ \tilde{v}_y = \dfrac{v_y}{V_\infty}$	$\dfrac{\partial \tilde{v}_x}{\partial \tilde{x}} + \dfrac{\partial \tilde{v}_y}{\partial \tilde{y}} = 0$
x momentum	$v_x \dfrac{\partial \rho v_x}{\partial x} + v_y \dfrac{\partial \rho v_x}{\partial x} = \nu \dfrac{\partial^2 \rho v_x}{\partial y^2}$	$\tilde{\rho v}_x = \dfrac{\rho v_x - \rho v_{x,s}}{\rho v_{x,\infty} - \rho v_{x,s}}$	$\tilde{v}_x \dfrac{\partial \tilde{\rho v}_x}{\partial \tilde{x}} + \tilde{v}_y \dfrac{\partial \tilde{\rho v}_x}{\partial \tilde{y}} = \dfrac{1}{Re} \dfrac{\partial^2 \tilde{\rho v}_x}{\partial \tilde{y}^2}$
Heat	$v_x \dfrac{\partial T}{\partial x} + v_y \dfrac{\partial T}{\partial y} = \alpha \dfrac{\partial^2 T}{\partial y^2}$	$\tilde{T} = \dfrac{T - T_s}{T_\infty - T_s}$	$\tilde{v}_x \dfrac{\partial \tilde{T}}{\partial \tilde{x}} + \tilde{v}_y \dfrac{\partial \tilde{T}}{\partial \tilde{y}} = \dfrac{1}{Re\,Pr} \dfrac{\partial^2 \tilde{T}}{\partial \tilde{y}^2}$
Mass	$v_x \dfrac{\partial C_A}{\partial x} + v_y \dfrac{\partial C_A}{\partial y} = D_A \dfrac{\partial^2 C_A}{\partial y^2}$	$\tilde{C}_A = \dfrac{C_A - C_{A,s}}{C_{A,\infty} - C_{A,s}}$	$\tilde{v}_x \dfrac{\partial \tilde{C}_A}{\partial \tilde{x}} + \tilde{v}_y \dfrac{\partial \tilde{C}_A}{\partial \tilde{y}} = \dfrac{1}{Re\,Sc} \dfrac{\partial^2 \tilde{C}_A}{\partial \tilde{y}^2}$

we find that the boundary conditions become

$$\tilde{y} = 0, \qquad \tilde{\theta} = 0,$$
$$\tilde{y} \Rightarrow \infty, \qquad \tilde{\theta} = 1,$$
$$\tilde{x} = 0, \qquad \tilde{\theta} = 1.$$

The equations for x-momentum, heat, and mass transfer can all be written in terms of $\tilde{\theta}$, with the generic boundary conditions (listed in Table 6.3 as the last row):

$$\tilde{v}_x \frac{\partial \tilde{\theta}}{\partial \tilde{x}} + \tilde{v}_y \frac{\partial \tilde{\theta}}{\partial \tilde{y}} = \frac{1}{Re} \frac{\partial^2 \tilde{\theta}}{\partial \tilde{y}^2}. \tag{6.2.9}$$

The solution of this equation has been given previously and is shown in Figure 6.2.

Of interest here is the solution for the gradient of the field evaluated at the surface of the plate. This was given for the velocity field in Eq. (6.2.9). In dimensionless form, it becomes

$$\left.\frac{\partial \tilde{\theta}}{\partial \tilde{y}}\right|_{\tilde{y}=0} = \frac{0.332}{\tilde{x}} Re_x^{1/2}, \tag{6.2.10}$$

where $\tilde{x}$ is the dimensionless distance along the plate. As before, the average value over the length of the plate (L) is calculated as

$$\int_0^1 \frac{\partial \tilde{\theta}}{\partial \tilde{y}} d\tilde{x} = 0.664\, Re_L^{1/2}. \tag{6.2.11}$$

Table 6.3. *Boundary conditions for the boundary-layer equations*

	$y = 0$	$\tilde{y} = 0$	$y \Rightarrow \infty$	$\tilde{y} \Rightarrow \infty$	$x = 0$	$\tilde{x} = 0$
Continuity	$v_y = 0$ $v_x = 0$	$\tilde{v}_y = 0$ $\tilde{v}_x = 0$	$v_x = V_\infty$	$\tilde{v}_x = 1$	$v_x = V_\infty$	$\tilde{v}_x = 1$
x momentum	$\rho v_x = \rho v_{x,s}$	$\tilde{\rho v}_x = 0$	$\rho v_x = \rho v_{x,\infty}$	$\tilde{\rho v}_x = 1$	$\rho v_x = \rho v_{x,\infty}$	$\tilde{\rho v}_x = 1$
Heat	$T = T_s$	$\tilde{T} = 0$	$T = T_\infty$	$\tilde{T} = 1$	$T = T_\infty$	$\tilde{T} = 1$
Mass	$C_A = C_{As}$	$\tilde{C}_A = 0$	$C_A = C_{A\infty}$	$\tilde{C}_A = 1$	$C_A = C_{A\infty}$	$\tilde{C}_A = 1$
Dimensionless		$\tilde{\theta} = 0$		$\tilde{\theta} = 1$		$\tilde{\theta} = 1$

Table 6.4. *Results for the laminar boundary layer* ($Sc = 1$, $Pr = 1$)

Transport	Flux	Transport coefficient	Result based on local position x	Result based on total length L
x momentum	$\tau_{yx} = -\mu \dfrac{dv_x}{dy}$	$\dfrac{f}{2} = \dfrac{-\tau_{yx}}{\rho V_\infty^2}$	$\dfrac{f_x}{2} = \dfrac{0.332}{\sqrt{Re_x}}$	$\dfrac{f_L}{2} = \dfrac{0.664}{\sqrt{Re_L}}$
Heat	$q_y = -k \dfrac{dT}{dy}$	$h = \dfrac{q_y}{\Delta T}$	$Nu_x = \dfrac{h_x x}{k} = 0.332\sqrt{Re_x}$	$Nu_L = \dfrac{hL}{k} = 0.664\sqrt{Re_L}$
Mass	$J_{Ay} = -D_A \dfrac{dC_A}{dy}$	$k_m = \dfrac{J_{Ay}}{\Delta C_A}$	$Sh_x = \dfrac{k_{m,x} x}{D_A} = 0.332\sqrt{Re_x}$	$Sh_L = \dfrac{k_m L}{D_A} = 0.664\sqrt{Re_L}$

Given this solution, we can determine the corresponding transport coefficient defined in Table 6.4 by following the procedure outlined at the start of this subsection: The transport coefficients are given by equating the flux calculated from the solution of the differential transport equations to that given by the empirical equations for the transport phenomenon of interest. The procedure has already been illustrated for the friction coefficient in Subsection 6.2.1 so only the results are given here.

The results in the last column of Table 6.4 show that, for $Pr = 1$ and $Sc = 1$, we can write

$$\frac{f}{2} Re = Nu = Sh. \tag{6.2.12}$$

This result is commonly referred to as the *Reynolds analogy* (Reynolds, 1874). This analysis supports the hypothesis stated at the beginning of this chapter, namely, that the dominant resistance to transport in a flowing fluid occurs in a thin boundary layer at the interface or surface. The Reynolds analogy arises because the boundary layer is shown to be identical for momentum, heat, and mass transfer for the simple case of laminar flow over a flat plate for a fluid with constant physical properties and $Sc = Pr = 1$. Note the value of this analogy: Given a measurement for a specific geometry for a mass transfer coefficient, heat transfer coefficient, or friction coefficient, any of the other transport coefficients can be determined by calculation. Thus analogies are very useful for obtaining estimates of transport coefficients for technically feasible design problems, such as those considered in Chapters 3 and 4. Further, the results in Table 6.4 also provide a method for reporting experimental results in terms of dimensionless groups, thus providing a method for correlating data across a variety of experimental conditions. Although the Reynolds analogy is limited to laminar boundary layers with $Pr = Sc = 1$, it serves as a basis for understanding more robust analogies as discussed next.

EXAMPLE 6.2 COOLING FAN. A fan blows over a computer chip. Estimate the surface temperature of the computer chip if the square chip is 1 cm in length, generates 1 W, and all of the heat must be removed by convection from the upper surface. The fan blows dry room air at about 1 m/s velocity.

SOLUTION. An energy balance at the surface of the chip operating at steady state is equivalent to balancing the flux of heat from the chip to that convected away by the flow. This becomes $Q = ha(T_s - T_{air})$. The value of the heat transfer coefficient can be estimated from the results shown in Table 6.4:

$$Nu_L = \frac{hL}{k} = 0.664\sqrt{Re_L},$$

$$Re_L = \frac{(1 \text{ m/s})(0.01 \text{ m})}{10^{-5} \text{ m}^2/\text{s}} = 1000,$$

$$Nu_L = 0.664\sqrt{500} = 21,$$

$$\therefore h = \frac{21 \times 0.026 \text{ W/m K}}{0.01 \text{ m}} = 55 \text{ W/m}^2 \text{ K}.$$

Thus the temperature at the surface can be estimated as

$$T_s = T_{air} + \frac{Q}{ha} = 25 + \frac{1}{55 \times 10^{-4}} = 207\,°\text{C}.$$

Clearly the fan is insufficient to cool the chip. This demonstrates why heat sinks are used to remove heat from electrical components by conduction into a larger metal part, often containing fins, which is then cooled by a fan.

6.2.3 Effects of Material Properties: The Chilton–Colburn Analogy

As shown in the previous section, the boundary-layers for laminar flow over a flat plate for x-momentum, heat, and mass transfer coincide if $Pr = Sc = 1$. For real materials, these boundary-layers will be of similar shape, but of different extent, as sketched in Figure 6.4. These differences arise because the diffusivities of momentum, energy, and species (i.e., ν, α, D_A) differ from one another in real materials. Solutions of the boundary layer equations by Pohlhausen (1921) explicitly include these differences through the ratios of properties in the form of Pr and Sc numbers. These solutions demonstrate that the boundary-layer thicknesses $\delta(x)$ for the three transport processes are related by

$$\left(\frac{\delta}{\delta_H}\right)^{\frac{1}{3}} = Pr = \frac{\nu}{\alpha}, \qquad \left(\frac{\delta}{\delta_D}\right)^{\frac{1}{3}} = Sc = \frac{\nu}{D_A}.$$

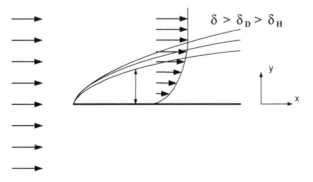

$$\delta > \delta_D > \delta_H$$

Figure 6.4. Boundary layers for varying $1 > Sc > Pr$.

The final result of Pohlhausen's analysis becomes

$$\mathrm{Nu_x} = 0.332\,\mathrm{Pr}^{\frac{1}{3}}\mathrm{Re_x^{\frac{1}{2}}}, \qquad \mathrm{Nu_L} = 2\mathrm{Nu_x},$$
$$\mathrm{Sh_x} = 0.332\,\mathrm{Sc}^{\frac{1}{3}}\mathrm{Re_x^{\frac{1}{2}}}, \qquad \mathrm{Sh_L} = 2\mathrm{Sh_x}. \qquad (6.2.13)$$

Clearly, the Reynolds analogy, Eq. (6.2.12), no longer holds. However, Colburn (1933) and Chilton and Colburn (1934) recognized that Pohlhausen's solution still satisfied the basic concepts recognized by Reynolds, namely, (1) dominant resistance in a boundary layer, and (2) similarity of the shape of the boundary layer for all three transport processes. Consequently, the Reynolds analogy can be extended by use of the Pohlhausen solution as follows. The *Stanton* number is defined as

$$\mathrm{St} = \frac{h}{\rho\hat{C}_p V_\infty} = \left(\frac{hL}{k}\right)\left(\frac{k}{\rho\hat{C}_p}\frac{1}{\nu}\right)\left(\frac{\nu}{V_\infty L}\right) = \frac{\mathrm{Nu}}{\mathrm{Pr\,Re}}. \qquad (6.2.14)$$

Colburn and then Chilton and Colburn further defined the "j factors," denoted as

$$j_H = \frac{h}{\rho\hat{C}_p V_\infty}(\mathrm{Pr})^{\frac{2}{3}} = \frac{\mathrm{Nu}}{\mathrm{Re\,Pr}}(\mathrm{Pr})^{\frac{2}{3}} = \frac{\mathrm{Nu}}{\mathrm{Re}\,(\mathrm{Pr})^{\frac{1}{3}}} = \mathrm{St}\,(\mathrm{Pr})^{\frac{2}{3}},$$
$$j_D = \frac{K_m}{V_\infty}(\mathrm{Sc})^{\frac{2}{3}} = \frac{\mathrm{Sh}}{\mathrm{Re\,Sc}}(\mathrm{Sc})^{\frac{2}{3}} = \frac{\mathrm{Sh}}{\mathrm{Re}\,(\mathrm{Sc})^{\frac{1}{3}}}. \qquad (6.2.15)$$

With this definition, the Polhausen solution for the laminar boundary layer can be written in the following compact form:

$$j_D = j_H = \frac{f}{2}. \qquad (6.2.16)$$

This relation is known as the *Chilton–Colburn analogy* and provides a simple and accurate relationship among the three transport processes.

EXAMPLE 6.3 WET-BULB PSYCHROMETER. Consider a psychrometer, which is a device used to measure the humidity of the air. It consists of two thermometers termed "wet" and "dry" bulbs. The wet bulb is covered with a wick wet with water. On rapidly spinning the thermometers in the air, the wet bulb cools because of evaporation. Determine the relationship between the wet-bulb temperature and the humidity of the air (see Figure E6.1).

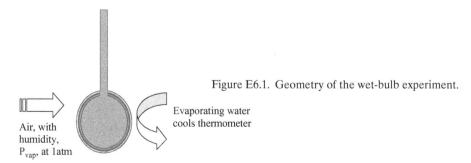

Figure E6.1. Geometry of the wet-bulb experiment.

Air, with humidity, P_{vap}, at 1 atm

Evaporating water cools thermometer

SOLUTION. This problem appears to be much more complex than what we have treated to date in that the flow past the wet bulb facilitates evaporation of the water, which results in a heat loss from the bulb. Consequently, all three transport mechanisms are coupled. However, we can start the analysis by writing a rate expression that expresses this coupling between the rates of heat and mass transfer, which are also related by the analogies derived in this section.

The physical phenomenon of interest is evaporative cooling. We ignore the relatively minor contributions from thermal conduction and convection in the air. The flux of heat from the surface of the wet bulb is given by an energy balance written at the surface of the wet bulb. The flux of heat entering the bulb is balanced by the energy required for vaporizing the evaporating water at steady state:

$$q_r = h(T_{air} - T_{bulb}) = \Delta \underline{H}^{vap} J_{w,r} = \Delta \underline{H}^{vap} k_m (C_{w,bulb} - C_{w,air}).$$

This equation can be rearranged to yield the following relationship:

$$C_{w,air} = C_{w,bulb} - \frac{h}{\Delta \underline{H}^{vap} k_m}(T_{air} - T_{bulb}).$$

Thus the water concentration in the air is seen to depend on the temperature difference of the wet-bulb thermometer with the surrounding air, the ratio of heat to mass transfer coefficients, and the concentration of water at the surface of the bulb. Notice that we require only the ratio of transport coefficients and do not need to know them individually. This ratio can be obtained directly from the Chilton–Colburn analogy as

$$j_D = j_H,$$

$$\frac{h}{\rho V_\infty \hat{C}_p}(Pr)^{\frac{2}{3}} = \frac{k_m}{V_\infty}(Sc)^{\frac{2}{3}},$$

$$\therefore \quad \frac{h}{k_m} = \rho \hat{C}_p \left(\frac{Sc}{Pr}\right)^{\frac{2}{3}} = \rho \hat{C}_p (Le)^{\frac{2}{3}}.$$

The velocity at which the psychrometer is spun, V_∞, does not enter into the final relationship. This is because the ratio of the transport coefficients will depend on only the relative size of the boundary layers, as given by the Lewis number Le. Substitution into the model equation leads to

$$C_{w,air} - C_{w,bulb} - \frac{\rho \hat{C}_p}{\Delta \underline{H}^{vap}}(Le)^{\frac{2}{3}}(T_{air} - T_{bulb}).$$

This result shows that knowledge of the thermodynamic properties of water and measurements of the "dry-" and "wet-" bulb thermometers will immediately yield the molar concentration of water in the air.

As an example application, consider the following data (corresponding to a very hot summer day that is not humid). The measurements are as follows:

$$T_{air} = 35 \ °C \quad T_{bulb} = 15 \ °C.$$

To calculate the relative humidity, physical property data are required. Note that the physical properties are estimated at the film temperature,

$$T_{film} = (T_{air} + T_{bulb})/2 = 25 \ °C,$$

for this problem. Also note that the density and heat capacity are to be evaluated at the film concentration as well, but, as we don't yet know the water content, we make the first calculation by using the properties of dry air (again at the film temperature). The required data are found in the *Chemical Engineers' Handbook* (1997) and the ideal-gas law.

Physical properties and constants
for wet-bulb example

$\Delta H_{water,25\,°C}^{vap}$	44 kJ/mol
$\hat{C}_{p,air,25\,°C}$	1 J/g K
$\rho_{air,25\,°C}$	1.2 kg/m^3
$P_{vap}(25\,°C)$	3169 Pa
R	8.314 J/mol K
$Sc_{air,25\,°C}$	0.61
$Pr_{air,25\,°C}$	0.72

Plugging in the physical property data and using the ideal-gas law, $C = P/RT$, yields

$$P_{bulb} = 3169(Pa)$$
$$- (8.314\,J/mol\,K \times 298\,K)\,\frac{(1.2\,kg/m^3)(1\,J/g\,K)(1000\,g/kg)}{44 \times 10^3\,J/mol}\left(\frac{0.61}{0.72}\right)^{\frac{2}{3}}$$
$$\times (308 - 288)\,(K)$$
$$= 3169\,Pa - 1224\,Pa = 1944\,Pa,$$

% relative humidity $= 100\% \times 1944\,Pa\,/3169\;Pa = 61\%$.

Note that this summer day is relatively dry by most standards. To improve the calculation, we could use the physical properties of air with the average water content. Note that, even at the surface, the mole fraction of water vapor is given by the relative pressures as $y_A = P_{vap}\,/P = 3169\,Pa/100\,000\,Pa \approx 0.032$, so in fact the assumption of the properties of dry air was indeed valid given the accuracy of the physical property data.

6.2.4 Turbulent Boundary Layers

Above $Re_x = 3 \times 10^5$ the boundary layer over the flat plate becomes turbulent (whereas in circular pipes, the transition Reynolds number is ~ 2300). Prandtl recognized that a boundary layer still exists in turbulent flow, a concept that was later improved on by von Kármán. Treatment of turbulent boundary layers is beyond the scope of this text, but the governing equation (Navier–Stokes equation) is again taken as the starting point for the analysis. The velocities in a turbulent flow, however, are time varying and chaotic. Yet the average properties of the flow can be treated largely similar to those of the laminar boundary layer. Without going into any mathematical detail, the same procedure is employed, i.e., determine the velocity profile and calculate the friction factor by equating the x-momentum flux calculated from Newton's constitutive equation for viscosity to the definition of the friction factor. Then, solving for the thermal and concentration profiles and equating to the appropriate constitutive equation of Fourier or Fick (Table 1.4), respectively, to the corresponding rate expression (Table 1.5) yields solutions for the heat and mass transfer coefficients for the turbulent boundary layer. The resulting *Prandtl analogy* for turbulent boundary layers is

$$\text{St} = \frac{\text{Nu}}{\text{Re Pr}} = \frac{f/2}{1 + 5\sqrt{f/2}(\text{Pr} - 1)},$$

$$\frac{\text{Sh}}{\text{Re Sc}} = \frac{f/2}{1 + 5\sqrt{f/2}(\text{Sc} - 1)}. \tag{6.2.17}$$

Thus knowledge of the friction factor for turbulent flow over a flat plate enables calculating the corresponding heat and mass transfer coefficients for given fluid properties.

6.3   Transport Correlations for Specific Geometries

In the previous section we note that the solutions presented here for specific boundary layers over a flat plate are themselves of limited value. What is of particular significance are the following observations:

- The resistance to mass, momentum, and heat transfer is located in a thin boundary layer where the fluid contacts an interface or surface.
- Correlations for Nusselt and Sherwood numbers depend on the strength and type of flow through the Reynolds number and material properties through the Prandtl and Schmidt numbers, respectively.
- Analogies among the rates of mass, momentum, and heat transfer arise from the mathematical similarity of the underlying transport equations and the resulting solutions for the species mass, velocity, and temperature profiles.

Thus the analysis justifies the hypothesis of the significance of the boundary layer in controlling the rate of transport in flowing systems. For geometries more complex than those of the flat boundary layer, we may not be able to derive analytical solutions for the rate of transport. However, the underlying structure will contain the same dimensionless groups, which is of value in correlating experimental data. Further, an underlying analogy between the transport properties should exist. The general form of this analogy is often taken to be that of the Chilton–Colburn analogy, Eq. (6.2.16). In this section we examine the validity of the extension of the analysis of the flat boundary layer to more complex flows, both internal and external to solid objects. The following discussion draws from experimental data analyzed in Sherwood and Pigford (1952).

To apply the Chilton–Colburn analogy relation to more complex geometries, we recognize that (a) the skin friction is separated from the additional contributions to the pressure drop because of deformation of streamlines, and (b) the mass transfer rate is not overly large such that the flux at the surface is dominated by diffusion. Further, although the result was derived for developing flow over a flat plate, we hypothesize that it will also hold for laminar and turbulent flow both external and internal to objects of much more complex geometry and for fully developed flow, such as turbulent flow in a pipe.

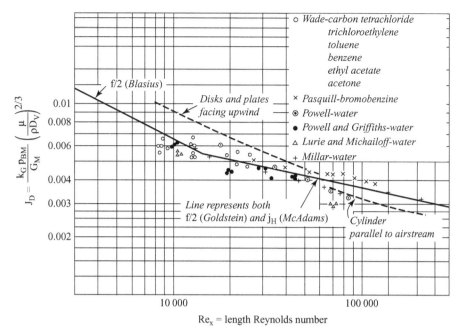

Figure 6.5. Evaporative mass transfer from flow over a plate. (Sherwood TK, Pigford RL. *Absorption and Extraction*, 1952. Copyright McGraw-Hill Companies, Inc. Reproduced with permission.)

Figure 6.5, which is adapted from Sherwood and Pigford (1952), illustrates experimental data for various liquids vaporizing from a flat plate with flow as a function of the Reynolds number based on the length of the plate. The friction factor is shown for laminar (Re < 10 000) and turbulent flow. The cited work of McAdams (1954) for heat transfer illustrates the agreement between j_H and f/2, leading to the same line correlating data for heat transfer to, and friction arising from, flow over a flat plate. Note that the results are also compared with data for flow parallel to a wetted cylinder as well as disks. The agreement supports the validity of extending the Chilton–Colburn analogy beyond the limiting case of the laminar flow over a flat plate that was proven mathematically in the previous section.

The next extension is to external flows orthogonal to bodies such as cylinders, as might be found as the fluid flow on the shell side of a shell-and-tube exchanger, or in a boiler. Figure 6.6 shows data from a number of different authors for heat transfer to and from forced gas flowing normal to a single cylinder. The value in using dimensionless groups to compare data from different cylinders, flow rates, and gases is apparent. The data are well represented by the following correlation:

$$\mathrm{Nu} = \left(0.35 + 0.47\mathrm{Re}^{0.52}\right)\mathrm{Pr}^{0.3}, \qquad 0.1 < \mathrm{Re} < 10^3. \qquad (6.3.1)$$

The physical properties are to be evaluated at the film temperature. Although the data shown are for gases, the preceding correlation is noted to work well for liquid with Prandtl numbers up to 1240 (McAdams, 1954).

The rate of mass transfer for evaporation from a wetted cylinder as well as adsorption of water from air is compared with known friction factors and heat transfer correlations of data by McAdams for flow orthogonal to cylinders in Figure 6.7. For the

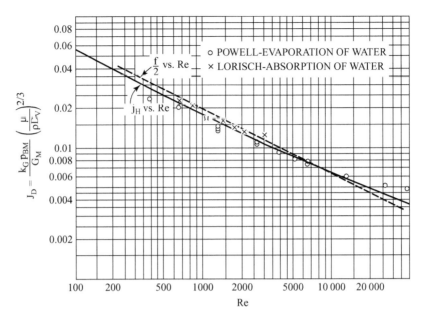

Figure 6.6. Mass transfer to a cylinder normal to air flow. (Sherwood TK, Pigford RL. *Absorption and Extraction*, 1952. Copyright McGraw-Hill Companies, Inc. Reproduced with permission.)

comparison, the skin friction is calculated by subtracting the measured pressure drop from the measured total drag on the cylinder. As noted by Sherwood and Pigford, "The agreement between j_D, j_H, and $f/2$ is quite remarkable." Consequently, there is clear experimental evidence that the analogy holds for turbulent flow orthogonal to cylinders.

Another model convective geometry that has received considerable attention because of its relevance for mass transfer from tablets, granular beds, packed towers,

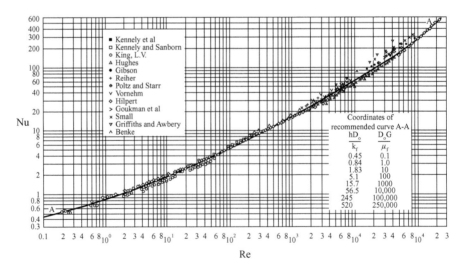

Figure 6.7. Heat transfer to and from a single cylinder, normal forced convection. (McAdams WH. *Heat Transmission*, 3rd ed., 1954. Copyright McGraw-Hill Companies, Inc. Reproduced with permission.)

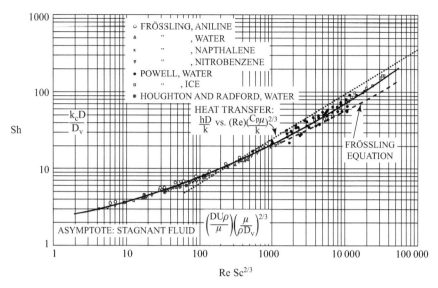

Figure 6.8. Mass transfer from flow over a sphere. (Sherwood TK, Pigford RL. *Absorption and Extraction*, 1952. Copyright McGraw-Hill Companies, Inc. Reproduced with permission.)

bubbles, and drops, as well as reactors with catalyst pellets is that of flow past a single sphere. Figure 6.8 shows results compared with the Frössling equation, which is a correlation of the said author's data:

$$\text{Sh} = \frac{k_m D}{D_{AB}} = 2.0\left(1 + 0.276\,\text{Re}^{\frac{1}{2}}\,\text{Sc}^{\frac{1}{3}}\right). \tag{6.3.2}$$

As seen in Figure 6.8, the comparison is good between the heat transfer correlation and the mass transfer data, and the data are well correlated by Frössling's correlation.

Note that the characteristic length appearing in the Reynolds number and Sherwood and Nusselt numbers is the diameter of the cylinder (there is no longer a "local" mass transfer coefficient and Reynolds number as in the developing boundary layer). For flow around objects such as cylinders and spheres, as well as flow inside of pipes, the characteristic length (such as given by the wetted perimeter) is used to form an average value for the transport coefficient based on the area available for heat or mass transfer.

The previous examples are all flows external to objects. Of significant practical importance for the design of tubular exchangers are correlations for turbulent heat transfer inside pipes. Friction factors for flow inside pipes are well established as pressure-drop measurements are relatively straightforward to implement, and heat transfer measurements for flow inside pipes requires measurements of only flow rates and temperatures (as shown in Chapter 3); hence, there are significant data sets available for correlation. The functional form for heat transfer correlations for flow in pipes follows from the results of Section 6.2, namely,

$$\text{Nu} = a\,\text{Re}^b\,\text{Pr}^c. \tag{6.3.3}$$

The coefficients (a, b, c) are fit to experiment, but generally fall within ranges close to those derived in Section 6.2, as shown in Table 6.5. As an example of the validity of

Table 6.5. Heat transfer correlation parameters for Eq. (6.3.3). (adapted from McAdams, 1954)

Author	Date	a	b	c	Bulk, film, or surface viscosity	Heating or cooling
Colburn	1933	0.023	0.8	0.333	B	H & C
Dittus & Boelter	1930	0.0243	0.8	0.4	B	H
Dittus & Boelter	1930	0.0265	0.8	0.3	B	C
Sieder & Tate	1936	$0.027\left(\dfrac{\mu}{\mu_s}\right)^{0.14}$	0.8	0.333	B & S	H & C

these correlations, Figure 6.9 shows data for gases inside pipes as compared against the Reynolds analogy [Eq. (6.2.12), line B] and the Prandtl analogy [Eq. (6.2.17), Pr = 0.74, line A]. The Prandtl analogy is 1.09 times the Reynolds analogy and better represents the data.

Mass transfer experiments have been performed for flow inside wetted wall columns. The latter are well represented by the relationship presented in Sherwood and Pigford (1952):

$$\frac{k_M D}{D_{AB}} \frac{p_{Bm}}{P} = 0.023 \, Re^{0.83} Sc^{0.44} \tag{6.3.4}$$

where the ratio (p_{Bm}/P) is the ratio of the mean partial pressure of the carrier gas to the total pressure. This correlation is valid for Reynolds numbers from 2000 to

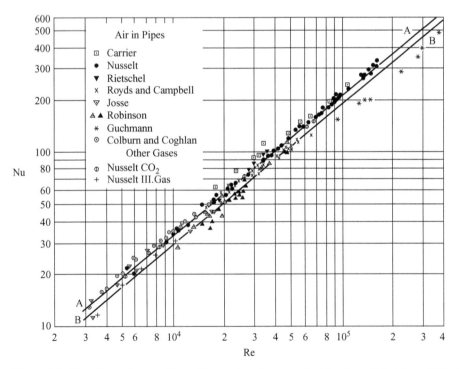

Figure 6.9. Nusselt number vs. Reynolds number for gases inside pipes. (McAdams WH. *Heat Transmission*, 3rd ed., 1954. Copyright McGraw-Hill Companies, Inc. Reproduced with permission.)

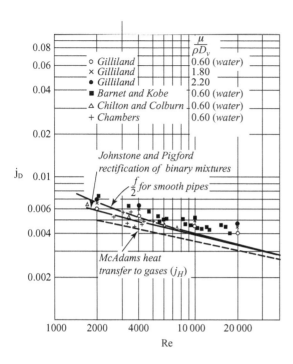

Figure 6.10. Heat and mass transfer in a wetted wall. (Sherwood TK, Pigford RL. *Absorption and Extraction*, 1952. Copyright McGraw-Hill Companies, Inc. Reproduced with permission.)

35 000 and Schmidt numbers from 0.6 to 2.5. Gilliland's equation can be rewritten as

$$j_D = 0.023 \text{Re}^{-0.17} \text{Sc}^{0.11}. \tag{6.3.5}$$

Comparison is shown in Figure 6.10, which shows reasonable agreement among experimental mass transfer data, the friction factor for internal pipe flow, McAdams correlation for heat transfer inside pipes,

$$j_H = 0.023 \text{Re}^{-0.20} \text{Pr}^{0.07}, \tag{6.3.6}$$

and the empirical equation developed by Johnstone and Pigford for mass transfer:

$$j_D = 0.033 \text{Re}^{-0.23}. \tag{6.3.7}$$

The fluid's material properties vary with temperature, and so the choice of where to evaluate the material properties must be specified. Logical choices are the film temperature or the bulk or surface values. It may also be relevant whether the fluid is being heated or cooled. The most significant property variation with temperature is the fluid's viscosity, which is often exponential in temperature. McAdams (1954) compares experiments and shows that the data correlate well with the bulk temperature. Of significance is the variation of the Nusselt number with Prandtl number, reproduced in Figure 6.11, which shows data at a fixed Reynolds number of 10^4 with variation in Prandtl number. Use of the film temperature is found to be less successful in correlating the data.

These examples are not exhaustive, but are used to illustrate the concept that the analogies have validity for geometries for which no exact derivation exists, including

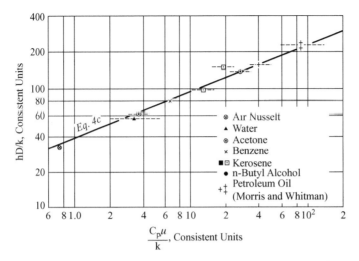

Figure 6.11. Dependence of Nusselt number on Prandtl number. (McAdams WH. *Heat Transmission*, 3rd ed., 1954. Copyright McGraw-Hill Companies, Inc. Reproduced with permission.)

both internal and external flows as well as laminar, transition, and turbulent flows. Notice that the functional form of the specific correlations (as well as the definition of the characteristic length) varies with the geometry as well as the type of flow (laminar or turbulent). Additional correlations of friction, heat and mass transfer coefficients can be found in the references given at the end of this chapter.

EXAMPLE 6.4 TUBULAR HEAT EXCHANGER. In Section 3.5 we considered the design of a tubular heat exchanger. In performing the technically feasible design, we assumed a value for the overall heat transfer coefficient of 1500 W/m^2 K. We are now in a position to determine a better estimate for the heat transfer coefficient for the specific fluid flow rates and pipe sizes specified in the technically feasible design.

SOLUTION. Table E3.1 summarizes the possible choices of pipe diameter, fluid velocities and Reynold's numbers, Consider the use of a 2.0-in. nominal pipe. The Reynolds number is 80848. The process stream is being cooled and the properties are that of water. For water, Pr = 2.4 at 80 °C and increases to 6 by 25 °C. This is due, as noted, to the strong dependence of viscosity on temperature. For purposes here, we consider the value at 80 °C. The resistance to heat transfer for the fluid being cooled inside the pipe can be estimated by the Dittus–Boelter relationship, Eq. (6.3.3). Thus,

$$Nu_1 = 0.0265 Re^{0.8} \, Pr^{0.3},$$
$$Nu_1(80 \text{ °C}) = 0.0265 \, (80848)^{0.8} \, (2.4)^{0.3} = 290,$$
$$\therefore h_1 = 290 \times 0.6/5.25 \times 10^{-2} = 3322.$$

The annular region is within a 3-in. nominal pipe. The estimation of the heat transfer coefficient is based on the hydraulic diameter, which is four times the cross-sectional area divided by the wetted perimeter, which becomes $D_2 - D_1 = 1.8$ cm for the annulus. The Reynolds number for this flow is 10088. The fluid is heating and has the properties of water. Here, the Dittus–Boelter equation is used for heating, as

$$Nu_2 = 0.0243 Re^{0.8} Pr^{0.4},$$

$$Nu_2(80\,°C) = 0.0243\,(10088)^{0.8}\,(6.0)^{0.4} = 79,$$

$$\therefore h_2 = 79 \times 0.6/1.8 \times 10^{-2} = 2647.$$

The overall heat transfer coefficient can now be estimated from Equation (5.4.12) as

$$\therefore U_{out} = \cfrac{1}{\cfrac{1}{h_{out}} + \cfrac{r_{out}\ln(r_{out}/r_{in})}{k_{wall}} + \cfrac{r_{out}}{h_{in}r_{in}}},$$

$$U_{out} = \cfrac{1}{\cfrac{1}{2647} + \cfrac{0.0603\ln(6.03/5.25)}{2 \times 16} + \cfrac{6.03}{3322 \times 5.25}}$$

$$U_{out} = \cfrac{1}{3.8 \times 10^{-3} + 2.6 \times 10^{-4} + 3.5 \times 10^{-4}} = 1016\ \text{W/m}^2\ \text{K}.$$

Thus we find the estimated heat transfer coefficient is slightly higher than the lowest value of $U = 1.0\ \text{kW/m}^2$ K we selected and considerably lower than the value we assumed of $1.5\ \text{kW/m}^2$ K. This will increase the required length of the design in Section 3.5 from 36 to 61 m. The lower value is due to the low flow rate in the annulus, which yields a low Reynold's number.

EXAMPLE 6.5 CHILTON–COLBURN ANALOGY. Experiments and analyses exploring the validity of heat and mass transfer correlations for flow inside a heated pipe were reported by Friend and Metzner (1958). In one experiment, they varied the Prandtl number over a broad range by use of fructose solutions (corn syrup). The data from the manuscript are plotted in Figure E6.2. Do the data support the Chilton–Colburn analogy between heat transfer and friction factors?

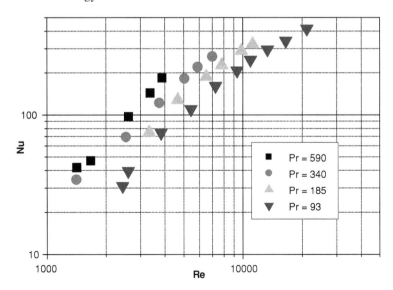

Figure E6.2. Heat transfer data (Friend and Metzner, 1958).

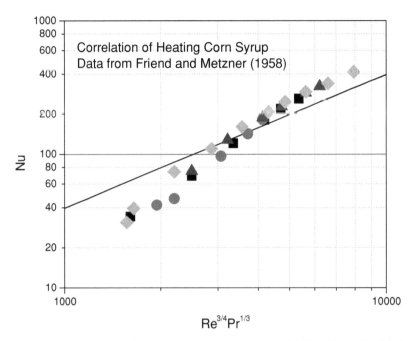

Figure E6.3. Comparing heating data of Friend and Metzner (1958) with Blasius friction factor.

SOLUTION. The friction factor for flow through a pipe for $2100 < \text{Re} < 10^5$ can be accurately represented by the Blasius formula: $f = \frac{0.0791}{\text{Re}^{1/4}}$. The Chilton–Colburn analogy, Eq. (6.2.16) can be used to transform this into a correlation for heat transfer, as

$$j_H = \frac{f}{2}$$

$$\frac{\text{Nu}}{\text{Re}\,(\text{Pr})^{\frac{1}{3}}} = \frac{1}{2}\frac{0.0791}{\text{Re}^{1/4}}$$

$$\text{Nu} = 0.0396\,\text{Re}^{\frac{3}{4}}\,\text{Pr}^{\frac{1}{3}}$$

The experimental data are plotted as suggested by the correlation, which shows that the dependence on Reynolds number is not a simple power law (see Figure E6.3). Indeed, Friend and Metzner (1958) show this and demonstrate that the power-law index of 0.33 for the Prandtl number is also incorrect. Empirically, it is better represented with a power-law exponent of 0.4. This illustrates the approximate nature of the correlations and analogies. A detailed discussion of the comparison between theories and the experiments can be found in the original article. However, a better analogy is to use the Prandtl analogy, Eq. (6.2.17). This is left as a homework assignment.

In closing this section, it is important to note that in these experiments the area available for mass transfer is known (or in the case of the wetted wall column, approximately known). In cases of direct fluid–fluid contacting, as often found in mass transfer operations, the area for transport is itself defined by the fluid motions and hence, correlations for mass transfer and area generation should be considered together. Such cases are discussed in more detail in the next section, as well as in Chapter 7.

6.4   Models for Estimating Transport Coefficients in Fluid–Fluid Systems

6.4.1 Film Theory

We noted after Eq. (6.2.7) that the heat transfer through the boundary layer could be approximated simply as the driving force divided by the thickness of the boundary layer (δ_T). This is formally equivalent to the simple conduction and diffusion problems investigated in the previous chapter in the absence of convection, in which the transport was across a fixed thickness of quiescent, homogeneous material. This concept is extended to more complex physical situations whereby it is assumed that the entire resistance to heat or mass transfer occurs in a fictitious boundary layer. Under this assumption, the heat and mass transfer coefficients (the latter for *equimolar counterdiffusion* or conditions for which the net molar velocity is small) become

$$\mathrm{Nu}_\delta = 1 \quad \rightarrow \quad h = \frac{k}{\delta_H},$$

$$\mathrm{Sh}_\delta = 1 \quad \rightarrow \quad k_m = \frac{D_A}{\delta_D}. \qquad (6.4.1)$$

The limitations of this model are that the film thickness is generally not known. Indeed, this model is of limited applicability, and, more often, penetration theory is employed to correlate and estimate heat and mass transfer coefficients when convection is present.

   An estimation of the film thickness can be obtained from boundary-layer theory. The effects of physical properties on the boundary layer thickness can be included by use of Pohlhausen's result (1921):

$$\delta = \frac{5L}{\sqrt{\mathrm{Re}_L}} = \mathrm{Pr}^{\frac{1}{3}}\delta_H = \mathrm{Sc}^{\frac{1}{3}}\delta_D. \qquad (6.4.2)$$

Replacing the thermal and mass transfer boundary layers in expressions for h and k_m yields

$$h = \frac{k}{5L}\mathrm{Re}^{\frac{1}{2}}\mathrm{Pr}^{\frac{1}{3}} \quad \text{or} \quad \mathrm{Nu} = \frac{hL}{k} = 0.2\mathrm{Re}^{\frac{1}{2}}\mathrm{Pr}^{\frac{1}{3}},$$

$$k_m = \frac{D_A}{5L}\mathrm{Re}^{\frac{1}{2}}\mathrm{Sc}^{\frac{1}{3}} \quad \text{or} \quad \mathrm{Sh} = \frac{k_m L}{D_A} = 0.2\mathrm{Re}^{\frac{1}{2}}\mathrm{Sc}^{\frac{1}{3}}. \qquad (6.4.3)$$

These correlations approximate the exact solutions of the laminar boundary layer, Eqs. (6.2.13).

6.4.2 Penetration Theory

Penetration theory stems from the solution of the unsteady or transient heat and mass transfer diffusion into a body, which was already solved in Section 5.8. One estimates the penetration of heat or species A into the fluid as it flows along the boundary layer by estimating the diffusion of heat or mass into a fluid element sliding along

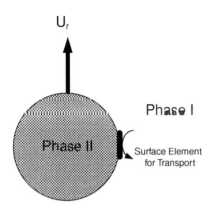

Figure 6.12. Bubble rising through a liquid phase.

the boundary surface. The velocity and distance yield a contact time for penetration. From the assumption of film theory and the solution for the laminar boundary layer equation (for $Pr = 1$, $Sc = 1$), the transport coefficients can be estimated in penetration theory as

$$h \approx \frac{k}{\delta_T} = \frac{k\sqrt{Re_L}}{5L} = \frac{k}{5\sqrt{\nu t}} \times \sqrt{Pr} \propto k\rho\sqrt{\frac{\alpha}{t}},$$

$$k_m \approx \frac{D_A}{\delta_D} = \frac{D_A\sqrt{Re_L}}{5L} = \frac{D_A}{5\sqrt{\nu t}} \times \sqrt{Sc} \propto \sqrt{\frac{D_A}{t}}. \qquad (6.4.4)$$

In the preceding equations, the characteristic penetration time is taken as $t = L/V_\infty$. The final expression on the right highlights only the important dependence on the physical properties (the constants have been dropped). The transport coefficients are assumed to depend on the square root of the molecular diffusivities, which is in contrast to film theory, in which they are linearly related. For turbulent transport, the time t is taken to be the *surface renewal time*, which has to do with the time a fluid element as part of a turbulent eddy spends at the surface, i.e., the contact time during which mass and heat transfer can take place. In the following, we investigate the application of these concepts to heat and mass transfer between two phases.

As an example of penetration theory, consider the problem of a gas bubble (Phase II) rising through a liquid (Phase I) in a mass contactor (Figure 6.12). Such a problem arises in the design of semibatch or continuous mass contactors in which a gas phase rich in one component is rising through a mixed liquid lean in that component (mixed–plug). Species A is being transferred from the gas phase to the liquid phase. If we consider as a control volume a small surface element, it is possible to locally neglect the curvature of the interfacial area. It is of no importance whether the liquid moves along the bubble at velocity u_r or the bubble rises at this velocity. It is the relative velocity of the bubble versus the continuous phase that is important.

We develop the penetration theory by considering a small surface element between a well-mixed Phase I and a Phase II, with relative motion between the phases. Locally, we can assume the velocity field to be a simple, constant velocity in the vertical direction, as shown in Figure 6.13. In the following, we consider mass transfer resistance only in the liquid phase (I).

If the boundary layer is thin relative to the size of the bubble, the velocity field is simplified to a velocity only in the x direction (i.e., along the surface of the bubble

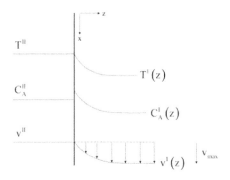

Figure 6.13. Local boundary layer for penetration theory analysis.

in the opposite direction of the bubble motion). This is shown in Figure 6.13, where the frame of reference is the bubble as stationary and the fluid moving. Further, the physical properties are assumed to be constant during the process. The differential transport equations derived in Section 6.1 simplify to the following balances, with their respective boundary conditions:

$$
\begin{array}{ll}
\textit{Energy} & \textit{Species A} \\[6pt]
v_x \dfrac{\partial T^I}{\partial x} = \alpha \dfrac{\partial^2 T^I}{\partial z^2}, & v_x \dfrac{\partial C_A^I}{\partial x} = D_A \dfrac{\partial^2 C_A^I}{\partial z^2}, \\[10pt]
x = 0, \quad T^I = T_b^I, & x = 0, \quad C_A^I = C_{A,b}^I, \\[4pt]
z = 0, \quad T^I = T^{II}, & z = 0, \quad C_A^I = C_{A,i}^I = M C_A^{II}, \\[4pt]
z = \infty, \quad T^I = T_b^I, & z = \infty, \quad C_A^I = C_{A,b}^I,
\end{array}
\tag{6.4.5}
$$

where b denotes the bulk property.

To simplify the analysis, one can assume that the bulk velocity in the x direction is constant and equal to the maximum velocity $V_{max} \approx U_r$. The maximum velocity is chosen as this is the velocity at the interface and it will not vary greatly over the range of penetration of the temperature or concentration profile.

Similar to the solution method employed in the boundary-layer theory (Section 6.2), one can find a similarity solution that reduces the partial differential equations to ordinary differential equations. The new variable is defined as

$$
\eta = \frac{z}{\sqrt{4Yx/V_{max}}},
\tag{6.4.6}
$$

where Y is the thermal diffusivity α for the energy conservation equation and the mass diffusivity D_A for the species mass conservation equation.

The final ordinary differential equations describing this process become

$$
\begin{array}{ll}
\dfrac{d^2 T^I}{d\eta^2} + 2\eta \dfrac{dT^I}{d\eta} = 0, & \dfrac{d^2 C_A^I}{d\eta^2} + 2\eta \dfrac{dC_A^I}{d\eta} = 0, \\[10pt]
\eta = 0, \quad T^I = T^{II}, & \eta = 0, \quad C_A^I = M C_A^{II}, \\[4pt]
\eta = \infty, \quad T^I = T_b^I, & \eta = \infty, \quad C_A^I = C_{A,b}^I.
\end{array}
\tag{6.4.7}
$$

Note that two of the previous boundary conditions (at $x = 0$ and $z = \infty$) result in the same $\eta = \infty$ boundary condition. These equations can be made dimensionless by defining the following scaled variables:

$$\tilde{T} = \frac{T^I - T^{II}}{T_b^I - T^{II}}, \qquad \tilde{C}_A = \frac{C_A^I - MC_A^{II}}{C_{A,b}^I - MC_A^{II}}. \tag{6.4.8}$$

The dimensionless equations are

$$\begin{aligned}
&\frac{d^2\tilde{T}}{d\eta^2} + 2\eta\frac{d\tilde{T}}{d\eta} = 0, \qquad &\frac{d^2\tilde{C}_A}{d\eta^2} + 2\eta\frac{d\tilde{C}_A}{d\eta} = 0, \\
&\eta = 0, \quad \tilde{T} = 0, \qquad &\eta = 0, \quad \tilde{C}_A = 0, \\
&\eta = \infty, \quad \tilde{T} = 1, \qquad &\eta = \infty, \quad \tilde{C}_A = 1,
\end{aligned} \tag{6.4.9}$$

Inspection shows that the equations and boundary conditions for both heat and mass transfer are identical when expressed in terms of the dimensionless variables. Thus we expect a single solution to describe both temperature and concentration fields, further emphasizing the analogies between heat conduction and diffusion. Integrating twice gives the temperature and concentration profile as an error function:

$$\tilde{T} = \tilde{C}_A = \mathrm{erf}(\eta) = \frac{2}{\sqrt{\pi}} \int_0^\eta e^{-s^2} ds. \tag{6.4.10}$$

The profile of the solution is essentially the same as the penetration solution derived in Section 5.8 for the transient heat conduction problem [Figure 5.16, which is $1 - \mathrm{erf}(x)$].

Fourier's and Fick's constitutive equations are now used to obtain expressions for the fluxes of heat and species A, respectively, at the boundary ($z = \eta = 0$). Using the mathematical result that $(d/dx)\mathrm{erf}(x) = (2/\sqrt{\pi})e^{-x^2}$, the fluxes at the interface become

$$q_z = -k \left.\frac{dT^I}{dz}\right|_{z=0} = \sqrt{\frac{k\rho\hat{C}_p V_{max}}{\pi x}} \left(T^{II} - T_b^I\right), \tag{6.4.11}$$

$$N_{A,z} \approx J_{A,z} = -D_A \left.\frac{dC_A^I}{dz}\right|_{z=0} = \sqrt{\frac{D_A V_{max}}{\pi x}} \left(MC_A^{II} - C_{A,b}^I\right).$$

One can immediately see that the flux will depend on the vertical position x along the boundary. Comparing expressions (6.4.11) with the constitutive equations for heat and mass transfer, we can obtain expressions for h and k_m:

$$\begin{aligned}
&q_z = h\left(T^{II} - T_b^I\right) \qquad &N_{A,z} = k_m\left(MC_A^{II} - C_{A,b}^I\right), \\
&\therefore h = \sqrt{\frac{k\rho\hat{C}_p V_{max}}{\pi x}}, \qquad &\therefore k_m = \sqrt{\frac{D_A V_{max}}{\pi x}}.
\end{aligned} \tag{6.4.12}$$

These values are dependent on the position x along the interface. Often we are interested only in average values for a given length (L) of interface. Averaging these results over the region $0 \le x \le L$ yields

$$h = \frac{1}{L}\int_0^L \sqrt{\frac{k\rho\hat{C}_p V_{max}}{\pi x}}\, dx = 2\sqrt{\frac{k\rho\hat{C}_p V_{max}}{\pi L}}, \tag{6.4.13}$$

$$k_m = \frac{1}{L}\int_0^L \sqrt{\frac{D_A V_{max}}{\pi x}}\, dx = 2\sqrt{\frac{D_A V_{max}}{\pi L}}.$$

These expressions were originally derived by Robert L. Pigford [(1917–1988), Ph.D. Thesis, University of Illinois, Urbana-Champagne, 1941] to describe the rate of mass transfer in a wetted wall column. In that problem, Phase I refers to a vapor or gas phase rich in species A and the transport of species A into a flowing liquid film (Phase II) is being analyzed. These relations have broad applicability, as will be shown now.

This expression can be used to estimate the mass transfer coefficient for the design of a plug–mixed semibatch mass contactor. L/V_{max} can be regarded as the total contact time between the two phases, and this is the residence time for the gas phase in the contactor. For rising bubbles in a mass contactor, the contact time for mass transfer can be approximated as the time it takes for the bubble to rise a single diameter d_B/V_{max}. Once it has risen this distance, it will be in contact with new liquid. The velocity profile in the boundary layer can be calculated to be a parabolic profile such that the average velocity would be $\overline{V} = 2/3 V_{max}$.

Rearranging the expressions gives then

$$Nu = \frac{hL}{k} = \sqrt{\frac{6}{\pi}}\sqrt{\frac{L\overline{V}}{\nu}}\sqrt{\frac{\nu}{\alpha}} = \sqrt{\frac{6}{\pi}} Re^{0.5} Pr^{0.5}, \qquad (6.4.14)$$

$$Sh = \frac{k_m L}{D_A} = \sqrt{\frac{6}{\pi}}\sqrt{\frac{L\overline{V}}{\nu}}\sqrt{\frac{\nu}{D_A}} = \sqrt{\frac{6}{\pi}} Re^{0.5} Sc^{0.5}.$$

Penetration theory thus suggests that Nu and Sh depend on the square root of the Reynolds number and the square root of the Prandtl and Schmidt number, respectively, which agrees with experiments.

EXAMPLE 6.6 MASS TRANSFER TO A RISING BUBBLE. Consider the problem of estimating the mass transfer coefficient for a bubble rising in a fluid in which methyl chloride is being adsorbed from the air bubble into a sodium hydroxide solution in a mass contactor. Penetration theory is to be used to obtain an estimate for the mass transfer coefficient k_m for the methyl chloride in the liquid phase. Any reaction of the methyl chloride with the sodium hydroxide will be ignored for this estimate (the reaction rate will be assumed to be slow for the particular conditions).

SOLUTION. For the bubble, V_{max} becomes the bubble rise velocity U_r and L becomes the bubble diameter d. This substitution is made because the water flows over the bubble and the contact time between bubble and water is roughly d_B/U_r. With the assumption that the bubble diameter in the mass contactor is 5 mm, the bubble rise velocity (for air in water) is taken to be 0.235 m/s. The diffusivity of a typical species we might be interested in, methyl chloride, in the sodium hydroxide solution can be estimated to be 4.4×10^{-9} m/s. The penetration depth is generally defined as the depth into the liquid phase (Phase I) where the dimensionless temperature or concentration in Eq. (6.4.10) decreases to 99% of its original value. This corresponds to $\eta = 2$, which yields $z \sim 0.02$ mm. Clearly, the boundary layer is much thinner than the 5-mm bubble diameter, such that neglecting the bubble curvature is acceptable.

The mass transfer coefficient can be estimated with these values from the penetration theory as

$$k_m = 2\sqrt{\frac{D_A V_{max}}{\pi L}} = 2\sqrt{\frac{(4.4 \times 10^{-9} \text{ m}^2/\text{s})(0.235 \text{ m/s})}{\pi(5 \times 10^{-3} \text{ m})}} = 5 \times 10^{-4} \text{ m/s}.$$

We can compare this value with that obtained from Figure 6.8, which is applicable for flow over a sphere. The Reynolds and Schmidt numbers are calculated to be

$$\text{Re} = \frac{\rho \, d_B U_r}{\mu} \approx \frac{(1000 \text{ kg/m}^3)(0.005 \text{ m})(0.235 \text{ m/s})}{0.001 \text{ Pa s}} = 1175,$$

$$\text{Sc} = \frac{\mu}{\rho \, D_{AB}} \approx \frac{0.001 \text{ Pa s}}{(1000 \text{ kg/m}^3)(4.4 \times 10^{-9} \text{ m}^2/\text{s})} = 227.$$

Therefore, the ordinate in Figure 6.8 is 44 000 and the Sherwood number is $\text{Sh} \sim$ 200, such that

$$k_m \sim 200 \frac{D_{AB}}{d_B} = 200 \frac{4.4 \times 10^{-9} \text{ m}^2/\text{s}}{0.005 \text{ m}} = 1.8 \times 10^{-4} \text{ m/s}.$$

The comparison in the previous example suggests that the penetration theory overestimates the rate of mass transfer under the stated conditions by a factor of about 3. A possible source of error is the assumption that the contact time is just the bubble velocity divided by the bubble diameter. In the next subsection, a different approach is taken to use penetration theory in turbulent flows.

6.4.3 Surface-Renewal Theory

Surface-renewal theory is an attempt to estimate mass transfer in a turbulent flow. The concept starts by considering fluid elements of Phase I that contact the interface with Phase II. Because of the turbulent fluid motion, these surface elements will arrive at the interface and be replaced regularly, as depicted in Figure 6.14. During the time of contact of the surface element at the interface, penetration theory is used to describe the heat and mass transfer between phases. The residence time distribution of the surface element at the surface is assumed to follow a Poisson process, which is due to the random nature of turbulence. The average residence time τ is then used directly with the results of penetration theory, Eqs. (6.4.4), to yield equations for the transport coefficients:

$$h = \sqrt{\frac{k \rho \hat{C}_p}{\tau}}, \qquad k_m = \sqrt{\frac{D_A}{\tau}}. \tag{6.4.15}$$

Because surface renewal theory uses the results from penetration theory, the same square-root dependence of h and k_m on the diffusivities is found. The difficulty, however, arises with the average residence time τ. This residence time is unknown and has to be either estimated from detailed, and complex, turbulent fluid mechanics calculations, or it has to be experimentally determined by comparison with experimental data.

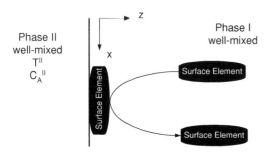

Figure 6.14. Surface-renewal model.

6.4.4 Interphase Mass Transfer

So far we have considered mass transfer within one phase, whereas most problems of interest involve transfer between two phases. This is shown in Figure 6.15, where there is a boundary layer in both phases in contact at the interface. An example would be the removal of acetone from water in a mass contactor with the organic solvent TCE. The resistances to mass transfer can occur in both phases. The starting point for the analysis is to write the flux in terms of transfer coefficients defined in each phase. Concentrations at the interface, which is taken to be infinitesimally thin, are denoted by the subscript i.

The flux of species A being transported between phases is

$$N_A = N_A^{I} = N_A^{II}, \qquad (6.4.16)$$

where we use the total flux to account for any convection or for the mass or mole average velocity when the amount of A being transferred is not small. Each flux can be written in terms of driving forces and the associated mass transfer coefficient as

$$\begin{aligned}
N_A &= K_m \left(C_A^{I} - M C_A^{II} \right), \\
N_A^{I} &= k_m^{I} \left(C_A^{I} - C_{A,i}^{I} \right), \qquad (6.4.17) \\
N_A^{II} &= k_m^{II} \left(C_{A,i}^{II} - C_A^{II} \right).
\end{aligned}$$

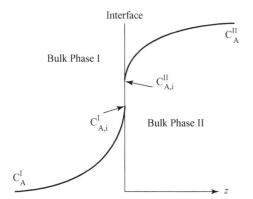

Figure 6.15. Interphase mass transfer geometry.

The overall mass transfer coefficient can then be calculated in terms of the individual mass transfer coefficients in each phase. The interface is assumed to be in chemical, mechanical, and thermal equilibrium:

$$C_A^I - C_{A,i}^I = \frac{N_A}{k_m^I},$$

$$MC_{A,i}^{II} - MC_A^{II} = \frac{MN_A}{k_m^{II}},$$

$$C_{A,i}^I = MC_{A,i}^{II},$$

$$\therefore C_A^I - MC_A^{II} = N_A \left(\frac{1}{k_m^I} + \frac{M}{k_m^{II}} \right).$$

Therefore, a comparison with Eq. (6.4.17) yields an expression for the overall mass transfer coefficient between phases in terms of the individual mass transfer coefficients in each phase:

$$\frac{1}{K_M} = \left(\frac{1}{k_m^I} + \frac{M}{k_m^{II}} \right). \tag{6.4.18}$$

This result can be generalized for the case in which the chemical equilibrium relationship is not a simple linear result. However, it always assumes that the interface is at equilibrium. Clearly, if either phase has a substantially lower mass transfer coefficient than the other, when accounting for the equilibrium relationship, that phase will be controlling the overall rate of mass transfer and resistance in the other phase can be neglected. An example of such a system might be a gas–liquid contacting process, in which often mass transfer is limiting in the liquid side. Alternatively, if a fast reaction is present in one phase, the resistance to mass transfer in that phase may be negligible.

EXAMPLE 6.7 OXYGENATION OF WATER. The mass transfer coefficient estimated for the oxygenation of water in the previous example assumed the resistance to mass transfer was only on the liquid side. Given the sum-of-resistances model for the interphase mass transfer coefficient, is this reasonable?

SOLUTION. The following table shows some physical property data we can consider in the analysis:

Property at 25 °C	Air	Water
$\nu(m^2/s)$	$\sim 10^{-5}$	10^{-6}
$\alpha_2(m^2/s)$	2.2×10^{-5}	1.4×10^{-7}
$D_{O_2} (m^2/s)$	1.76×10^{-5}	1.8×10^{-9}

Thus the binary diffusivity of oxygen is significantly lower in water than in air. Consequently, if the boundary layers are similar, the dominant resistance will clearly be in the liquid phase. To show this, we estimate the typical film thickness for mass transfer into water is often of the order of ~ 0.1 mm or $\delta_m = 10^{-4}$ m. For the gas inside the air, the film thickness is expected to be comparable. Therefore we can use the measured diffusivity of oxygen in water from the preceding table to estimate the mass transfer coefficient in both phases as

$$k_m^I \approx \frac{D_{O_2/water}}{\delta_{m,water}} = \frac{1.8 \times 10^{-9} \text{ m}^2/\text{s}}{10^{-4} \text{ m}} = 1.8 \times 10^{-5} \text{ m/s},$$

$$k_m^{II} = \frac{D_{O_2/air}}{\delta_{m,air}} = \frac{1.76 \times 10^{-5} \text{ m}^2/\text{s}}{10^{-4} \text{ m}} = 0.18 \text{ m/s}.$$

The overall mass transfer coefficient can be calculated from Eq. (6.4.18) given the value of the partition coefficient as

$$\frac{1}{K_M} = \left(\frac{1}{k_m^I} + \frac{M}{k_m^{II}} \right).$$

The value of the partition coefficient is calculated from the known Henry's "Law" coefficient written in terms of the liquid side mole fraction of oxygen:

$$P_{O_2}^{II} = H y_{O_2}^I$$
$$H = 4.4 \times 10^4 \text{ bars}.$$

To put this into the form

$$C_{O_2}^I = M C_{O_2}^{II},$$

the following manipulation is performed:

$$C_{O_2}^I = \frac{P_{O_2}^{II}}{RT} = \frac{H}{RTC^I} C^I y_{O_2}^I = M^{-1} C_{O_2}^I,$$

$$M = \frac{RTC^I}{H} = \frac{(8.314 \text{ N m/mol K})(298 \text{ K})(55.6 \times 10^3 \text{ mol/m}^3)}{4.4 \times 10^4 \times 10^5 \text{ N/m}^2} = 0.031.$$

Then

$$K_m^I = \frac{1}{\dfrac{1}{k_m^I} + \dfrac{M}{k_m^{II}}} = \frac{1}{\dfrac{1}{1.8 \times 10^{-5} \text{ m/s}} + \dfrac{0.031}{0.18 \text{ m/s}}} = 1.8 \times 10^{-5} \text{ m/s}.$$

Clearly, the resistance to mass transfer is entirely on the liquid side.

6.5 Summary of Convective Transport Coefficient Estimations

This chapter presents an organized development of transport correlations, transport analogies, and additional methods of estimating transport coefficients. Table I (page xx in the Instructors' and Readers' Guide) provides a classification of the simplified equipment types considered in this book. Here, in this brief summary, we provide guidance on the choice of method applicable for each type.

6.5.1 Heat Exchangers

As most heat exchangers, tank type or tubular, mixed or plug flow, separate the fluids by a solid barrier, the overall heat transfer coefficient can be calculated by a sum-of-resistances method in which the local heat transfer coefficients are obtained from correlations of the form Nu(Re, Pr). There are differences, however, in considering stirred devices, such as tank-type exchangers, versus pressure-driven flow, such as jackets and tubular devices.

Tank Type (Mixed–Mixed)

Heat transfer from or to a well-mixed fluid inside the containing wall of a stirred tank, such as that illustrated in Figure 3.1(a), is governed by the convection established by the impeller. As such, correlations are often found for specific impeller geometries as a function of the impeller Reynolds number, which is defined on the diameter and rotation speed of the impeller or agitator as

$$\mathrm{Re} = \frac{L_p^2 N_r \rho}{\mu},$$

where N_r is the agitator rotation rate and L_p is the diameter of the agitator.

The Nusselt number, however, is typically defined on the tank diameter (D): $\mathrm{Nu} = (hD/k)$, and hence typical correlations take the form

$$\mathrm{Nu} = a\,\mathrm{Re}^b\mathrm{Pr}^{1/3}\left(\frac{\mu_b}{\mu_w}\right)^m. \tag{6.5.1}$$

In the preceding equation, the subscripts b and w refer to bulk and wall, respectively. Notice how the exponent for the Prandtl number is fixed to that calculated from boundary-layer theory—this is because the Prandtl number for liquids does not vary over a broad enough range to warrant independent correlation. The following example of a table of correlations coefficients (Table 6.6) is specific for particular agitator designs and ranges of operation, as noted.

The jacketed or shell side of a typical mixed-tank exchanger is actually of less importance as the flow rate of the utility stream is sufficiently high such that the dominant resistance comes from the stirred internal phase and possibly the wall of the vessel. Correlations for the jacket side can be found in typical heat transfer handbooks if required.

Table 6.6. Correlation coefficients for Eq. (6.5.1) (from *Chemical Engineers' Handbook*, 1997, Table 10-6)

Agitator	a	b	m	Re
Paddle, Chilton, IEC (1944)	0.36	2/3	0.21	$300\text{--}3\times10^5$
Pitch blade turbine, Uhl, Chem. Eng. Prog. Sym. Sr. (1945)	0.53	2/3	0.24	8–200
Disk flat blade turbine, Brooks, Chem. Eng. Prog. (1959)	0.54	2/3	0.14	$40\text{--}3\times10^5$
Propeller, Brown, Trans. Inst. Chem. Eng. (1947)	0.54	2/3	0.14	2×10^3 (1 pt)
Anchor, Uhl	1	1/2	0.18	1–300
Anchor, Uhl	0.36	2/3	0.18	$300\text{--}4\times10^4$
Helical ribbon, Glutz, J. Appl. Chem. USSR (1966)	0.633	1/2	0.18	$8\text{--}1\times10^5$

EXAMPLE 6.8 USE OF IMPELLER HEAT TRANSFER CORRELATIONS. Estimate the heat transfer coefficient for the following conditions corresponding to typical mixed tank-type exchangers considered in Chapter 3.

System Properties

Tank (Stainless)	$D = 1$ m
	$k = 16$ W/m K
	$L_{wall} = 1$ cm
	Area $= 1$ m^2
Agitator	0.3-m-diameter paddle
	100 rpm
Fluid (water)	$\mu = 0.89 \times 10^{-3}$ Pa s (25 °C)
	$\mu = 0.54 \times 10^{-3}$ Pa s (50 °C)
	$\rho = 1000$ kg/m^3
	$k = 0.6$ W/m K
	$Pr = 0.7$
	$\hat{C}_p = 4.2$ J/mol K
	$M_w = 18$ g/mol
Tank temperature	25 °C
Wall temperature	50 °C

SOLUTION. The following calculations can be performed to estimate the local heat transfer coefficient between the tank and the tank wall under the specified conditions. First, the Reynolds number based on the agitator properties is estimated as

$$Re = \frac{L_p^2 N_r \rho}{\mu} = \frac{0.3^2 \, (100/60) \, 1000}{0.89 \times 10^{-3}}.$$
$$= 1.7 \times 10^5$$

For the paddle-type agitator in this range of conditions, the coefficients for correlation equation (6.5.1) are read from Table 6.6 as

$$Nu = 0.36 Re^{2/3} \, Pr^{1/3} \left(\frac{\mu_b}{\mu_w} \right)^{0.21}.$$

Using the property values supplied, we calculate the local heat transfer coefficient as

$$Nu = (1.7 \times 10^5)^{2/3} (0.7)^{1/3} \left(\frac{0.89}{0.54} \right)^{0.21} = 1100,$$

$$h = Nu \frac{k}{D} = 1100 \frac{0.6}{1} = 660 \text{ W/m}^2 \text{ K}.$$

This value is reasonable as judged from the values discussed for this tank in Chapter 3.

Tank Type (Mixed–Plug)

Internal plug flow inside a pipe in a tank is equivalent to that of a plug-flow tubular exchanger, which can be modeled by correlations of the form of (6.3.6), as subsequently discussed. As the coils are generally located at or near the wall of the tank to avoid the impeller, the impeller correlations can be applied to obtain estimates for the heat transfer coefficient from the exterior of the pipe to the stirred fluid. Again, the dominant resistance to heat transfer is usually in the stirred interior fluid.

Tubular (Cocurrent and Countercurrent, Plug–Plug)

Turbulent flow in pipes, annular regions, and shells can be estimated by correlations of the form of Eq. (6.3.3), namely,

$$Nu = a Re^b Pr^c. \tag{6.3.3}$$

As noted in Table 6.5, the coefficients depend on the geometry, range of Reynolds numbers (and sometime Prandtl numbers), and whether the fluid in question is being heated or cooled. The choices of conditions to evaluate the material properties are also specified along with the correlation. Manufacturers of equipment and the handbooks in the reference section have numerous correlations applicable for more complex, specific geometries, such as multipass shell-and-tube and plate-and-frame exchangers. The use of such correlations has been shown in Chapter 3 and is further illustrated in the design example in Chapter 8.

6.5.2 Mass Contactors

Mass contactors, unlike heat exchangers, usually involve direct contacting of the fluids involved, and hence the area as well as the rate of mass transfer depends on the microscale fluid mechanics. This sufficiently complicates the analysis such that few, if any, correlations of mass transfer coefficients exist for most mass contactors of industrial interest.

However, some of the results presented in this chapter are of relevance for specific operations. For example, mass transfer to or from a solid object, such as leaching of minerals or solutes from solids (e.g., decaffeination of coffee beans), or the evaporation of volatiles from solids in airstreams (drying) can be calculated from the correlations presented in Section 6.3 or in the cited references. Failing that, the use of a transport analogy, such as the Chilton–Colburn analogy, Eq. (6.2.16), enables using correlations developed for the analogous geometry but for heat transfer or fluid friction to be converted to an appropriate correlation for mass transfer:

$$j_D = j_H = \frac{f}{2}. \tag{6.2.16}$$

Here we rely on the central hypothesis that the same boundary-layer mechanics governs all three types of transport considered in this equation. Hence correlations developed for heat transfer, which are much more prevalent and for which more experimental data are generally available, can be converted to mass transfer correlations. It is important, however, to have the same geometry and range of Reynolds numbers in constructing the analogy, and the rate of mass transfer must be sufficiently small such that diffusion-induced convection effects are not dominant.

We next consider the cases in which both streams are fluids and in direct contact. The interphase mass transfer modeling of Section 6.4 applies to relate the overall mass transfer coefficient to the local coefficients within each fluid phase:

$$\frac{1}{K_M} = \left(\frac{1}{k_m^I} + \frac{M}{k_m^{II}} \right). \qquad (6.4.18)$$

However, the local mass transfer coefficients are not readily obtained, except perhaps by experiment. Because the area and mass transfer coefficients are determined simultaneously by the fluid mechanics of the device, estimation of these properties is the subject of the next chapter.

Tank Type (Mixed–Mixed)
This may be the most difficult case. As will be shown in Chapter 7, the importance of area generation in the mixing zone near the impeller of a stirred tank dominates the rate of mass transfer.

Tank Type (Mixed–Plug)
The most common scenario is a gas plume rising in a mixed liquid tank, as described in detail in Chapter 4. The resistance is therefore usually dominated by the liquid side. As shown in Section 6.4, the penetration model can be used to estimate the mass transfer coefficient. A convenient framework for correlating and estimating mass transfer to rising bubbles in turbulent conditions is given by the surface-renewal theory:

$$k_m = \sqrt{\frac{D_A}{\tau}} \qquad (6.4.15)$$

Estimation of the mean contact time τ will also be considered further in the next chapter as it is related to the bubble rise velocity and fluid motions induced in the liquid by the bubble rise velocity, which themselves depend on the bubble size and hence area for mass transfer.

Tubular (Cocurrent and Countercurrent, Plug–Plug)
Common examples are the stripping of a volatile from a liquid by a gas stream, or the absorption or scrubbing of a contaminant from a gas stream by a liquid. These operations are nearly always countercurrent and run in a packed column. As the area for contact depends on the gas and liquid flow rates, as well as on the type of packing and geometry of the column, typically one finds correlations for the product of the mass transfer coefficient times the area—that is, there is little attempt to independently estimate the rate of mass transfer from the area for transfer. This poses significant problems in scale-up and technically feasible design. In the following chapter, we will demonstrate, however, that it is possible to estimate or even measure the area independently, and hence determine mass transfer rates from experiment. Further, the penetration and surface renewal models can once again be used to help correlate and scale up mass transfer coefficients.

Nomenclature

C	Concentration (mol/m^3)
$\hat{C}_p$	Heat capacity (kJ/kg K)
D	Diffusivity (m^2/s)
f	Coefficient of skin friction
F	Force (N)
ΔH^{vap}	Heat of vaporization (kJ)
h	Local heat transfer coefficient (W/m^2 K)
J	Diffusivity flux (kg/m^2 s)
j_D	Colburn J factor for mass transfer, Sh/ReSc$^{1/3}$
j_H	Colburn J factor for heat transfer, Nu/RePr$^{1/3}$
k_m	Local mass transfer coefficient (m/s)
k	Thermal conductivity (W/m K)
Le	Lewis number
L	Distance (m)
L_p	Diameter of agitator (m)
N_A	Total mass flux of species A (kg/m^2 s)
M	Distribution coefficient
p_A	Partial pressure of species A (N/m^2)
P	Total pressure (N/m^2)
Nu	Nusselt number
Pr	Prandtl number
q	Heat flux (W/m^2)
Re	Reynolds number
Sc	Schmidt number
Sh	Sherwood number
T	Temperature (K)
U	Overall heat transfer coefficient (W/m^2 K)
v	Velocity (m/s)
x	x direction
y	y direction
z	z direction

Greek

α	Thermal diffusivity (m^2/s)
δ	Thickness (m)
μ	Viscosity (Pa-s)
ν	Kinematic viscosity (m^2/s)
ρ	Density (kg/m^3)
τ	Residence time (s)
$\tau_{i,j}$	Shear stress (Pa)

Subscripts

∞	Bulk

A	Species A
S	Surface
x	x direction

REFERENCES

Aris R. *Vectors, Tensors, and the Basic Equations of Fluid Mechanics*. New York: Dover, 1990.

Bird RB, Stewart WE, Lightfoot EN. *Transport Phenomena* (2nd ed.). New York: Wiley, 2002.

Chemical Engineers' Handbook (7th ed.). New York: McGraw-Hill Professional, 1997.

Chilton TH, Colburn AP. Mass transfer (absorption) coefficients—Prediction from data on heat transfer and fluid friction. Ind. Eng. Chem. 1934; 26: 1183.

Colburn AP. A method of correlating forced convection heat transfer data and a comparison with fluid friction. Trans. AIChE J. 1933; 29: 174.

Friend WL, Metzner, AB. Turbulent heat transfer inside tubes and the analogy among heat, mass, and momentum transer. AIChE J. 1958; 4: 393–402.

McAdams, WH. *Heat Transmission* (3rd ed.). New York: McGraw-Hill, 1954.

Pohlhausen E. The heat exchange between solid bodies and liquids with little heat conduction. Zeitschrift für Angewandte Math. und Mechanik 1921; 1: 115–21.

Reynolds O. On the extent and action of the heating surface on steam boilers. Proc. Lit. Philos. Soc. Manchester 1874; 14: 7.

Reynolds O. An experimental investigation of the circumstances which determine whether the motion of water shall be direct or sinuous, and of the law of resistance in parallel channels. Philos. Trans. 1883; 174: 935–82.

Schlichting H. *Boundary-Layer Theory* (7th ed.). New York: McGraw-Hill, 1979.

Sherwood TK, Pigford RL. *Adsorption and Extraction*. New York: McGraw-Hill, 1952.

Welty J, Wicks CE, Wilson RE, Rorrer GL. *Fundamentals of Momentum, Heat, and Mass Transfer* (4th ed.). New York: Wiley, 2001.

The HTRI also provides resources and links to the manufacturers of heat exchanger equipment, www.htri-net.com.

PROBLEMS

6.1 Consider a catalytic isothermal CFSTR reactor, in which a first-order A→B reaction takes place in porous catalyst pellets. The feed concentration and flow rate are 1 M and 1 L/min, respectively. The reactor volume is 20 L, the reaction constant in the catalyst (based on the pore volume) is 0.002 s^{-1}, and the effective reagent diffusivity is 1.3×10^{-9} m^2/s. The reactor contains 13 000 pellets of 5-mm diameter. Finally, the catalyst manufacturer claims that the porosity of the catalyst is 0.54, and tortuosity is close to 1.

   a. Write the model equations for the catalyst. Explain your assumptions, and include the necessary boundary conditions.
   b. Determine the model equation for the reactor.
   c. Compute the Thiele modulus.
   d. Calculate the conversion achieved in the reactor.

e. How would the achieved conversion change if the diameter of the catalyst pellets was changed to 2 mm, provided that the catalyst mass remained the same? What about 10 mm? Comment on the effects that you observe.

6.2 Consider the heat transfer to the surface of a rocket nozzle. The combustion gases flow out at $1100\,°C$ and the heat flux to the surface is $10\,kW/m^2$. The current operating temperature of the nozzle is $1000\,°C$.

a. The maximum surface temperature of the nozzle material is to be constrained by material properties to be $600\,°C$. What is the heat duty of the coolant required (per square meter of surface area)?

b. It is suggested as an alternative that one can reduce the nozzle length L by a factor of two. To maintain rocket performance, the exhaust velocity must be increased by a factor of two. What is the new heat duty?

6.3 A heat exchanger is being designed for use with radioactive fluids, and so prototype testing has been requested to validate the design. The fluid is to be cooled. As the experiments are complicated and expensive because of the radioactivity, a nonradioactive simulant has been developed. Over the temperature ranges of interest, the simulant has the same thermal diffusivity as the radioactive fluid, but a viscosity that is twice as large, twice the conductivity, but also twice the heat capacity. The two fluids have the same density. A shell-and-tube heat exchanger is to be designed with water on the shell side.

a. Tests indicate that the dominant heat transfer resistance is on the tube side. If the heat transfer coefficient is measured to be $100\,W/m^2\,K$ and the Re of the simulant is 50 000, what can one expect for the heat transfer coefficient of the radioactive fluid pumped through the tube side with the same flow rate as the simulant?

b. The prototype is deemed successful and is to be scaled up. For safety and maintenance purposes, only one exchanger can be used. However, in the scale-up, the pressure drop in the tube side cannot be increased or decreased. Discuss the methods of scaling from the laboratory prototype to the working exchanger. Be specific on what the dimensions and flow rates of the full-scale exchanger will have relative to the prototype, and identify any possible limitations to the scale-up.

6.4 Determine which phase limits the rate of oxygen transfer during oxygen removal from water by pure nitrogen. Nitrogen at 1.05 bar is bubbled into water at $20\,°C$. Use the *penetration model* to determine the relative rate of mass transfer in each phase. Which phase has the greatest resistance to mass transfer?

Data:

Property	Nitrogen	Water
D_{AB} (m^2/s)	2×10^{-5}	2.1×10^{-9}
Oxygen solubility (mol/L)		1.4×10^{-3} at $p_{O_2} = 1$ bar
Concentration of solvents	1.1 bar (gas)	55.5 mol/L

6.5 There is saturated steam at 1 atm flowing through a duct at a high velocity. You decide to use this stream to heat water. The water that is being heated starts out at $15\,°C$ and needs to be heated to a final temperature of $75\,°C$. The expected resistance that is due to fouling on the inside and outside of the pipe is $1.7 \times 10^{-4}\,m^2\,K/W$ and $2.3 \times 10^{-4}\,m^2\,K/W$, respectively. After purchasing a 5-cm-ID thin-walled copper pipe, having a roughness of

0.026 mm, you want to design a heat exchanger that can handle a flow rate of 10 L/min. *Hint: Because the heat transfer coefficient of condensing steam is high, the resistance to heat transfer in the steam is negligible.*

a. Determine the overall heat transfer coefficient using two different correlations that are applicable to the flow conditions. Correlations listed in Bird et al. (2002) are given below.

b. Use the estimated heat transfer coefficients from the two correlations to evaluate the uncertainty of the calculations that you performed.

c. What is the required pipe length?

d. Plot the required pipe length as a function of pipe diameter, varying the diameter from 1 to 30 cm.

e. Plot the pressure drop through the heat exchanger as a function of the pipe diameter.

Bennett and Myers:

$$\text{Nu} = 1.62 \left(\frac{4 \times \dot{m} \times \hat{C}_p}{\pi \times k \times L} \right)^{1/3} \text{(laminar)}.$$

Sieder and Tate:

$$\text{Nu} = 1.86 \times \text{Re}^{1/3} \times \text{Pr}^{1/3} \times \left(\frac{D}{L} \right)^{1/3} \times \left(\frac{\mu}{\mu_s} \right)^{0.14} \text{(laminar)}.$$

Dittus and Boelter:

$$\text{Nu} = 0.023 \times \text{Re}^{0.8} \times \text{Pr}^{0.4} \text{(turbulent)}.$$

Colburn:

$$\text{Nu} = 0.023 \times \text{Re}^{0.8} \times \text{Pr}^{1/3} \text{(turbulent)}.$$

Note: The transition from laminar flow is $\text{Re} = 2100$ for pipe flow, with $\text{Re} > 10\,000$ required for fully turbulent flow. The intermediate regime is denoted the "transition regime" and requires interpolation. The properties are evaluated at the mean bulk temperature unless noted otherwise.

6.6 Reconsider Problem 6.5. There is saturated steam flowing through a duct at a high velocity. You decide to use this steam to heat water. The water that is being heated starts out at 15°C and needs to be heated to a final temperature of 75 °C. The expected resistance that is due to fouling on the inside and outside of the pipe is 1.7×10^{-4} m² K/W and 2.3×10^{-4} m² K/W, respectively. After purchasing a 5-m-long (5-cm-ID) thin-walled copper pipe, having a roughness of 0.026 mm, you want to design a heat exchanger that can handle a flow rate of 10 L/min. *Hint: Because the heat transfer coefficient of condensing steam is high, the resistance to heat transfer in the steam is negligible.*

a. Determine the overall heat transfer coefficient using the pressure drop through the pipe.

b. What is the temperature of the saturated steam necessary to handle the load?

6.7 Rewrite the correlations for heat transfer presented in Problem 6.5 in the form of the Colburn j factor for heat transfer. Then, using the Chilton–Colburn analogy, determine the corresponding correlations for mass transfer written in terms of the Sherwood number. Plot the Colburn correlation for the Sherwood number (assume $\text{Sc} = 1$) for Reynolds numbers between 1000 and 100 000. Be careful to indicate the range of expected validity of the correlations.

6.8 The following correlation has been presented for heat transfer coefficients for water and air in an annulus:

$$\frac{h_i}{\hat{C}_p \rho v} = 0.023 \left(\frac{D_o}{D_i} \right)^{0.5} \left(\frac{(D_o - D_i) \rho v}{\mu} \right)^{-0.2} \left(\frac{\hat{C}_p \mu}{k} \right)^{-2/3} ,$$

where D_i and D_o are the inner and outer diameters of the annulus, respectively, and h_i is the heat transfer coefficient for the inner surface at D_i. Consider now the rate of mass transfer of naphthalene into air, as measured from a naphthalene rod of 2-in. diameter, which is concentric with a brass tube that has an ID of 3 in. Air flows through the annulus between the rod and the tube at a mass velocity of 31 ft./s, at 1 atm and 32 °F. The Schmidt number for the system is 2.57, and the density and viscosity of the gas are 0.0807 lb_m/ft^3 and 1.176×10^{-5} lb_m/ft s, respectively. What is the gas-phase mass transfer coefficient under these conditions?

6.9 A sphere is suspended in a quiescent fluid. The Biot number is small such that the sphere can be considered at uniform temperature. Heat flows out of the sphere by conduction.

a. Show that the Nusselt number for this problem is Nu $= 2$. Compute the temperature (as a function of time) of a 1-cm-diameter lead sphere with initial temperature of 50 °C if the surrounding air temperature is 25 °C (*Note*: Ignore any natural convection in the air.)

b. For a constant applied flow, Ranz and Marshall (Chem. Eng. Prog. 1952; 48: 141) recommend the following correlation:

$$Nu = 2 + 0.60 Re^{1/2} \, Pr^{1/3}.$$

Compute the temperature of the lead sphere if the air is moving at 1 cm/s per second and compare with the result just obtained.

c. Compute the Biot number for this problem. At what flow rate range does the assumption of a lumped temperature analysis fail for the sphere?

6.10 Compute the effects of *wind chill* by considering a model for a human as a cylinder of 30-cm diameter and 1.8 m high, with a surface temperature of 30 °C.

a. Estimate the rate of heat loss from the cylinder in the absence of wind, but assume a typical human motion of 3 mph (you can neglect the top and bottom of the cylinder).

b. Calculate the rate of heat loss in a 20-mph wind at 5 °C.

c. What is the rate of heat loss if the surface is covered by a thin film of water (at the same temperature) that evaporates and the relative humidity of the air is 75%?

6.11 Consider fully developed laminar flow between parallel plates in which a Newtonian fluid is being heated by the walls that are maintained at a constant wall flux.

a. If the distance between the walls is 2H, derive and solve the differential energy balance corresponding to this flow situation to obtain solutions for both the temperature distribution between the walls and the average temperature in the duct, $\bar{T}$. It may be assumed that the molecular energy transport in the z direction can be neglected and the axial temperature gradient, dT/dz, is constant at $\Delta T/L$. Note that it is necessary to solve for the velocity profile between the plates.

b. Use the solution for the temperature profile obtained in part a to obtain a numerical value for the Nusselt number if the latter is based on a characteristic dimension of H and having a driving force of $T_w - \bar{T}$.

6.12 Redo the previous problem if the flow is between parallel plates and the plates are composed of solid salt. You can assume that the water flowing between the salt plates does not penetrate the walls and the amount of salt dissolution is small. In part a compute the concentration profile and average salt concentration, and in part b compute the Sherwood number.

6.13 You and a friend are camping and you would like to boil water to cook some vegetables. However, you forgot to bring a pot to cook in. The only thing you brought that holds water is a 6-L plastic water bucket. Because the bucket cannot be placed in the campfire you suggest that the water in the bucket can be heated by hot stones taken from your campfire. Your friend claims that it would take hours to boil water by using stones. Being a good chemical engineering student, you have brought all of your textbooks with you camping and in one of the textbooks are the following experimental data for the dissolution of salt in water.

   Experiment: 30 spherical salt tablets are placed in 6000 mL of water and stirred at 100 rpm. The concentration of salt in the water phase is measured as a function of time. The density of salt is 2.16 g/cm^3, the saturation concentration of salt is 0.360 g/cm^3, the total mass of tablets put in water is 19.2 g, and the diffusivity of salt in water is 2.2×10^{-5} cm^2/s. Water's thermal conductivity, heat capacity, and viscosity are 600 W/m K, 4.18 kJ/kg K, and 630×10^{-6} N s/m^2, respectively.

Time (s)	$1000C_A^I$ (g/cm^3)
0	0
15	0.30
30	0.35
45	0.64
60	0.89
75	1.08
90	1.10
105	1.24
120	1.40
135	1.49
150	1.68
165	1.76
195	2.06
200	2.14
240	2.31
270	2.43

   You plan to boil the water by holding one spherical stone 10 cm in diameter inside the bucket with metal tongs and stirring at 100 rpm. You may assume that the stone has a high heat capacity so that the temperature of the stone is constant at 450 °C (this is of course an approximation) and the initial temperature of the water is 25 °C. Determine how long it will take to boil the bucket of water.

6.14 Mass transfer coefficients can be estimated (Nadkarni VM, Russell TWF. Mass transfer to naturally flowing streams. I&EC, Proc. Des. and Dev. 1973; 12: 414) in gas–liquid turbulent multiphase flows typical of waste water treatment by using penetration theory and assuming that the contact time is $\tau = \bar{l}/|\bar{u}|$, where $\bar{l} = \left(v_L^3/\varepsilon\right)^{\frac{1}{4}}$ and $|\bar{u}| = (v_L\varepsilon)^{\frac{1}{4}}$ Here, ε is the viscous dissipation per unit mass of fluid.

   a. Derive an expression for the mass transfer coefficient in terms of the liquid velocity and the energy dissipation.

   b. Derive a relationship for the Sherwood number in terms of Reynolds and Schmidt numbers based on this idea and that fact that, for pipe flow, $\varepsilon = 2f\,(v_L)^3/d_{pipe}$, and the Blasius correlation for friction in turbulent flow (see Example 6.5).

6.15 Derive Eqs. (6.4.15), surface-renewal theory, as follows. Take as a system the interface in Figure 6.14 and consider an equation for the probability that a surface element remains in contact with a surface E(t). Application of the word statement of the conservation equation, Figure 1.6, leads to the following: {*Probability that a surface element is at the surface at time t + Δt*} = {*Probability that a surface element is at the surface at time t*} − {*rate of removal of surface elements*}Δt. If the rate of removal can be described by a first-order rate constant, k_1, which has units of inverse seconds, show that the residence time distribution follows an exponential distribution: $E(t) = Ae^{-k_1 t}$. Further, show that the constant of integration can be determined from the normalization condition: $\int_0^\infty E(t)dt = 1$. Compute the average residence time, τ, and show that $\tau = (1/k_1)$, such that $E(t) = (1/\tau)\exp(-t/\tau)$. Finally, use the flux expressions derived for penetration theory, Eqs. (6.4.11), and this residence time distribution to calculate the fluxes of heat and mass of A, and arrive at the final expressions for the transport coefficients.

6.16 Gas adsorption into a liquid, such as a removal of CO_2, can be enhanced by adsorption into a reactive liquid, such as ethanolamine solutions. Consider the wetted wall column analyzed in Subsection 6.4.2. Assume that A, which is dilute, now reacts in the liquid with a first-order reaction that depends on the concentration of A in the liquid. Derive an expression for the rate of mass transfer analogous to Eqs. (6.4.13) that includes the reaction rate. What new dimensionless group characterizes the effect of reaction on the rate of mass transfer? Derive an expression for the enhancement of the rate of mass transfer (rate with reaction divided by rate in the absence of reaction).

6.17 A hot process stream with properties similar to those of water with a flow rate of 100 L/min must be cooled from 100 to 80 °C by use of cooling water available at 5 °C. The cooling water must be returned at a temperature no greater than 30 °C. The heat exchanger will be constructed of schedule 40S stainless steel pipes, 3 in. for the inner pipe (through which the process stream will flow) and 5 in. for the outer pipe (through which the cooling water will flow) and is completely insulated from the surroundings. Data for the thermal properties of water as well as for the pipe are given in the following tables below (from Welty et al., 2001).

Thermal properties of water:

T (°C)	ρ (kg/m³)	$\hat{C}_p$ (J/kg K)	k (W/m K)	μ (cP)
0	999.3	4226	0.558	1.794
20	998.2	4182	0.597	0.993
40	992.2	4175	0.633	0.658
60	983.2	4181	0.658	0.472
80	971.8	4194	0.673	0.352
100	958.4	4211	0.682	0.278

Thermal properties of stainless steel:

T ($°C$)	ρ (kg/m^3)	$\hat{C}_p$ ($J/kg\ K$)	k ($W/m\ K$)
20	7820	4226	16.0
100	7820	4226	17.3

a. Given the process conditions and information in the preceding table, which heat transfer correlation would be most applicable to this process?
b. Compute the minimum flow rate of cooling water required to carry out this operation. Using this flow rate, determine the maximum length of heat exchanger required.
c. Compute and plot the cooling water flow rate as well as the length of exchanger required as a function of the outlet cooling water temperature in the range of 10–30 $°C$. Your calculations should be plotted in a similar manner to that shown in Figure E3.6.
d. How would you choose a flow rate and exchanger length for the final design, and what factors would be important?

APPENDIX A: DERIVATION OF THE TRANSPORT EQUATIONS

Derivations of the differential transport equations for momentum, heat, and mass transfer follow a common paradigm, which starts with the differential control volume given in Figure 6.A1 and the word statement of conservation, Figure 1.6. In the following, we derive simplified equations relevant for the analysis of the model problems to be considered in this chapter. The interested student is referred to references cited in this chapter for further information and examples of their use (Bird et al., 2002).

Fluxes of mass, momentum, and energy enter and leave the differential control volume through the six orthogonal faces, taken to be aligned with the Cartesian coordinate axes. Further, stresses act on the control volume, performing work on the fluid inside, and body forces, such as gravity, act on the mass inside the control volume.

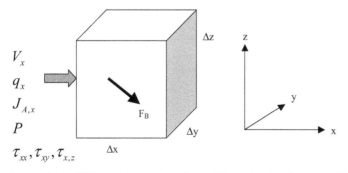

Figure 6.A1. Differential control volume. (Note that the fluxes and forces are indicated for only the surface normal to x—there will be appropriate terms acting on the surface at x + Δx, as well as the surfaces normal to the y and z directions. Also, F_B refers to a body force that, for example, includes gravity.)

Using the word statement introduced in Figure 1.6, the conservation laws for mass, momentum (in the x, y, z directions), energy, and species mass are derived for this microscopic control volume. For each conserved quantity, the appropriate constitutive equation (Newton, Fourier, or Fick; see Table 1.4) is required for providing a closed equation for the field of interest (fluid velocity, temperature, concentration).

Mass Conservation—Continuity Equation

From the word statement of the conservation law (Figure 1.6), the conservation of mass is expressed as follows. The mass contained in the control volume is given by the density ρ times the volume, which is given by $\Delta x \Delta y \Delta z$. Mass enters and leaves the control volume because of convection, given by the fluxes of mass through the faces of the control volume. The mass flux is given by the density times the velocities (v_x, v_y, v_z). The equation for mass conservation becomes

$$
\begin{aligned}
\rho|_{t+\Delta t}\Delta x \Delta y \Delta z \quad &= \rho|_t \Delta x \Delta y \Delta z \\
+ \rho v_x|_x \Delta y \Delta z \Delta t \quad &- \rho v_x|_{x+\Delta x}\Delta y \Delta z \Delta t \\
+ \rho v_y|_y \Delta x \Delta z \Delta t \quad &- \rho v_y|_{y+\Delta y}\Delta x \Delta z \Delta t \\
+ \rho v_z|_z \Delta x \Delta y \Delta t \quad &- \rho v_z|_{z+\Delta z}\Delta x \Delta y \Delta t.
\end{aligned}
\tag{6.A1}
$$

Dividing the preceding equation by $\Delta x \Delta y \Delta z \Delta t$ and taking the limit as $\Delta x, \Delta y, \Delta z, \Delta t \to 0$ enables writing the conservation of mass in terms of derivatives as

$$
\frac{\partial \rho}{\partial t} + \frac{\partial \rho v_x}{\partial x} + \frac{\partial \rho v_y}{\partial y} + \frac{\partial \rho v_z}{\partial z} = 0.
\tag{6.A2}
$$

For the problems that follow, we assume that the density of the fluid is constant. Consequently, Eq. (6.A2) can be written as

$$
\frac{\partial v_x}{\partial x} + \frac{\partial v_y}{\partial y} + \frac{\partial v_z}{\partial z} = 0.
\tag{6.A3}
$$

This equation is known as the *equation of continuity*, and it expresses in mathematical form the concept of conservation of mass for a constant density fluid.

Momentum

Consider the flux of x momentum, which, for the boundary-layer problems of interest, yields the relevant velocity profile for analysis of the heat and mass transfer from the plate. Using the statement of conservation of momentum, we can construct the linear momentum equation by accounting for all of the changes in x momentum in the control volume, the fluxes of x momentum in and out of the control volume, as well as all surface and body forces that will increase or decrease the x momentum of the control volume.

One important point to recall is the effect of the stresses acting on the control volume surfaces and their ability to generate momentum. The definition of the components of the stress tensor (τ) in terms of forces acting on the surfaces enables proper account for this term in the momentum equations, namely,

$$\tau = \begin{bmatrix} \tau_{xx} & \tau_{xy} & \tau_{xz} \\ \tau_{yx} & \tau_{yy} & \tau_{yz} \\ \tau_{zx} & \tau_{zy} & \tau_{zz} \end{bmatrix},$$

$$\tau_{ij} = \frac{\text{force j direction}}{\text{area normal i direction}} = \text{flux of j momentum in i direction}.$$

The stress just defined is the *deviatoric* stress, which does not include the isotropic pressure, which is considered separately. Consequently, stress × area × time increment will yield the change in momentum that is due to shear and normal stresses.

The terms affecting the x momentum are as follows:

Accumulation:

$$\Delta x \Delta y \Delta z [(\rho v_x)_{t+\Delta t} - (\rho v_x)_t]$$

= flux in − out:

$$\Delta y \Delta z \Delta t \{[(\rho v_x) v_x]_x - [(\rho v_x) v_x]_{x+\Delta x}\}$$
$$+ \Delta x \Delta z \Delta t \{[(\rho v_x) v_y]_y - [(\rho v_x) v_y]_{y+\Delta y}\}$$
$$+ \Delta x \Delta y \Delta t \{[(\rho v_x) v_z]_z - [(\rho v_x) v_z]_{z+\Delta z}\}$$

+ generation by surface forces:

$$+ \Delta y \Delta z \Delta t \left[(\tau_{xx})_x - (\tau_{xx})_{x+\Delta x}\right]$$
$$+ \Delta x \Delta z \Delta t \left[(\tau_{yx})_y - (\tau_{yx})_{y+\Delta y}\right]$$
$$+ \Delta x \Delta y \Delta t \left[(\tau_{zx})_z - (\tau_{zx})_{z+\Delta z}\right]$$
$$+ \Delta y \Delta z \Delta t \left[(P)_x - (P)_{x+\Delta x}\right]$$

+ generation by body forces:

$$+ \Delta x \Delta y \Delta z \Delta t \rho\, F_{B,x}.$$

Dividing by $\Delta x \Delta y \Delta z \Delta t$ and taking the differential limit yields

$$\frac{\partial}{\partial t}(\rho v_x) = -\frac{\partial}{\partial x}(\rho v_x v_x) - \frac{\partial}{\partial y}(\rho v_x v_y) - \frac{\partial}{\partial z}(\rho v_x v_z) - \frac{\partial}{\partial x}(\tau_{xx})$$
$$-\frac{\partial}{\partial y}(\tau_{yx}) - \frac{\partial}{\partial z}(\tau_{zx}) - \frac{\partial P}{\partial x} + \rho\, F_{B,x}. \qquad (6.A4)$$

This equation can be written in more compact form by defining the *substantial* derivative as

$$\frac{D}{Dt} = \frac{\partial}{\partial t} + v_x \frac{\partial}{\partial x} + v_y \frac{\partial}{\partial y} + v_z \frac{\partial}{\partial z} = \frac{\partial}{\partial t} + \mathbf{v} \cdot \nabla \qquad (6.A5)$$

where in the last equality vector notation has been employed. Then the linear momentum equation can be written as

$$\frac{D}{Dt}(\rho v_x) + (\rho v_x)\left[\frac{\partial v_x}{\partial x} + \frac{\partial v_y}{\partial y} + \frac{\partial v_z}{\partial z}\right] = -\frac{\partial}{\partial x}(\tau_{xx}) \tag{6.A6}$$

$$-\frac{\partial}{\partial y}(\tau_{yx}) - \frac{\partial}{\partial z}(\tau_{zx}) - \frac{\partial P}{\partial x} + \rho F_{B,x}.$$

This equation can be further simplified by subtracting the continuity equation (conservation of mass) multiplied by the x velocity,

$$v_x\left(\frac{D}{Dt}(\rho) + (\rho)\left[\frac{\partial v_x}{\partial x} + \frac{\partial v_y}{\partial y} + \frac{\partial v_z}{\partial z}\right]\right) = 0, \tag{6.A7}$$

to yield

$$\rho\frac{Dv_x}{Dt} = -\frac{\partial}{\partial x}(\tau_{xx}) - \frac{\partial}{\partial y}(\tau_{yx}) - \frac{\partial}{\partial z}(\tau_{zx}) - \frac{\partial P}{\partial x} + \rho F_{B,x}. \tag{6.A8}$$

This equation can be expressed more compactly in vector notation (note that there are similar equations for the y and z momentum):

$$\rho\frac{Dv}{Dt} = -\nabla\cdot\boldsymbol{\tau} - \nabla P + \rho\mathbf{F_B}. \tag{6.A9}$$

To proceed, a constitutive equation is required for relating the flux of momentum (shear stress) to the velocity field. Here we employ Newton's constitutive equation for viscosity as (where we only consider a shear viscosity) $\tau_{yx} = -\mu(dv_x/dy)$ to yield

$$\rho\frac{\partial v_x}{\partial t} + \rho v_x\frac{\partial v_x}{\partial x} + \rho v_y\frac{\partial v_x}{\partial y} + \rho v_z\frac{\partial v_x}{\partial z}$$

$$= \mu\left(\frac{\partial^2 v_x}{\partial x^2} + \frac{\partial^2 v_x}{\partial y^2} + \frac{\partial^2 v_x}{\partial z^2}\right) - \frac{\partial P}{\partial x} + \rho F_{B,x}, \tag{6.A10}$$

which, generalized to all three coordinates in vector notation, is $\rho(Dv/Dt) = \mu\nabla^2 v - \nabla P + \rho\mathbf{F_B}$. The result is known as the *Navier–Stokes* equation, and it is commonly used to solve problems in Newtonian fluid mechanics.

Energy

The energy equation is written for the sum of the internal and kinetic energy, per mass:

$$\hat{E} = \hat{U} + \frac{1}{2}v^2. \tag{6.A11}$$

Construction of the conservation law for energy proceeds in a manner similar to that used in the earlier chapters of this book (see Figure 1.5), with the added complication that forces acting on the boundary of the control volume must also be included in the analysis. The volume is given by $\Delta x\Delta y\Delta z$ and rates of in and out flows of energy are given by conductive fluxes ($\mathbf{q}$) times the area through which the flux enters times the time increment Δt. In addition to heat conduction, the flow of fluid in and out of the control volume also changes the energy content. Finally, pressure and shear stresses perform work on the system (PV work) and therefore change the energy content. The terms are as follows:

Accumulation:

$$\Delta x \Delta y \Delta z [(\rho \hat{E})_{t+\Delta t} - (\rho \hat{E})_t]$$

= flux in – out:

$$\begin{aligned} = {} & \Delta y \Delta z \Delta t [(q_x)_x - (q_x)_{x+\Delta x}] \\ & + \Delta x \Delta z \Delta t [(q_y)_y - (q_y)_{y+\Delta y}] \\ & + \Delta x \Delta y \Delta t [(q_z)_z - (q_z)_{z+\Delta z}] \end{aligned}$$

+ convection in – out:

$$\begin{aligned} & + \Delta y \Delta z \Delta t [(\rho \hat{E} v_x)_x - (\rho \hat{E} v_x)_{x+\Delta x}] \\ & + \Delta x \Delta z \Delta t [(\rho \hat{E} v_y)_y - (\rho \hat{E} v_y)_{y+\Delta y}] \\ & + \Delta x \Delta y \Delta t [(\rho \hat{E} v_z)_z - (\rho \hat{E} v_z)_{z+\Delta z}] \end{aligned}$$

+ work by pressure:

$$+ \Delta y \Delta z \Delta t \left[(P v_x)_x - (P v_x)_{x+\Delta x} \right]$$

+ work by shear on surface:

$$\begin{aligned} & + \Delta y \Delta z \Delta t \left[(\tau_{xx} v_x)_x - (\tau_{xx} v_x)_{x+\Delta x} \right] \\ & + \Delta x \Delta z \Delta t \left[(\tau_{yx} v_x)_y - (\tau_{yx} v_x)_{y+\Delta y} \right] \\ & + \Delta x \Delta y \Delta t \left[(\tau_{zx} v_x)_z - (\tau_{zx} v_x)_{z+\Delta z} \right] \\ & + \Delta y \Delta z \Delta t \left[(\tau_{xy} v_y)_x - (\tau_{xy} v_y)_{x+\Delta x} \right] \\ & + \Delta x \Delta z \Delta t \left[(\tau_{yy} v_y)_y - (\tau_{yy} v_y)_{y+\Delta y} \right] \\ & + \Delta x \Delta y \Delta t \left[(\tau_{zy} v_y)_z - (\tau_{zy} v_y)_{z+\Delta z} \right] \\ & + \Delta y \Delta z \Delta t \left[(\tau_{xz} v_z)_x - (\tau_{xz} v_z)_{x+\Delta x} \right] \\ & + \Delta x \Delta z \Delta t \left[(\tau_{yz} v_z)_y - (\tau_{yz} v_z)_{y+\Delta y} \right] \\ & + \Delta x \Delta y \Delta t \left[(\tau_{zz} v_z)_z - (\tau_{zz} v_z)_{z+\Delta z} \right] \end{aligned}$$

– work by body forces:

$$-\Delta x \Delta y \Delta z \Delta t \rho \left(F_{B,x} v_x + F_{B,y} v_y + F_{B,z} v_z \right).$$

Taking the differential limit and converting to vector notation for compactness (the next appendix includes a discussion of vector symbols and operators) yields

$$\frac{\partial (\rho \hat{E})}{\partial t} + \nabla \cdot (\rho \hat{E} \mathbf{v}) = -\nabla \cdot \mathbf{q} - \nabla \cdot (P \mathbf{v}) + \rho \mathbf{F_B} \cdot \mathbf{v} - \nabla \cdot (\boldsymbol{\tau} \cdot \mathbf{v}). \qquad (6.\text{A}12)$$

The following manipulations are performed to reduce the equation to a simpler form:

- define the enthalpy per unit mass: $\hat{H} = \hat{U} + P/\rho$,
- subtract the velocity times linear momentum equation (6.A9),
- substitute $d\hat{H} = \hat{C}_p dT + \left[\frac{1}{\rho} - T \left(\frac{\partial \frac{1}{\rho}}{\partial T} \right)_P \right] dP$,
- substitute Newton's constitutive equation for viscosity,
- substitute Fourier's constitutive equation for heat conduction.

The end result is the energy equation, which, under the further assumption of *constant physical properties*, becomes

$$\rho \hat{C}_p \frac{DT}{Dt} = k\nabla^2 T + \mu \Phi_v. \tag{6.A13}$$

The last term is the viscous dissipation, where $\Phi = \boldsymbol{\tau} \cdot \nabla \mathbf{v}$. With some algebraic manipulations, Eq. (6.A13) can be written in the following dimensionless form:

$$\frac{D\tilde{T}}{D\tilde{t}} = \frac{1}{\mathrm{Re\,Pr}} \nabla^2 \tilde{T} + \frac{\mathrm{Br}}{\mathrm{Re\,Pr}} \Phi_v. \tag{6.A14}$$

The following dimensionless group, known as the *Brinkman* number, is identified in the analysis in addition to Re and Pr:

$$\mathrm{Br} = \frac{\mu V_x^2}{k\Delta T} = \frac{\text{viscous dissipation}}{\text{heat conduction}}.$$

Species Mass

The derivation of the species mass equation follows that in Section 6.1. The resulting balance equation in vector notation is

$$\frac{DC_A}{Dt} = -\nabla \cdot \mathbf{J}_A + r_A. \tag{6.A15}$$

Applying Fick's constitutive equation and assuming a constant diffusivity leads to

$$\frac{DC_A}{Dt} = D_A \nabla^2 C_A + r_A. \tag{6.A16}$$

Scaling the equation in the absence of reaction leads to the result,

$$\frac{D\tilde{C}_A}{D\tilde{t}} = \frac{1}{\mathrm{Re\,Sc}} \nabla^2 \tilde{C}_A, \tag{6.A17}$$

where $\tilde{C}_A = C_A/\Delta C_A$

The presence of chemical reaction leads to additional dimensionless groups. If the reaction is first order or pseudo-first order and takes the form of $r_A = -k_{rxn} C_A$, for example, the scaled equation becomes

$$\frac{D\tilde{C}_A}{D\tilde{t}} = \frac{1}{\mathrm{Re\,Sc}} \nabla^2 \tilde{C}_A - \frac{\phi^2}{\mathrm{Re\,Sc}} \tilde{C}_A \tag{6.A18}$$

where the new dimensionless group is known as the Thiele Modulus; $\phi = \sqrt{k_{rxn}L^2/D_A}$ gives the relative rate of reaction to diffusion for a homogeneous, first-order reaction. Clearly when the Thiele modulus is large, reaction dominates, and vice versa. Finally, we note that in the presence of *surface reaction* the reaction rate does not appear in the balance equation, but rather, enters through the boundary conditions. For the case in which the surface reaction is pseudo-first order, making the boundary conditions dimensionless leads to the identification of the second *Damköhler number* as $Da^{II} = [(k_{rxn}^{surface}L)/D_A]$, where the rate of surface reaction is per unit area (not per unit volume as for homogeneous reactions).

APPENDIX B: VECTOR NOTATION

In Cartesian coordinates, the vector operators used in the previous appendix are

$$\nabla = \hat{\delta}_x \frac{\partial}{\partial x} + \hat{\delta}_y \frac{\partial}{\partial y} + \hat{\delta}_z \frac{\partial}{\partial z},$$

$$\nabla s = \hat{\delta}_x \frac{\partial s}{\partial x} + \hat{\delta}_y \frac{\partial s}{\partial y} + \hat{\delta}_z \frac{\partial s}{\partial z},$$

$$\nabla^2 s = \frac{\partial^2 s}{\partial x^2} + \frac{\partial^2 s}{\partial y^2} + \frac{\partial^2 s}{\partial z^2}, \tag{6.B1}$$

$$\nabla \cdot \mathbf{g} = \frac{\partial g_x}{\partial x} + \frac{\partial g_y}{\partial y} + \frac{\partial g_z}{\partial z},$$

$$\nabla \times \mathbf{g} = \hat{\delta}_x \left(\frac{\partial g_z}{\partial y} - \frac{\partial g_y}{\partial z} \right) + \hat{\delta}_y \left(\frac{\partial g_x}{\partial z} - \frac{\partial g_z}{\partial x} \right) + \hat{\delta}_z \left(\frac{\partial g_y}{\partial x} - \frac{\partial g_x}{\partial y} \right).$$

In curvilinear coordinates, the operator and the first few operations are as follows. Cylindrical:

$$\nabla = \hat{\delta}_r \frac{\partial}{\partial r} + \hat{\delta}_\theta \frac{1}{r} \frac{\partial}{\partial \theta} + \hat{\delta}_z \frac{\partial}{\partial z},$$

$$\nabla s = \hat{\delta}_r \frac{\partial s}{\partial r} + \hat{\delta}_\theta \frac{1}{r} \frac{\partial s}{\partial \theta} + \hat{\delta}_z \frac{\partial s}{\partial z},$$

$$\nabla^2 s = \frac{1}{r} \frac{\partial \left(r \frac{\partial s}{\partial r} \right)}{\partial r} + \frac{1}{r^2} \frac{\partial^2 s}{\partial \theta^2} + \frac{\partial^2 s}{\partial z^2}, \tag{6.B2}$$

$$\nabla \cdot \mathbf{g} = \frac{1}{r} \frac{\partial r g_r}{\partial r} + \frac{1}{r} \frac{\partial g_\theta}{\partial \theta} + \frac{\partial g_z}{\partial z},$$

$$\nabla \times \mathbf{g} = \hat{\delta}_r \left(\frac{1}{r} \frac{\partial g_z}{\partial \theta} - \frac{\partial g_\theta}{\partial z} \right) + \hat{\delta}_\theta \left(\frac{\partial g_r}{\partial z} - \frac{\partial g_z}{\partial r} \right) + \hat{\delta}_z \left(\frac{1}{r} \frac{\partial r g_\theta}{\partial r} - \frac{1}{r} \frac{\partial g_r}{\partial \theta} \right).$$

Spherical:

$$\nabla = \hat{\delta}_r \frac{\partial}{\partial r} + \hat{\delta}_\theta \frac{1}{r} \frac{\partial}{\partial \theta} + \hat{\delta}_\phi \frac{1}{r \sin \theta} \frac{\partial}{\partial \phi},$$

$$\nabla s = \hat{\delta}_r \frac{\partial s}{\partial r} + \hat{\delta}_\theta \frac{1}{r} \frac{\partial s}{\partial \theta} + \hat{\delta}_\phi \frac{1}{r \sin \theta} \frac{\partial s}{\partial \phi},$$

$$\nabla^2 s = \frac{1}{r^2} \frac{\partial \left(r^2 \frac{\partial s}{\partial r} \right)}{\partial r} + \frac{1}{r^2 \sin \theta} \frac{\partial \left(\sin \theta \frac{\partial s}{\partial \theta} \right)}{\partial \theta} + \frac{1}{r^2 \sin^2 \theta} \frac{\partial^2 s}{\partial \phi^2},$$

$$\nabla \cdot \mathbf{g} = \frac{1}{r^2} \frac{\partial (r^2 g_r)}{\partial r} + \frac{1}{r \sin \theta} \frac{\partial (\sin \theta g_\theta)}{\partial \theta} + \frac{1}{r \sin \theta} \frac{\partial g_\phi}{\partial \phi}, \tag{6.B3}$$

$$\nabla \times \mathbf{g} = \hat{\delta}_r \left[\frac{1}{r \sin \theta} \frac{\partial (\sin \theta g_\phi)}{\partial \theta} - \frac{1}{r \sin \theta} \frac{\partial g_\theta}{\partial \phi} \right],$$

$$+ \hat{\delta}_\theta \left(\frac{1}{r \sin \theta} \frac{\partial g_r}{\partial \phi} - \frac{1}{r} \frac{\partial r g_\phi}{\partial r} \right) + \hat{\delta}_z \left(\frac{1}{r} \frac{\partial r g_\theta}{\partial r} - \frac{1}{r} \frac{\partial g_r}{\partial \theta} \right).$$

With regards to the tensor operations, we give only the index form of the general result:

$$\nabla \cdot \tau = \sum_k \hat{\delta}_k \left(\sum_i \frac{\partial}{\partial x_i} \tau_{ik} \right). \tag{6.B4}$$

Additional information about vector and tensor notation can be found in the work by Aris (1990) given in the references.

7 Estimation of the Mass Transfer Coefficient and Interfacial Area in Fluid–Fluid Mass Contactors

In our study of mass transfer, we noted that both technically feasible analysis and design were complicated by the need to estimate two parameters in the rate expression: the mass transfer coefficient (K_m) and the interfacial area (a). In Chapter 6 theoretical and correlative methods for estimating mass transfer coefficients were discussed for those cases in which transfer of some species was limited by a well-defined boundary layer, such as at a solid or liquid surface. Correlations were identified relating the Sherwood number to the Reynolds and Schmidt numbers, with the functionality specific to the geometry and type of flow (i.e., laminar or turbulent). Complications arise in applying these developments when estimating K_m in liquid–liquid or gas–liquid mass contactors. We need estimates of bubble or drop size and some knowledge of the fluid motions in the vicinity of the bubbles or drops to calculate the Reynolds and Sherwood numbers. Furthermore, as shown in Section 6.5, resistances to mass transfer may occur in both phases, necessitating knowledge of the fluid motions inside the bubbles or drops. In this chapter we examine methods for estimating these quantities. This is an active research area requiring Level VI two-phase fluid motion modeling and experiment and is not understood nearly as well as its single-phase counterpart. For this reason, most handbooks and textbooks alike provide only empirical correlations of the product K_ma specific to particular process equipment and do not address the problem of interfacial area determination independently from the prediction of mass transfer coefficients. We attempt to simplify the existing work in this field so it can be used for analysis of existing contactors and, with some care, for mass contactor design.

We first consider estimation of bubble and drop size as this is necessary for estimation of area and K_m. A solid–fluid system is the easiest to analyze for particle size and provides a structure for the more complicated problem of liquid–liquid or liquid–gas system in which the bubble or drop is created in the mixing process. An example of such as a system was treated in Chapter 4 with the salt-tablet experiment. This example was simple as all the tablets were taken to be the same size. With any solid system a distribution of particle sizes can be obtained with some experimental effort before the solid particles are introduced into a mass contactor. This may be as simple as classification through sieving, by using different size screens, or as complex as using light-scattering experiments. Such data can be depicted graphically in a histogram or fit to a probability distribution function

(PDF) model such as the normalized log-normal distribution function shown in the following:

$$f(x) = \frac{1}{\sqrt{2\pi}\, x \ln(\sigma_{LN})} \exp\left\{ -\frac{1}{2}\left[\frac{\ln(x/d_{50})}{\ln(\sigma_{LN})} \right]^2 \right\}, \qquad (7.0.1)$$

where $f(x)$ describes the fraction or probability of all particles of a particular size, x. In this case, d is a measure of the diameter, d_{50} is the statistical median of the diameter, and σ_{LN} is the log-normal standard deviation. The total area can be calculated if the shape of every particle is defined. This is easy if one has spherical particles, and one can use the shape factor as described for the salt-tablet experiment in Chapter 3 for shapes other than spherical.

The Sauter mean diameter d_{32} defined in the following equations can be used to estimate interfacial areas in multiphase systems:

$$d_{32} = \sum_{i=1}^{N_t} d_i^3 \bigg/ \sum_{i=1}^{N_t} d_i^2,$$

$$d_{32} = 6\, V_d / a_d.$$

In the above, V_d is the particle volume, a_d is the particle area, and d_{32} is related to the log-normal distribution as follows:

$$d_{32} = d_{50} \exp\left[5/2 \ln (\sigma_{LN})^2\right].$$

d_{32} can be obtained for solid particles for any given distribution. More information about characterizing particle size distributions and methods of measurement can be found in the monograph by Allen (1990).

It is much more difficult to estimate d_{32} in liquid–liquid or liquid–gas systems because the drops and bubbles are created within the mass contactor. The drop and bubble sizes in either a tank-type or tubular contactor depend on the level of turbulence in the device. In a tank-type vessel the turbulence is created by mechanical agitation or in some cases by the power input of one of the phases. In large aeration basins the sparged gas creates small bubbles in the rising plume of gas and liquid over the orifices delivering the gas. This plume of bubbles mixes the basin by circulating water in the vessel. In a tubular contactor the pump supplies the energy to create turbulence in a pipe, with a diameter selected to have high fluid velocities. Turbulence in tubular systems can also be induced by putting internals in the pipe; such devices are called static mixers, such as that shown in Figure 7.6 in Section 7.3.

In semibatch contactors, as discussed in Chapter 4, we need to consider only solid–fluid or gas–liquid systems. In a solid–fluid semibatch contactor, the particle size can be determined as previously discussed because the solids have to be in the contactor before the fluid is introduced. In a gas–liquid semibatch contactor, the gas is sparged into the contactor and can be the sole means by which the contactor is mixed. The bubble size at the point of introduction must be determined, and the bubble size in the contactor, almost always smaller than the initial bubble size, must also be determined. Figure 7.1 shows the initial bubble size and the bubble size in the

Figure 7.1. (top) Air–water in a 12 m × 12 m × 6 m tank showing initial bubble size (Otera, 1983) and (bottom) bubble plumes. (Photo from Otero Z, Tilton JN, Russell TWF. Some observations on flow patterns. Int. J. Multiphase Flow 1985; 11: 584. Copyright Elsevier. Reproduced with permission.)

plume. When the mass contactor is equipped with a motor and one or more impellers, it is more complicated to estimate interfacial area.

In continuous tank-type or tubular liquid–liquid or liquid–gas mass contactors both phases are continuously introduced into the contactor. Drop and bubble sizes have to be experimentally determined or estimated within the contactor as it operates. The initial-size bubble or drop is formed at the point where the liquid or gas is introduced. This bubble or drop size may be larger or smaller than the bubble or drop size that can be attained in the turbulent fluid motions of the contactor. The sizes of bubbles formed at the orifice in gas spargers are discussed by Bhavaraju et al. (1978) and by Tilton and Russell (1982).

In all two-phase contactors there are "active" regions in which the turbulent level is sufficient to break up drops or bubbles and there are neutral regions in which the drops or bubbles are only transported and coalescence may occur. For a drop or bubble to break up, it must be in an active region of the contactor for a sufficient period of time to allow breakup to occur. To properly determine drop or bubble size as a function of the complex fluid motions in any contactor, population balance models would need to be developed that describe changes in drop or bubble size because of breakage or coalescence. It is easy to derive such balances, but the determination of constitutive equations needed to describe breakage and coalescence is an active research area. In addition we require more detailed descriptions of fluid motion within the contactor to derive pragmatically useful balances in which the active and neutral regions are delineated.

Maximum drop or bubble sizes can be estimated based on a force balance analysis, and in the next section a method for so doing is discussed. This will be useful for both technically feasible analysis and design problems.

7.1   Estimation of Bubble and Drop Size

Work originally developed by Kolmogorov and Hinze in the 1940s and 1950s utilized a force balance on a bubble or drop to determine the maximum stable bubble size. The maximum bubble size can then be related to a mean bubble size, which in turn can be used to estimate the interfacial area. We first discuss the assumptions in developing an expression for estimating drop and bubble size and then show how the resulting model compares with experimental data.

The approach developed by Kolmogorov (1949) and Hinze (1955) postulated that a critical drop size could be determined by a balance of forces on a drop. In a turbulent field, the local pressure fluctuations outside the drop are mostly responsible for its deformation. The surface-tension force opposes this deformation. A drop in which these two forces balance exactly is the maximum stable size. This maximum stable size, d_{max}, is then characterized by a dimensionless quantity called the *Weber* number:

$$\text{We}_{crit} = \frac{\tau}{\sigma/d_{max}}. \tag{7.1.1}$$

In this equation, τ is a stress induced by pressure fluctuations and σ is the interfacial tension between the continuous and dispersed phases. This is to be distinguished from the surface tension of a fluid, which is usually measured against air. Thus, the Weber number is a dimensionless group that represents a ratio between an inertial force (in this case resulting from a turbulent flow field) and the force that is due to interfacial tension.

Later work by Levich (1962) proposed a slightly different force balance in which the internal pressure of the bubble was balanced against the capillary pressure of the deformed bubble. The result of this analysis is a modified Weber number:

$$\text{We}'_{crit} = \frac{\tau}{\sigma/d_{max}} \left(\frac{\rho^{II}}{\rho^{I}} \right)^{1/3}. \tag{7.1.2}$$

The dispersed- and continuous-phase densities are represented by ρ^{II} and ρ^{I}, respectively, following the convention we established in Chapter 4. Throughout our discussion, we distinguish between the properties of these phases by using superscripts II and I. Note that, for gas–liquid systems, the densities differ by roughly three orders of magnitude, and thus these two critical Weber number will differ by about one order of magnitude. We will examine the consequences of this result later.

In an isotropic (homogenous) field, the pressure fluctuation stress is

$$\tau = \rho^{I}\overline{v^2} = \rho^{I}\left[2(\varepsilon\, d_{max})^{2/3}\right]. \tag{7.1.3}$$

Here, ε refers to the specific power input, which has units of power per mass. Making these substitutions into the originally proposed Weber numbers, followed by some rearrangement to obtain dimensionless groups, yields the following equations:

$$d_{max} = \left(\frac{We_{crit}}{2}\right)^{3/5}\left(\frac{\sigma}{\rho^{I}}\right)^{0.6}\varepsilon^{-0.4}, \tag{7.1.4}$$

$$d_{max} = \left(\frac{We'_{crit}}{2}\right)^{3/5}\left(\frac{\sigma}{\rho^{I}}\right)^{0.6}\left(\frac{\rho^{I}}{\rho^{II}}\right)^{0.2}\varepsilon^{-0.4} \tag{7.1.5}$$

We can see from this analysis that both theories predict the same dependence on the energy per unit mass; ε and d_{max} will vary proportionally with $\varepsilon^{-0.4}$.

Some ρ and σ values are presented in Table 7.1.

It is sometimes more convenient to express the power per mass in terms of power per volume that can be readily found by dividing ε by the density of the continuous phase, ρ^{I}.

Equations (7.1.4) and (7.1.5) require that the phase that forms bubbles or drops be in the "active" volume of the contactor for a long enough period of time that breakup will occur. This time has not been investigated extensively, but is probably within an order of magnitude of 1 s.

One must also be able to estimate the power per mass, ε, correctly in any analysis of existing equipment or to develop a technically feasible design. The power input

Table 7.1. Density and interfacial tension against H_2O of some liquids at 25 °C (adapted from Murphy NF, Lastovica JE, Fallis JG. Correction of interfacial tension of two-phase three-component systems. Ind. Eng. Chem. 1957; 49: 1035)

Compound name	Formula	T (°C)	ρ (kg/m³)	σ (m N/m)
Benzene	C_6H_6	25	869.3	28.7
Carbon Tetrachloride	CCl_4	25	1576.3	26.5
n-Hexane	C_6H_{14}	25	658.0	50.0
Toluene	C_7H_8	25	859.8	27.2
(1,1,2)-TCE	$C_2H_3Cl_3$	25	1422.7	27.1
1-Butanol	$C_4H_{12}O$	27	805.6	1.7
Cyclohexanol	$C_6H_{12}O$	25	936.2	3.1
Furfural	$C_5H_4O_2$	25	1150.8	4.3
Methyl ethyl ketone	C_4H_8O	27	797.6	2.4
Nitromethane	CH_3NO_2	25	1117.6	8.8

to any contactor can be determined quite readily for most situations, but the mass or volume over which it is effective is not so easily determined.

Our assumption that the control volume is well mixed will still be valid, but we need to expand our understanding of the fluid motions in mass contacting equipment by identifying three regions:

- an introduction region where bubble and drop sizes are set by the method of introduction,
- a neutral region where bubbles and drops are merely moved around, and
- an "active" region where bubbles and drops are created as previously described.

Figure 7.2 shows a conceptual sketch for these regions for tank-type and tubular systems.

The impact of these different regions for both tank-type and tubular mass contactors is discussed in the following sections.

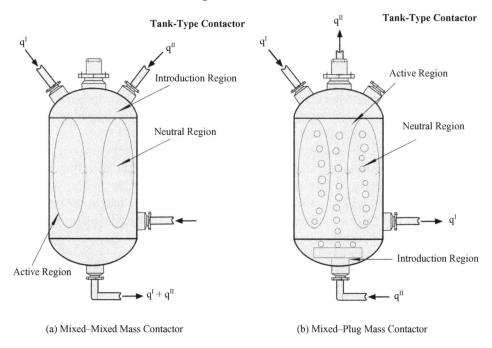

(a) Mixed–Mixed Mass Contactor   (b) Mixed–Plug Mass Contactor

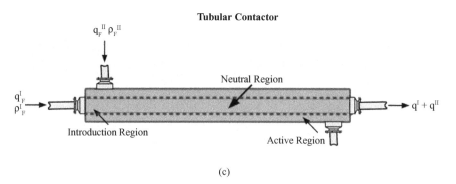

(c)

Figure 7.2. Introduction, neutral, and active regions of mass contacting equipment.

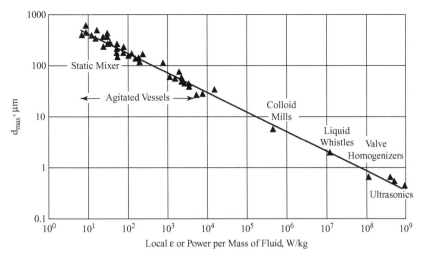

Figure 7.3. d_{max} dependence on local power (ε). (Paul et al., 2004. Reproduced with permission.)

Figure 7.3, presented originally by Davies (1987), shows how d_{max} varies with the "local" power input, (in watts per kilogram) over a wide range. The slope of the line in Figure 7.3 shows the $\varepsilon^{-0.4}$ dependence predicted by Eqs. (7.1.4) and (7.1.5). The "local" power input volume is that designated by the "active" region in Figure 7.2 and is calculated for agitated vessels at half of the swept out volume of the impeller [Eq. (7.2.1)].

7.2 Tank-Type Mass Contactors

To estimate interfacial areas in tank-type mass contactors we need to examine two kinds of fluid motions: mixed–mixed and mixed–plug. The estimations of area under these classifications are applicable to the batch, semibatch, and continuous mass contactors requiring the Level IV analysis described in Chapter 4.

7.2.1 Mixed–Mixed Interfacial Area Estimation

Mixed–mixed systems are either liquid–liquid or solid–fluid. Liquid–gas systems that are mechanically agitated are sometimes treated as mixed–mixed to estimate area. The particle size distribution for solid–fluid systems can be determined before the solid is placed into the contactor.

Mechanically mixed liquid–liquid systems are very common and a subject of considerable research today. The mechanical mixing produces very complex flow patterns in any vessel and a wide range of energy dissipation rates in the "active" region of the vessel. Although it is a matter of some debate, the "active" region has been defined as one half of the swept-out volume of the impeller:

$$V_{active} = \frac{\pi L_p^2}{4} h', \qquad (7.2.1)$$

where L_p is the impeller diameter and h' is the height of the impeller blade (Davies, 1987).

Even if this definition is a reasonable representation of reality there remains the issue of how long any drop needs to remain in the active region to be reduced in size. Despite these uncertainties we can get estimates of d_{32} that will allow technically feasible analysis and design to be completed. Most drops will travel through the active region many times, further complicating any analysis.

Equation (7.1.4) or (7.1.5) can be used to obtain d_{max} if one can determine ε. d_{32} is proportional to d_{max}, and if d_{32} data are obtained experimentally then one can develop a correlation for d_{32}. For dilute low-viscosity liquid–liquid systems the following equation for d_{32} has been obtained by fitting data for 14 different liquid–liquid pairs in tanks agitated with a Rushton turbine impeller (Paul et al., 2004):

$$\frac{d_{32}}{L_p} = 0.053\,(We_I)^{-3/5}. \tag{7.2.2}$$

The Weber number in this expression is based on the impeller design and its rotational speed:

$$We_I = \frac{\rho^I N^2 L_p^3}{\sigma}, \tag{7.2.3}$$

where N is the impeller speed (in inverse seconds), L_p is the impeller diameter (in meters), and σ is the surface tension (in newtons per meter).

As pointed out in Subsection 4.3.3.2, the "optimal" drop size can be determined only by a careful analysis of the liquid–liquid separator design.

EXAMPLE 7.1. A 1000-L liquid–liquid mass contactor is equipped with a Rushton turbine of diameter $L_p = 0.5$ m (both the tank height and diameter are 1.08 m). The dispersed and continuous phases both have a density of 1200 kg/m^3. The interfacial tension σ is 3.0×10^{-3} N/m. The mixer is a variable-speed motor and drive that is capable of impeller speeds between 40 and 70 rpm (0.67–1.17 s^{-1}). What range of drop sizes can be produced?

SOLUTION. The impeller Weber number is

$$We_I = \frac{(1200\ \text{kg/m}^3)(N^2\ \text{s}^{-2})(0.5^3\ \text{m}^3)}{3 \times 10^{-3}\ \text{kg m/s}^2\ \text{m}}$$

$$= 50\,000\ N^2,$$

$$\frac{d_{32}}{0.5} = 0.053\,(50\,000\ N^2)^{-3/5},$$

$$(N = 40\ \text{rpm}) = 0.67\ \text{s}^{-1},$$

$$d_{32} - 65\ \mu\text{m},$$

$$(N = 70\ \text{rpm}) = 1.17\ \text{s}^{-1},$$

$$d_{32} = 21\ \mu\text{m}.$$

With d_B and V^{II}, the volume of the dispersed phase is known, the interfacial area can be calculated from Eq. (4.3.39):

$$V^{II} = \frac{1}{6}\,a d_{32}$$

There are corrections for other impeller types and for viscous systems presented in the *Handbook of Industrial Mixing* (Paul et al., 2004) as is an extensive table listing over 20 correlations for d_{32} in mixed–mixed liquid–liquid stirred vessels. They must be used with extreme caution because a variety of experimental equipment and measuring techniques were used and several studies involve high fractions of the dispersed phase.

The problem of area estimation becomes much more complex if there is a system that has a high percentage of dispersed phase, if coalescence is suspected, or if surfactants are present.

7.2.2 Mixed–Mixed K_m Estimation

There are few, if any, reasonable procedures for determining K_m in mixed–mixed liquid–liquid mass contactors. The fundamental problem is that we do not have good knowledge of the structure of the boundary layers outside and inside the drop. The mass transfer rates in liquid–liquid contactors will be of the form developed in Chapter 6, Sh(Re, Sc). The impeller Reynolds number might be used in such correlations, but this has not been investigated for the reasons subsequently outlined. For the same reasons the penetration and surface-renewal theories (Section 6.4) have been difficult to apply.

Design and analysis of such contactors must be accompanied by experimental determination of the rate of mass transfer. An analysis of an experiment with the model equations from Chapter 4 along with measurements of concentration as function, of time produces a value for the product $K_m a$. The estimation of area follows from the procedure previously described, which then enables K_m to be estimated.

However, as shown in Chapter 4, such experiments are very difficult to perform because phase equilibrium in liquid–liquid systems is often achieved within 5–100 s. Consequently, most mixed–mixed liquid–liquid mass contactors and their associated separators closely approximate an equilibrium stage and the analysis described in Subsections 4.3.3.1 and 4.3.3.2 can be applied. This Chapter 4 analysis is critical to understanding the logic behind equilibrium stage operations prevalent in the chemical and petrochemical process industries.

Another conclusion that one might draw from the rapid approach to equilibrium is that a tank-type mass contactor may not be the best design. It might well be better to use a static mixer in a tubular cocurrent contactor that can also act as a separator if pipe diameter is varied downstream from the mixer.

7.2.3 Mixed–Plug Area Estimation

This area estimation applies to a semibatch or continuous contactor in which either a solid phase is mixed with a fluid phase in plug flow or a liquid phase is mixed with a gas phase in plug flow. Such contactors also act as separators. The fluid phase rises through the contactor because of the density difference and exits at the top. The solid phase can be treated as previously discussed, so area determination is straightforward.

With a mixed liquid and plug-flow gas the two-phase fluid mechanics required for estimating area is complex. One must be able to predict

- bubble size at the orifice,
- bubble size in the vessel,
- bubble rise velocity, and
- the liquid velocity in the contactor produced by the rising plume of gas bubbles.

Many of these quantities are equipment specific and depend on vessel dimensions and the gas sparger design. As shown in Subsection 4.3.3.3, the gas-phase plug-flow analysis is the same for both the semibatch and continuous contactors. Even when the contactor is mechanically agitated the gas passes from the sparger to the liquid surface in plug flow. It is difficult to entrain gas bubbles and circulate them throughout the contactor, even though some equipment manufacturers claim they can do so. The mixed–plug model equation predictions [Eqs. (4.3.44) and (4.3.47)] often give the best fit to experimental data.

Bubble sizes at the orifice for the gas flow rates used in laboratory-, pilot-, and commercial-scale mass contactors are difficult to measure accurately because of jet formation at the orifice and liquid turbulence at the point where the bubble is formed. Typical orifice-formed bubbles are shown in Figure 7.1.

There are numerous studies of extremely low gas rates where single bubbles are formed. The following equation is typical and is often recommended to calculate the orifice bubble diameter, d_{B0}:

$$d_{B0} = \left[\frac{6\sigma d_0}{g(\rho^I - \rho^{II})} \right]^{1/3}. \tag{7.2.4}$$

There is good agreement between this prediction and experiment but only at gas rates several orders of magnitude below that employed in any practical contactor. A summary of research carried on bubble formation is described by Bhavaraju et al. (1978). They developed the following equation based on photographic analysis of air–water systems and air–water–carbopol solutions (all quantities are expressed in SI units):

$$d_{B0} = \frac{3.15\, d_0^{0.05} (q^{II})^{0.32}\, \mu_L^{0.1}}{g^{0.21} (\rho^I)^{0.1}}. \tag{7.2.5}$$

For air–water contactors at 20 °C this reduces to

$$d_{B0} = 0.39\, (q^{II})^{0.32}. \tag{7.2.6}$$

By analyzing photographic data from a different system Otero (1983) developed the following prediction for air sparged into water:

$$d_{B0} = 0.66\, (q^{II})^{0.32}\, d_0^{0.12}. \tag{7.2.7}$$

There is no clear evidence that one of the preceding expressions is better than the other. Both give useful limits on the orifice bubble diameter that one might encounter in any technically feasible analysis or design problem. For air–water flows that are commonly encountered in commercial equipment bubble sizes can range from 0.035 to 0.15 m.

If the mass contactor under consideration has a liquid depth less than about a meter the orifice-formed bubble may persist throughout the liquid depth with little breakup. Above a liquid depth of a meter and a gas rate sufficient to supply enough power to the system, a liquid circulation is established. The liquid is entrained by the stream of rising gas bubbles and flows upward until it reaches the top of the liquid, where it flows radially outward until it encounters a vessel wall or the radially flowing liquid from another plume [Figure 4.10(a)]. The liquid then flows down and is recirculated. A single bubbling orifice can pump substantial amounts of water and is the most common means by which large aeration basins are mixed and supplied with oxygen. The turbulent flowing liquid breaks up the bubble stream into small bubbles, which can reach a limiting value of about 0.0045 m for water–air systems with sufficient power input from the gas stream.

An extensive experimental study of such flows was carried out by Otero (1983) and reported with design procedures for commercial scale contactors by Otero Keil and Russell (1987) and Tilton and Russell (1982). These references should be consulted to obtain more details on the gas–liquid fluid mechanics and their impact on design and operation of mixed–plug mass contactors.

The power input from the gas flow through the orifice that creates the bubbles and drives the liquid circulation is

$$\text{power input} = q_{AVG}^{II} g \rho^I h_L. \tag{7.2.8}$$

This is the power supplied by the gas in the two-phase gas–liquid plume, h_L is the clear-liquid height, g is the acceleration that is due to gravity, and q_{AVG}^{II} is the gas flow rate, calculated as follows:

$$q_{AVG}^{II} = q_{OR}^{II} p_s \left(\frac{\ln p_o / p_s}{p_o - p_s} \right),$$

p_o is the pressure at the orifice, p_s is the pressure at the liquid surface and q_{OR}^{II} is the gas flow rate evaluated at p_o.

The plume mass must be defined to determine ε, which is the power input from Eq. (7.2.8) divided by the plume mass. Above each orifice there is a circulation element that consists of a gas–liquid plume surrounded by a liquid annulus (Figure 7.4). Continuity requires that the mass of liquid flowing up in the plume of diameter d must equal the mass of liquid flowing down in the annulus of diameter D:

$$d^2 = 0.5 D^2. \tag{7.2.9}$$

Gas-sparged mass contactors can have a single plume or multiple plumes, depending on the size of orifice-formed bubbles and the orifice spacing in the sparger. If the distance between the orifices is about the same size as the orifice-formed bubble, d_{B0} given by Eq. (7.2.5), then there will likely be a single plume. With orifice spacing much greater than d_{B0}, multiple plumes will exist (Figure 7.1).

The mass of the plume required for calculating the power input per unit mass, ε, is, under these assumptions, given by

$$\rho^I h_L \pi d^2 / 4.$$

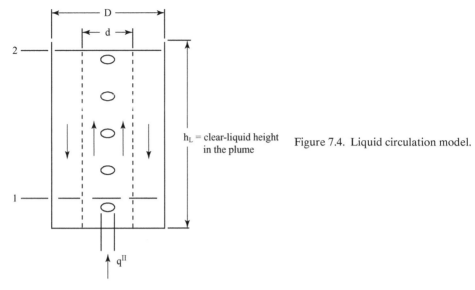

h_L = clear-liquid height in the plume

Figure 7.4. Liquid circulation model.

Then, from Eq. (7.2.8) we obtain

$$\varepsilon = \frac{4q_{AVG}^{II}g}{\pi d^2}.$$

Bhavaraju et al. (1978) and Tilton and Russell (1982) examined laboratory-scale gas–liquid data in sparged tanks to obtain the following expression based on Eqs. (7.1.4) and (7.2.9). We designate bubble diameters as d_B:

$$d_B = 0.53 \left(\frac{\sigma}{\rho^I}\right)^{0.6} \left(\frac{\mu^I}{\mu^{II}}\right)^{0.1} \varepsilon^{-0.4}. \qquad (7.2.10)$$

The viscosity ratio term in Eq. (7.2.10) is shown as the continuous-phase liquid viscosity divided by the dispersed-phase viscosity. For air–water systems Eq. (7.2.10) reduces to

$$d_B = (0.26 \times 10^{-2})\varepsilon^{-0.4}. \qquad (7.2.11)$$

The interfacial area per unit volume of the two-phase plume in mixed–plug vessels, a_v, can be estimated if one knows the percentage volume of the mass contactor occupied by the gas (often referred to as the gas holdup), $\phi = V^{II}/V^I$, and the bubble diameter in the plume d_B:

$$a_v = \frac{6\phi}{d_B}. \qquad (7.2.12)$$

The holdup can be estimated by use of the volumetric gas flow q^{II} to obtain the liquid velocity in the plume v^I and the bubble diameter d_B to obtain the bubble rise velocity u_r:

$$v^I = 0.79d^{-2/3}(g\,q^{II})^{1/3}\,h_L^{1/3}, \qquad (7.2.13)$$

$$u_r = \left(\frac{2\sigma}{\rho^I d_B} + \frac{g d_B}{2}\right)^{1/2}. \qquad (7.2.14)$$

These simplified forms of the equations for v^I derived by Tilton and Russell (1982) should provide a reasonable estimate for the liquid velocity for turbulent liquid flow in the plume, the situation most commonly encountered. The expression for u_r assumes a bubble Reynolds number much greater than 1.

The holdup ϕ can be estimated from an overall mass balance on the gas phase, assuming negligible mass transfer from the bubble in the plume because their time in the liquid phase is small:

$$\phi = \frac{q^{II}}{\frac{\pi D^2}{4}(v^I + u_r)} \left[\frac{p_0}{p_s + \rho^I g \left(\frac{h_L}{2} \right)} \right]. \tag{7.2.15}$$

The term in brackets accounts for the expansion of bubble volume. It assumes an average holdup at half the liquid depth. It is frequently small enough to be neglected. p_0 is the pressure at the orifice, and p_s is the pressure at the liquid surface.

7.2.4 Mixed–Plug K_m Estimation

The penetration theory developed in Chapter 6 can be employed to obtain values for K_m by use of Eqs. (6.4.13). With an estimate of bubble size and an estimate of bubble rise velocity from Subsection 7.2.3, an estimate of K_m can be obtained from Eqs. (6.4.13), where $V_{max} = u_r$ and $L = d_B$:

$$k_m = \sqrt{\frac{4}{\pi} D \frac{u_r}{d_B}}. \tag{7.2.16}$$

Equation (7.2.16) estimates the liquid side resistance k_m. In a gas–liquid system we can generally ignore the gas-phase resistance so that $K_m \approx k_m$.

k_m was calculated by using Equation (7.2.16) by Tilton and Russell (1982) who examined the experimental results of Miller (1974), Towell et al. (1965), and Alvarez-Cuenca and Nerenberg (1981).

A comparison of the model predictions with experiment is shown in Figure 7.5, which uses the preceding expression for k_m and the equations in Subsection 7.2.3.

EXAMPLE 7.2. The data obtained in the tank described in Example 4.2 need to be further analyzed so that the effect of air flow and tank geometry can be predicted for any tank-type contactor. The Level III analysis and experiment produced two values for $k_m a/V^I$, where V^I is the total volume of the mass contactor. We now need to estimate a so we can predict performance for other mass contactors.

SOLUTION. The power input to the plume, Eq. (7.2.8), is

$$\text{Power} = (3.9 \times 10^{-3} \text{ m}^3/\text{s})(9.8 \text{ m/s}^2)(1000 \text{ kg/m}^3)(6 \text{ m}) = 230 \text{ W}$$

The plume diameter and plume volume, Eq. (7.2.9), are

$$d^2 = 0.5 \, (1.4)^2,$$
$$d = 0.98 \text{ m}.$$

The plume liquid volume is

$$V = \frac{\pi (0.98)^2}{4} \times 6 = 4.5 \text{ m}^3.$$

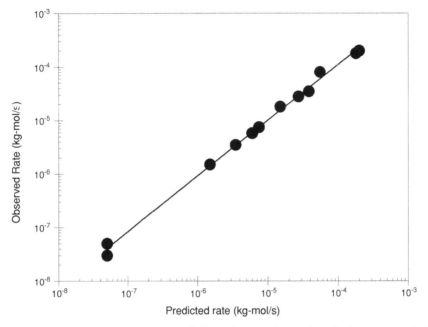

Figure 7.5. Comparison of model predictions with experimental results for mass transfer in sparged vessels (adapted from Tilton, 1981). Each • represents several data points.

The power per unit mass ε is

$$\varepsilon = \frac{230 \text{ W}}{(4.5 \text{ m}^3)(1000 \text{ kg/m}^3)} = 0.05 \frac{\text{W}}{\text{kg}}.$$

Because this is an air–water system the bubble diameter can be estimated from Eq. (7.2.11):

$$d_B = 0.26 \times 10^{-2} (0.05)^{-0.4}$$
$$= 0.0087 \text{ m } (0.34 \text{ in.}).$$

The liquid velocity in the plume is estimated with Eq. (7.2.13):

$$v^l = 0.79 (0.98)^{-2/3} (9.8 \times 3.9 \times 10^{-3})^{1/3} 6^{1/3}$$
$$= 0.47 \text{ m/s}.$$

The bubble rise velocity for air–water, Eq. (7.2.14), is

$$u_r = \left[\frac{1.45 \times 10^{-4}}{0.0087} + 4.9 (0.0087) \right]^{1/2} = 0.23 \frac{\text{m}}{\text{s}}.$$

The holdup, Eq. (7.2.15), is

$$\phi = \frac{3.9 \times 10^{-3}}{\dfrac{\pi (0.98)^2}{4} (0.47 + 0.23)} \left[\frac{1 \times 10^5 + 1000 (9.8) 6}{1 \times 10^5 + 1000 (9.8) 3} \right]$$
$$= 0.008.$$

The interfacial area per unit volume of liquid in the plume a_v is

$$a_v = \frac{6\Phi}{d_B} = \frac{6 \times 0.008}{0.0087} = 5.51 \text{ m}^{-1}.$$

For the tank, the total interfacial area is that of bubbles in the plume because there are no bubbles in the liquid that flow down in the annular region surrounding the plume:

$$a = 5.51 \times 4.5 = 24.8 \text{ m}^2.$$

Our experiment produced

$$\frac{K_m a}{V^I} = \frac{k_m a}{V^I} = 0.0007,$$

because

$$V^I = \frac{\pi}{4} (1.4)^2 \times 6$$

$$= 9.28 \text{ m}^3,$$

$$k_m = \frac{0.0007 \times 9.28}{24.8}$$

$$= 2.6 \times 10^{-4} \text{ m/s}.$$

Predicting k_m from Eq. (7.2.16) yields

$$k_m = \sqrt{\frac{4}{\pi} \frac{D}{d_B} u_r}.$$

For air–water this is

$$k_m = 4.79 \times 10^{-5} \bigg/ \left(\frac{u_r}{d_B}\right)^{1/2}$$

$$= 4.79 \times 10^{-5} \left(\frac{0.23}{0.0087}\right)^{1/2}$$

$$= 2.46 \times 10^{-4} \text{ m/s}.$$

This is rather good agreement with the experiment.

If we increase our gas flow rate, q^{II}, to 16.9×10^{-3} m^3/s, the power per unit now becomes 0.22 W/kg and the bubble size becomes 0.0047 m. The superficial liquid velocity is now 0.78 m/s, and the bubble rise velocity is 0.23 m/s. Holdup is increased to 0.027, giving an interfacial area per unit volume of 34.5 m^{-1} in the plume and a total area of 155 m^2. For the same K_m value, Eq. (4.3.15) shows that we can increase the oxygen concentration at 10 minutes from 4.8×10^{-3} kg/m^3 to 10.8×10^{-3} kg/m^3:

$$C_A^I = 11.5 + 9.8 e^{\frac{-K_m a t}{V^I}},$$

$$\frac{k_m a t}{V^I} = \frac{2.6 \times 10^{-4} \text{ (m/s)} \times 155 \text{ (m}^2) \times 10 \text{ (min)} \times 60 \text{ (s/min)} = 2.6}{9.28 \text{ (m}^3)},$$

$$C_A^I = 11.5 - 0.7 = 10.8 \times 10^{-3} \text{ kg/m}^3$$

7.3   Tubular Contactors

To estimate interfacial areas in continuously operating tubular mass contactors we need to consider cocurrent and countercurrent mass contactors separately. The required fluid motions in either type of tubular contactor set serious limits on the flow rates of the fluid phases and complicates the estimation of area. The details of fluid motions as functions of flow rates of the phases, fluid properties, and pipe sizes are beyond the scope of what we can cover in this book. Our aim therefore is to suggest approaches that may motivate the reader to consider tubular mass contactors and examine the literature on two-phase flow. The text by Govier and Aziz (1972) is well organized and contains material useful for technically feasible designs.

As noted in Chapter 4, cocurrent direct-contact tubular mass contactors can be either horizontal or vertical devices. The interfacial area is generated by the turbulent flow of the continuous phase and can be related to the pressure drop in the pipe.

Countercurrent tubular mass contactors are all vertical devices and are commonly referred to as towers or columns. They serve as both mass contactors, and separators, and almost all are gas–liquid systems. The interfacial area is generated by the countercurrent flow over column internals, either packing or plates.

7.3.1 Cocurrent Area Estimation

For either horizontal or vertical contactors the bubble or drop size depends on the power input to the contactor as shown in Eqs. (7.1.4) and (7.1.5). For cocurrent tubular mass contactors with no internals, the power input is the pressure drop in the pipe times the volumetric flow rate. We can express ε as follows:

$$\varepsilon = \Delta P q^{I}/[L_{pipe}(\pi D^{2}/4)\rho^{I}].$$

We can express ε in terms of a friction factor f:

$$\varepsilon = 2 f (v^{I})^{3}/d_{pipe}. \tag{7.3.1}$$

For low concentrations of the dispersed phase, the turbulent field will be similar to that of a pure liquid. Under these conditions, the Blasius correlation for the friction factor may be used as a first approximation (see Example 6.5):

$$f = 0.079 Re^{-0.25}. \tag{7.3.2}$$

Making these substitutions and again rearranging terms, one can write Eqs. (7.1.4) and (7.1.5) in a form with dimensionless groups:

$$\frac{d_{max}}{d_{pipe}} = 1.38 \, We_{crit}^{3/5} Re^{-1/2} Ca^{-3/5} \tag{7.3.3}$$

$$\left(\frac{d_{max}}{d_{pipe}}\right) = 1.38 \, We_{crit}'^{3/5} Re^{-1/2} Ca^{-3/5} \left(\frac{\rho^{I}}{\rho^{II}}\right)^{1/5}. \tag{7.3.4}$$

The latter of these expressions comes from Levich's proposed Weber number, whereas the former is derived from the Kolmogorov–Hinze theory. Aside from the critical Weber number, these equations contain two other nontrivial dimensionless groups: the Reynolds and capillary numbers. For dilute flows, these can be defined in terms of the properties of the continuous phase:

$$\text{Re} = \frac{\rho^I v^I d_{pipe}}{\mu^I}, \tag{7.3.5}$$

$$\text{Ca} = \frac{v^I \mu^I}{\sigma}. \tag{7.3.6}$$

The Reynolds number Re represents a ratio of inertial and viscous forces, and the capillary number Ca relates viscous forces to surface tension.

From inspection of the involved groups, it is clear that (7.3.3) and (7.3.4) will predict similar behavior for suspensions when the two phases are of comparable density. However, the similarity between these theories will not hold for liquid–gas flows, unless the critical Weber number is fundamentally different in these cases. The data and analysis that show that the Levich expression fits all the available data for both liquid–liquid and gas–liquid systems is presented in the appendix. d_{32} and d_{max} have been shown by Hesketh et al. (1986) to be related as follows:

$$d_{32} = 0.6 d_{max}.$$

The equation needed to estimate interfacial areas in cocurrent tubular contactors is therefore

$$\frac{d_{32}}{d_{pipe}} = 0.828 \times \text{Re}^{-1/2} \text{Ca}^{-3/5} \left(\frac{\rho^I}{\rho^{II}}\right)^{1/5}. \tag{7.3.7}$$

We may extract an estimate of interfacial area in a tubular mass contactor by assuming that the drop or bubble diameter is represented by d_{32}:

$$a = \frac{6 V^{II}}{d_{32}}. \tag{7.3.8}$$

The velocity of the dispersed phase may or may not be the same as that of the continuous phase. For liquid–liquid dispersions, pressure–volume effects are negligible and drops are likely to flow at the same speed as the bulk fluid. In the case of compressible gases, though, there may be considerable slippage between the phases.

Equation (7.3.8) has been derived for a restrictive set of conditions. The most relevant of these are as follows:

• Low dispersed-phase concentration. This requires that gas–liquid flows be in the bubble region. The liquid velocity to achieve this pattern needs to be at least 3 m/s.
• Continuous-phase turbulent flow that is well-characterized by the Blasius equation.
• Equilibrium bubble/drop distribution is achieved quickly relative to dispersed-phase retention time.

One can employ a static mixer with tubular cocurrent mass contactors. Figure 7.6 shows a typical static mixer. Such devices are useful for generating both bubbles and drops and one should consult manufacturers or the *Handbook of Industrial Mixing* (Paul et al., 2004) for design procedures and estimates of bubble size.

Figure 7.6. A KMSR model Kenics® static mixer. (Photo courtesy of Chemineer, Inc.)

7.3.2 Cocurrent K_m Estimation

Equation (7.2.16) can be used to get an estimate of K_m for tubular cocurrent mass contactors if bubble or drop diameter is known and the fluid velocity of the continuous phase is known.

If we can achieve a bubble size of 0.005 m in an air–water horizontal mass contactor the continuous-gas-phase velocity will be about 3 m/s. This will produce a value of $K_m = 11 \times 10^{-4}$ m/s if we can assume that the gas velocity is the quantity we need in Eq. (7.2.16).

7.3.3 Countercurrent Area Estimation

Tubular countercurrent mass contactors or towers all have some type of internals, trays, or packing. There are many different types of commercially available packing or trays available, each claiming a particular advantage for a given service. The area for mass transfer depends on the type of packing or the tray design and the complex fluid mechanics of the two-phase flows through the trays or over the packing. Most systems are gas–liquid, although if there is a great enough density difference liquid–liquid systems can be used for mass contacting in towers. There are several proprietary liquid–liquid tower contactors, most of which have some type of rotating internals.

In tray or plate towers the gas or vapor phase forms bubbles as it passes through the liquid on the tray. Liquid levels are of the order of some 4–8 cm, and the bubbles of gas are in contact with the liquid for very short times and do not break up as in deep-liquid situations. The mixed–plug analysis described in Subsection 7.2.2 is not applicable. Furthermore, froth forms above the highly agitated liquid, complicating even further any estimation of the interfacial area. Interfacial area a can be related to efficiency of a tray or plate in a similar way to that developed in Chapter 4, Subsection 4.3.3.2. Tray and packed-tower efficiency estimation is part of the extensive research programs carried out at Fractionation Research Inc. (F.R.I.) (www.fri.org): "F.R.I. is a non profit research consortium consisting of the leading engineering, chemical and petroleum companies in the world." It has been successfully operating since the early 1950s and has a number of research efforts underway today. We will not attempt to summarize the extensive research carried by F.R.I. or any of the many academic organizations who are active today as these are well beyond the scope of this book.

In a packed tower the liquid flows down over and through the packing, of which there are many proprietary devices (Figure 7.7), and contacts the upward flowing gas. The interfacial area for mass transfer depends in a complex way on the type of packing, the liquid distribution devices, and the flow rates of the liquid and the

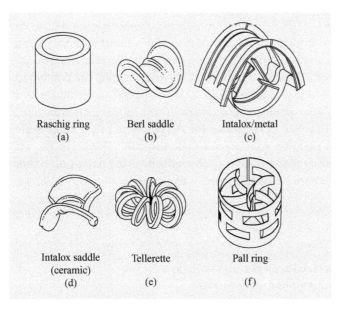

Figure 7.7. Representative random packings: Types (c), (e), and (f) are the through-flow variety. (*Chemical Engineers' Handbook*, 1997, Figure 4.45a. Reproduced with permission.)

gas. There are critical limits on the gas flow rate in particular; if it is too low and/or the liquid distributors are poorly designed, the liquid will not distribute itself evenly over the packing and liquid channels can develop, greatly reducing the area for mass transfer. If the gas rate is too high, liquid can be entrained and taken overhead out of the tower. This issue will be addressed in an example in Chapter 8.

There has been much experimental effort devoted to the performance of many different types of packing but to date there is no universal correlation that one can employ to determine area. Typical of the attempts to develop useful correlations is the following expression reported in *Chemical Engineers' Handbook* (1997) for interfacial area for 25-mm Rashig rings as a function of gas and liquid flow rates:

$$\frac{a_v}{a_w} = 0.54\,G'^{0.31}L'^{0.07}. \tag{7.3.9}$$

The external area of the packing a_w is available in tables for all different types of packing (*Chemical Engineers' Handbook*, 1997, Table 14-7b). The 25-mm Rashig ring has a surface area a_w of 190 m^2/m^3. The mass gas flow rate per tower cross-sectional area G' and the liquid flow rate L' both have units of kilograms per second times square meters. The correlation was obtained for L' between 4 and 17 kg/m^2 s.

There are many other correlations of varying degrees of complexity obtained by experiments in towers by use of the analysis presented in Subsection 4.4.3 to plan experiments and analyze the experimental data. Such experiments are costly to carry out because they must be done on commercial-scale towers. We do not know enough about the two-phase fluid mechanics and how to predict flow behavior for the packing and various fluid distributors to be confident of scaling up from laboratory-scale equipment.

7.3.4 Countercurrent K_m Estimation

This is a much more difficult problem because one is faced with very complex fluid motions in either a packed column or a tray column. Current practice is to collect experimental data on specific column internals to determine the combined quantity $K_m a$. This is almost always expressed in terms of height of a transfer unit defined by integrating and rearranging the countercurrent model equations, as shown in Subsection 4.4.3. However, correlations for the combined $K_m a$ make scale-up from laboratory or pilot plant to commercial scale difficult. Therefore we will explore methods for estimating mass transfer coefficients by using penetration theory in the following chapter, where we consider the technically feasible design of a gas–liquid packed column.

Nomenclature

a_v	Interfacial area per unit volume
Ca	Capillary number, $(v/\mu\sigma)$
D	Circulation element diameter (m)
d	Diameter (m), plume
d_{32}	Sauter diameter (m)
d_{50}	Median diameter, initial median diameter (m)
d_B	Bubble diameter (m)
d_{B0}	Bubble diameter at orifice (m)
d_{max}	Maximum stable diameter (m)
d_{pipe}	Diameter of pipe (m)
f	Fanning friction factor
g	Gravitational acceleration (m/s^2)
G'	Gas flux (kg/m^2 s)
h'	Height of impeller blade (m)
h_L	Clear-liquid height (m)
L_p	Impeller diameter (m)
L_{pipe}	Length of pipe (m)
L'	Liquid flux (kg/m^{2}s)
$P, \Delta P$	Pressure, pressure drop (Pa)
p_0	Pressure at orifice (N/m^2)
p_s	Pressure at surface (N/m^2)
q^I	Flow rate of continuous phase (m^3/s)
q^{II}	Flow rate of dispersed phase (m^3/s)
Re	Reynolds number
T	Temperature (K)
v^I	Velocity of continuous phase (m/s)
v^{II}	Velocity of dispersed phase (m/s)
V^I	Volume of continuous phase (m^3)
V^{II}	Volume of dispersed phase (m^3)
W	Impeller width (m)
We_{crit}	Critical Weber number
z	Axial dimension of pipe (m)

Greek Symbols

α	Shape factor
δ_{pipe}	Roughness of pipe (m)
ε	Power dissipation per unit mass (W/kg)
ϕ	Dispersed-phase volume fraction
μ^I	Viscosity of continuous phase (Pa s)
μ^{II}	Viscosity of dispersed phase (Pa s)
ρ^I	Density of continuous phase (kg/m^3)
ρ^{II}	Density of dispersed phase (kg/m^3)
σ	Interfacial tension (kg/s^2)
σ_{LN}	Log-normal standard deviation
τ	Stress (Pa)

REFERENCES

Allen T. *Particle Size Measurement* (4th ed.). New York: Chapman and Hall, 1990.

Alvarez-Cuenca M, Nerenberg M. Oxygen mass transfer in bubble columns working at large gas and liquid flow rates. AIChE J. 1981; 27: 66–73.

Andreussi P, Paglianti A, Silva F. Dispersed bubble flow in horizontal pipes. Chem. Eng. Sci. 1999; 54: 1101–7.

Bhavaraju SM, Russell TWF, Blanch HW. The design of gas-sparged devices for viscous liquid systems. AIChE J. 1978; 24: 454–66.

Calabrese RV, Chang TPK, Dang PT. Drop breakup in turbulent stirred-tank contactors. Part I: Effect of dispersed phase viscosity. AIChE J. 1986a; 32: 657–66.

Calabrese RV, Wang CY, Bryner NP. Drop breakup in turbulent stirred-tank contactors. Part III: Correlations for mean size and drop size distribution. AIChE J. 1986b; 32: 677–81.

Chemical Engineers' Handbook (7th ed.). New York: McGraw-Hill Professional, 1997.

Collins SB, Knudson JG. Drop-size distributions produced by turbulent pipe flow of immiscible liquids. AIChE J. 1970; 16: 1072–80.

Davies ST. A physical interpretation of drop sizes in homogenizers and agitated tanks, including the dispersion of viscous oils. Chem. Eng. Sci. 1987; 42: 1671–6.

Govier GW, Aziz K. *The Flow of Complex Mixtures in Pipes*. New York: Van Nostrand Reinhold, 1972.

Hesketh RP, Etchells AW, Russell TWF. Bubble breakage in pipeline flow. Chem. Eng. Sci. 1991; 46: 1–9.

Hesketh RP, Russell TWF, Etchells AW. Bubble size in horizontal pipelines. AIChE J. 1986; 33: 663–7.

Higbie R. The rate of adsorption of a pure gas into a still liquid during short periods of exposure. Trans. AIChE. 1935; 31: 365–89.

Hinze JO. Fundamentals of the hydrodynamic mechanism of splitting in dispersions processes. AIChE J. 1955; 1: 289–5.

Holmes TL. Fluid mechanics of horizontal bubble flow. Ph.D. dissertation, University of Delaware, 1973.

Holmes TL, Russell TWF. Horizontal bubble flow. Int. J. Multiphase Flow 1975; 2: 51–66.

Karabelas AJ. Droplet size spectra generated in turbulent pipe flow of dilute liquid/liquid dispersions. AIChE J. 1978; 24: 170–80.

Kolmogorov AN. On the breaking of drops in turbulent flow. Dokl. Akad. Nauk. SSSR. 1949; 66: 825–8.

Kubie J, Gardner GC. Drop sizes and drop dispersion in straight horizontal tubes and in helical coils. Chem. Eng. Sci. 1977; 32: 195–202.

Levich VG. *Physiochemical Hydrodynamics*. Englewood Cliffs, NJ: Prentice-Hall, 1962, p. 464.

Miller DN. Scale-up of agitated vessels gas–liquid mass transfer. AIChE J. 1974; 20: 445.

Otero Z. Liquid circulation and mass transfer in gas–liquid contactors. Ph.D. dissertation. University of Delaware, 1983.

Otero Z, Tilton JN, Russell TWF. Some observations on flow patterns in tank-type systems. Int. J. Multiphase Flow 1985; 11: 581–9.

Otero Keil, Z, Russell TWF. Design of commercial-scale gas–liquid contactors. AIChE J. 1987; 33: 488–6.

Paul EL, Atiemo-Obeng VA, Kresta SM, eds. *Handbook of Industrial Mixing: Science and Practice*. New York: Wiley-Interscience, 2004.

Razzaque MM, Afacan A, Liu S, Nandakumar K, Masliyah JH, Sanders RS. Bubble size in coalescence dominant regime of turbulent air–water flow through horizontal pipes. Int. J. Multiphase Flow 2003; 29: 1451–71.

Tilton JN. Modeling and computer aided design of gas–sparged vessels for mass transfer. M.Ch.E. thesis, University of Delaware, 1981.

Tilton JN, Russell TWF. Designing gas sparged vessels for mass transfer. Chem. Eng. November 29, 1982: 61–8.

Towell GD, Strand CP, Ackermann GH. Mixing and mass transfer in large diameter bubble columns. AIChE–IChemE Symposium Series 1965; 10: 97.

PROBLEMS

7.1 100 L/min of TCE is to be contacted with 100 L/min of water containing 5% by weight acetone. The droplet sizes are to be 2 mm, 5 mm, and 1 cm. Compare the energy requirements (as total kW) to contact these in

   a. a tank-type mass contactor,
   b. a cocurrent horizontal tubular contactor.

7.2 Oxygen is to be removed from water by contacting with nitrogen in a countercurrent mass contactor packed with 25-mm Rashig rings. The superficial liquid flow rate is 0.25 m/s and the packed area is 185 m^2/m^3. Estimate K_m assuming that the resistance to mass transfer is in the liquid phase by using the penetration theory and assuming the contact time is the time the liquid takes to flow across one Rashig ring.

7.3 Two mixers are being evaluated for use in a liquid–liquid extraction to be performed in a continuous-flow stirred tank. Both mixers are operated to predict a mean bubble size of 5 mm, but mixer one has a narrow size distribution (10% relative size distribution) whereas mixer two produces a broad size distribution (50% relative size distribution). Assuming the distributions are log-normal, compute the average bubble size that can be used to predict the area for transport (Sauter mean). Plot the two distributions and comment on possible advantages and disadvantages to having a broad size distribution (remember, the liquids have to be separated in a separator after the contactor).

7.4 A vendor of mixing equipment tells you that he can achieve the same bubble size as that obtained in the mass contactor described in Examples 4.2 and 7.2, $d_B = 0.0087$ m, by installing a Rushton impeller mixer in the mass contactor. The salesperson recommends using a motor capable of driving the mixer at 50 rpm and states that Eq. (7.2.2) applies. What diameter of the impeller would be needed? Would you purchase such a motor and mixer for this tank? What difficulties would you encounter if you did so?

APPENDIX: BUBBLE AND DROP BREAKAGE

We can resolve the discrepancy between Eqs. (7.1.4) and (7.1.5) by considering the breakup mechanism of a bubble or drop in relation to the Weber number. Hinze reports that a relatively simple mechanism is characteristic of breakup around the critical Weber number, whereas the mechanism becomes much more complex and even chaotic at larger Weber numbers. From this behavior, one can infer that the Weber number is characteristic of the breakage mechanism. Moreover, experiments by Collins and Knudson (1970) on liquid–liquid dispersions and by Holmes (1975) on an air–water system show a very similar mechanism of breakup—deformation into a dumbbell shape, followed by the breakage event. Taken together, these two observations indicate that *the critical Weber number should be nearly the same for liquid–liquid and gas–liquid dispersions*. Thus, for a theory to describe both drop and bubble sizes, it must yield a consistent value of the critical Weber number across different fluid combinations.

There are data available on the maximum stable drop size in pipelines for a number of different liquid–liquid combinations. Studies by Collins and Knudson (1970) involve three different organics in water, and Kubie and Gardner (1977) report data on water–n-butyl acetate and water–alcohol dispersions. Karabelas (1978), meanwhile, studied oil–water dispersions, by using both kerosene and transformer oil as the continuous phase. Study of gas–liquid dispersions, meanwhile, has been mostly limited to air–water systems. Holmes (1973), Hesketh et al. (1991), Andreussi et al. (1999), and Razzaque et al. (2003) are examples of literature available for gas–liquid systems.

Figures 7.A1 and 7.A2 show experimental data from some of these experiments, plotted according to the Kolmogorov–Hinze and Levich theories. There exist a large range of experimental conditions in these experiments. Information is summarized in Table 7.A1.

Figure 7.A1 is a plot of the scaled maximum diameter (d_{max}/d_{pipe}) versus the product $Re^{-0.5}Ca^{-0.6}$. There are two main clusters of data, which show two distinctly different linear trends. The first cluster, which contains all of the liquid–liquid dispersion data, shows that these systems can be well represented with a critical Weber number of the order of 1. The other cluster, however, contains air–water data and is well characterized by a critical Weber number of about 10.6. Two distinct critical Weber numbers indicate that the Kolmogorov–Hinze theory is inadequate.

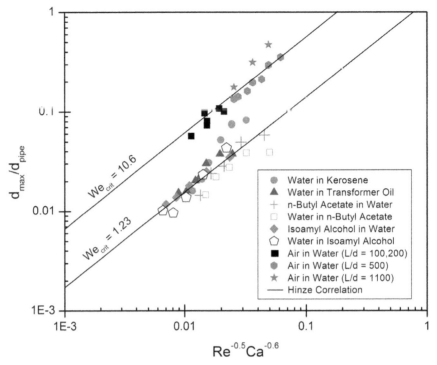

Figure 7.A1. Experimental data compared with Eq. (7.3.3).

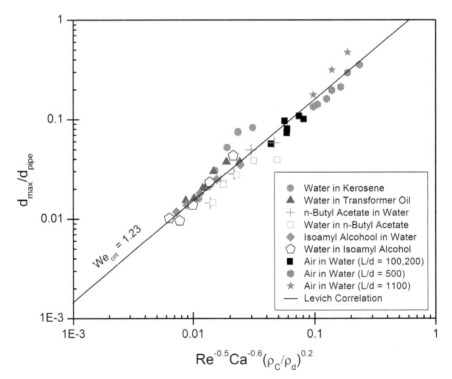

Figure 7.A2. Experimental data compared with Eq. (7.3.4).

Table 7.A1. Summary of experimental techniques and conditions

Investigator	Experimental Method	Continuous Phase	Dispersed Phase	ϕ_d (%)	Re	Ca	ρ_c/ρ_d	We_{crit}	We'_{crit}
Kubie & Gardner	Photography	Isoamyl alcohol	Water	Not available	[2500, 7600]	[0.84, 2.5]	0.83	1.4	1.5
Kubie & Gardner	Photography	Water	Isoamyl alcohol	Not available	[14000, 43000]	[0.18, 0.54]	1.21	1.3	1.2
Karabelas	Drop encapsulation & photography	Transformer oil	Water	0.20	[3400, 9300]	[0.54, 1.4]	0.89	1.5	1.6
Karabelas	Drop Encapsulation & Photography	Kerosene	Water	[0.16, 0.26]	[26000, 66000]	[0.06, 0.16]	0.80	2.9	3.1
Kubie & Gardner	Photography	n-Butyl Acetate	Water	Not available	[18000, 55000]	[0.04, 0.12]	0.89	0.9	0.9
Kubie & Gardner	Photography	Water	n-Butyl acetate	Not available	[14000, 43000]	[0.06, 0.18]	1.13	0.6	0.6
Holmes	Photography	Water	Air	0.1	[87000, 177000]	[0.04, 0.08]	~800	10.6	1.1
Razzaque et al.	Photography	Water	Air	0.15	[32000, 71000]	[0.02, 0.04]	~800	10.0	1.1

Figure 7.A2 is a plot consistent with the framework developed from the Levich force balance. In this case, the plot shows a single trend line that describes nearly all of the liquid–liquid and gas–liquid data. The entire set is well characterized by a Weber number of about 1.2, and the relative standard error of this regression is about 27.5%. For this reason, the Kolmogorov–Hinze result is typically used only for liquid–liquid systems or discarded entirely in favor of Levich's force balance.

8   Technically Feasible Design Case Studies

Figure 1.2 presents the logic leading to technically feasible analysis and design. In this chapter we illustrate the design process that follows from the analysis of existing equipment, experiment, and the development of model equations capable of predicting equipment performance. Design requests can come in the form of memos, but an ongoing dialogue between those requesting a design and those carrying out the design helps to properly define the problem. This is difficult to illustrate in a textbook but we will try to give some sense of the process in the case studies presented here.

Technically feasible heat exchanger and mass contactor design procedures were outlined in Sections 3.5 and 4.5. In this chapter we present case studies to illustrate how one can proceed to a technically feasible design. Recall that such a design must satisfy only the design criteria, i.e., the volume of a reactor that will produce the required amount of product, the heat exchanger configuration that will meet the heat load needed with the utilities available, or the mass contactor that will transfer the required amount of material from one phase to another given the flow rate of the material to be processed. Even for relatively simple situations, design is always an iterative process and requires one to make decisions that cannot be verified until more information is available and additional calculations are made. Such additional information may require one to quantitatively determine the effect of the design on the rest of a process or it may require one to expand an economic evaluation. A technically feasible design is always the first step in the process to develop an equipment design that can then be built, operated, and properly controlled to meet a process need.

There are three case studies presented in this chapter. The first develops a technically feasible design for double-pipe and shell-and-tube exchangers by extending Example 3.4. The second case examines a design for a countercurrent tubular mass contactor, more commonly known as a packed tower, which will be used to deoxygenate water by contacting a downward-flowing liquid with a flowing nitrogen stream. The third case study deals with a two-phase mass contactor and reactor, which is a commercially available pilot-scale bioreactor or fermentor described in Chapter 2, Figure 2.4. We expand on the semibatch reactor analysis in Chapter 2, where we determined which of several rate expressions best fit experimental data reported in the literature, to include an analysis of the oxygen transfer to the fermentor broth by air sparging. This analysis then allows us to develop a rational procedure for the scale-up of air-sparged bioreactors.

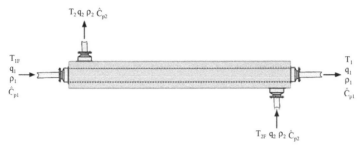

Figure 3.15. Countercurrent double-pipe exchanger.

8.1   Technically Feasible Design of a Heat Exchanger

This design problem is defined in Chapter 3 as Example 3.4. Recall that an exchanger is required to cool 100 L/min from 80 to 50 °C. There is available chilled water at 5 °C. The technically feasible design presented in Chapter 3 yielded a utility flow rate $q_2 = 46$ L/min with an exit temperature of $T_2 = 70$ °C. A double-pipe exchanger of 120 ft. of 2-in. pipe inside a 3-in. outer pipe was proposed with a U value of 1.5 kW/m² K as a first iteration. The transfer area a is 6.0 m². Such an exchanger is shown in Figure 3.15, reproduced here for convenience. This is a simple but effective first step to show how the design should proceed. The logic for exchanger design is presented in Section 3.5 and illustrated in Figure 8.1. The iterations to justify (or improve) our selection of U proceed as follows.

Step 7. Iteration to a Technically Feasible Design (Section 3.5). To use the correlations in Chapter 6 for determination of U we need both the Reynolds number and the Nusselt number. For the process fluid in the inner pipe the Reynolds number Re_1 is

$$Re_1 = \frac{D_{1,in}\rho_1 v_1}{\mu_1} = \frac{D_{1,in}\rho_1 \frac{q_1}{\pi D_{1,in}^2/4}}{\mu_1} = \frac{4\rho_1 q_1}{\pi D_{1,in}\mu_1}$$

The following calculation requires material properties, which vary with temperature in the exchanger. As a reasonable simplification we use approximate constant values for the heat capacity, density, and thermal conductivity as the variation of these parameters with temperature is within acceptable limits for this calculation. Further, we approximate both streams as water. The viscosity, however, varies strongly with temperature, so we use a value calculated at the bulk average temperature for each stream. A monograph that presents the viscosity of many liquids as a function of temperature is available in the *Chemical Engineers' Handbook* (1997), and some selected values for water are presented in Table 8.1.

$$Re_1 = \frac{4(1000 \text{ kg/m}^3)(1.67 \times 10^{-3} \text{ m}^3/\text{s})}{\pi (0.0526 \text{ m}) (5 \times 10^{-4} \text{ kg/m-s})} = 80848.$$

The physical properties are calculated at the average bulk temperature of 65 °C.

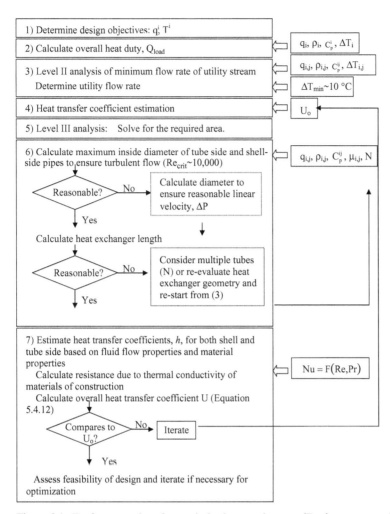

1) Determine design objectives: q_i^i, T^i

q_i, ρ_i, C_p^i, ΔT_i

2) Calculate overall heat duty, Q_{load}

$q_{i,j}$, $\rho_{i,j}$, C_p^{ij}, $\Delta T_{i,j}$

3) Level II analysis of minimum flow rate of utility stream
   Determine utility flow rate

$\Delta T_{min} \sim 10\ ^\circ C$

4) Heat transfer coefficient estimation

U_o

5) Level III analysis: Solve for the required area.

6) Calculate maximum inside diameter of tube side and shell-side pipes to ensure turbulent flow ($Re_{crit} \sim 10,000$)

$q_{i,j}$, $\rho_{i,j}$, C_p^{ij}, $\mu_{i,j}$, N

Reasonable? — No → Calculate diameter to ensure reasonable linear velocity, ΔP

Yes

Calculate heat exchanger length

Reasonable? — No → Consider multiple tubes (N) or re-evaluate heat exchanger geometry and re-start from (3)

Yes

7) Estimate heat transfer coefficients, h, for both shell and tube side based on fluid flow properties and material properties
   Calculate resistance due to thermal conductivity of materials of construction
   Calculate overall heat transfer coefficient U (Equation 5.4.12)

$Nu = F(Re, Pr)$

Compares to U_o? — No → Iterate

Yes

Assess feasibility of design and iterate if necessary for optimization

Figure 8.1. Design procedure for a tubular heat exchanger. (Design summary in Section 3.5.)

Because the liquid in the outer pipe flows in an annulus, the Reynolds number is defined with a hydraulic diameter D_h, defined by the wetted perimeter P_w, as

$$D_h = \frac{4A_c}{P_w},$$

$$D_h = \frac{4A_c}{P_w} = \frac{4\left(\dfrac{\pi D_{2,in}^2}{4} - \dfrac{\pi D_{1,out}^2}{4}\right)}{\pi D_{2,in} + \pi D_{1,out}} = D_{2,in} - D_{1,out}.$$

Table 8.1. Viscosity of water

T °(C)	μ (kg/m s) $\times 10^3$
10	1.31
20	1.02
40	0.72
60	0.55
80	0.44

The Reynolds number for the utility fluid, Re_2, also calculated with properties evaluated at the average bulk temperature, is

$$Re_2 = \frac{(D_{2,\text{in}} - D_{1,\text{out}}) \rho_2 \left[\dfrac{4q_2}{\pi \left(D_{2,\text{in}}^2 - D_{1,\text{out}}^2\right)} \right]}{\mu_2} = \frac{4\rho_2 q_2}{\pi \mu_2 (D_{2,\text{in}} + D_{1,\text{out}})},$$

$$Re_2 = \frac{4(1000 \text{ kg/m}^3)\,(7.67 \times 10^{-4} \text{ m}^3/\text{s})}{\pi(7 \times 10^{-4} \text{ kg/m s})\,(0.078 \text{ m} + 0.0603 \text{ m})} = 10088.$$

The Prandlt numbers, Pr_1 (65°C) and Pr_2 (40°C), need to be calculated for both streams:

$$Pr_1 = \frac{\hat{C}_p \mu_1}{k} = \frac{(4184 \text{ J/kg K})(5.0 \times 10^{-4} \text{ kg/m s})}{0.6 \text{ W/m K}} = 3.5,$$

$$Pr_2 = \frac{\hat{C}_p \mu_2}{k} = \frac{(4184 \text{ J/kg K})(7.0 \times 10^{-4} \text{ kg/m s})}{0.6 \text{ W/m K}} = 4.9.$$

Using the Colburn correlation from Chapter 6, Eq. (6.3.3), and Table 6.5, we find

$$Nu = 0.023 Re^{0.8} Pr^{0.33},$$
$$Nu_1 = 0.023 \, (80854)^{0.8} \, (3.5)^{0.33} = 293,$$
$$Nu_2 = 0.023 \, (10088)^{0.8} \, (4.9)^{0.33} = 62.$$

With the Nusselt number, $Nu = hD/k$ and k for water (Table 5.1) known, the local heat transfer coefficients h can be calculated:

$$h_1 = \frac{Nu_1 k}{D_1} = \frac{293(0.6 \text{ W/m K})}{0.0526 \, [\text{m}]} = 3342 \text{ W/m}^2 \text{ K},$$

$$h_2 = \frac{Nu_2 k}{D_2 - D_1} = \frac{62(0.6 \text{ W/m K})}{0.0779 \text{ m} - 0.0603 \text{ m}} = 2120 \, \frac{\text{W}}{\text{m}^2 \text{ K}}.$$

The overall heat transfer coefficient for a double-pipe heat exchanger was developed in Chapter 5, Eq. (5.4.12):

$$U = \frac{1}{\dfrac{1}{h_{\text{out}}} + \dfrac{r_{\text{out}}(\ln r_{\text{out}}/r_{\text{in}})}{k_{\text{wall}}} + \dfrac{1}{h_{\text{in}}}\dfrac{r_{\text{out}}}{r_{\text{in}}}}. \quad (5.4.12)$$

For our problem $r_{1,\text{out}} = 0.030$ m, $r_{1,\text{in}} = 0.026$ m, and $k = 16$ W/m K:

$$U = \frac{1}{\dfrac{1}{2120} + \dfrac{0.030 \ln\left(\dfrac{6.03}{5.26}\right)}{16} + \dfrac{6.03}{3342 \times 5.26}}$$

$$= 0.93 \, \frac{\text{kW}}{\text{m}^2 \text{ K}}.$$

For our assumed value of 1.0 kW/m² K, we then need 180 ft. for the design. Heat exchangers are most frequently encountered as part of some process, and optimal designs are difficult to define without much more information on the process requiring the exchanger and the uncertainties arising for any number of reasons.

Even without additional information there are difficulties in manufacture that make the double-pipe exchanger a poor choice. If we elect to use 10-ft. pipe sections, 18 will be required, and they will have to be stacked one on top of each other or some other configuration requiring return bends for both the inner and outer pipe. It makes more sense to use a number of smaller pipes inside one shell.

This configuration is a countercurrent shell-and-tube heat exchanger; an example is shown in Figure 3.10. In this day and age this is best done by going out on performance and having a reputable firm with access to the HTRI procedures do this. We illustrate the process here in its simplest form. Typical shell-and-tube exchangers have tube sizes ranging from 0.25 to 1.25 in., with 0.75 and 1 in. being the most common. Standard tube lengths are 8, 19, 12, and 20 ft., with 20 ft. being the most widely used. We base our initial design on a velocity of 1 m/s on the tube side, which contains the process fluid. Our required area is 9 m² based on a heat transfer coefficient of 1.0 kW/m² K.

The number of pipes, N, can be determined from the area, length, and diameters:

$$a = LN\pi D_i;$$

where

$$L = 20 \text{ ft.} = 6.1 \text{ m,}$$

$$N = \frac{a}{L\pi D_i}.$$

With a velocity of 1 m/s, the tube inner diameter can be determined given the required flow rate and number of tubes:

$$v = \frac{4q_1}{\pi D_{1,in}^2 N},$$

Substituting for N and rearranging yields

$$D_{1,in} = \frac{4q_1 L}{v_1 a} = \frac{4(1.667 \times 10^{-3} \text{ m}^3/\text{s})(6.1 \text{ m})}{(1 \text{ m/s})9.0 \text{ m}^2} = 0.0045 \text{ m} = 0.177 \text{ in.}$$

We can select a 1/4 27 BWG tubing, which has an inner diameter of 0.218 in. (0.0055 m) and an outer diameter of 0.25 in. (Table 8.2):

$$N = \frac{a}{\pi L D_{1,in}} = \frac{9.0 \text{ m}^2}{(6.1 \text{ m})\pi 0.00557 \text{ m}} = 84.$$

Design of the shell can be complicated and is beyond the scope of this book. To illustrate our technically feasible design it is sufficient to specify a single-pass shell with no internal baffles. We select a shell diameter sufficient to contain the tube bundle. The tubes in a shell-and-tube heat exchanger are supported in a tube bundle that can be removed from the shell for inspection and cleaning.

We can obtain the shell diameter through which the utility fluid flows in a simple fashion by assuming that the tubes occupy one third of the cross-sectional area of the shell. This is a conservative assumption that avoids having to specify the tube layout in detail.

Table 8.2. *Pipe schedule for heat exchanger piping*

D_{nom} (in.)	BWG	D_{in} [in. (cm)]	x_{wall} [in. (cm)]
¼	22	0.194 (0.493)	0.028 (0.071)
¼	27	0.218 (0.554)	0.016 (0.041)
½	16	0.370 (0.940)	0.065 (0.165)
½	18	0.402 (1.021)	0.049 (0.124)
¾	15	0.606 (1.539)	0.072 (0.183)
¾	16	0.620 (1.575)	0.065 (0.165)
1	16	0.870 (2.210)	0.065 (0.165)
1	18	0.902 (2.291)	0.049 (0.124)
1 ¼	13	1.060 (2.692)	0.095 (0.241)
1 ½	14	1.334 (3.388)	0.083 (0.211)
1 ½	16	1.370 (3.480)	0.065 (0.165)

BWG 1/4-in. tube external cross-sectional area $= \pi (0.25)^2/4 = 0.05$ in^2

Total tube cross-sectional area $= 84 (0.05) = 4.2$ in^2 (0.0027 m^2)

Shell cross-sectional area $= 3 (4.2) = 12.6$ in^2

Shell diameter $= (12.6(4)/\pi)^{0.5} = 4.0$ in. (10.2 cm)

Examining Table E3.1 we find that a 4-in. Schedule 40 pipe has an internal diameter of 4.0 in. and a cross-sectional area of 12.6 in^2 (0.0081 m^2). This will accommodate the tube bundle.

Step 4 (repeated; Section 3.5). To estimate the heat transfer coefficient for the shell we need a velocity and a hydraulic diameter D_h to obtain a Reynolds number. We estimate the liquid velocity in the shell by dividing the volumetric flow rate, 7.6×10^{-4} m^3/s, by the cross-sectional area of the shell, 0.0081 m^2, minus that of the tubes, 0.0018 m^2:

$$v_2 = \frac{(7.6 \times 10^{-4} \text{ m}^3/\text{s})}{(0.0081 - 0.0023) \text{ m}^2}$$

$$= 0.13 \text{ m/s}.$$

Because there will be some flow in the shell perpendicular to the tube, this velocity is an approximation that could be improved at the expense of some detailed fluid mechanic modeling in the shell coupled with experimentation on commercial-scale equipment. Many correlations have been proposed; we illustrate the process by using the simplest.

The cross-sectional area of the shell available for fluid flow multiplied by 4 divided by the wetted perimeter, consisting of the shell wetted perimeter and the wetted perimeter of the multiple tubes in the bundle, can be used to calculate D_h:

$$D_h = \frac{4 (0.0081 - 0.0023) \text{ m}^2}{\pi (4 + 84 \times 0.23) \text{ in. } (2.54 \times 10^{-2} \text{ m/in.})}$$

$$= 0.012 \text{ m}.$$

The shell-side Reynolds number is

$$Re_2 = \frac{0.018 \text{ m } (1000 \text{ kg/m}^3)(0.13 \text{ m/s})}{7 \times 10^{-4} \text{ kg/m s}}$$
$$= 2328.$$

The velocity in each tube is 0.83 m/s and the tube-side Reynolds number is

$$Re_1 = \frac{(0.25 \text{ in.})(2.54 \times 10^{-2} \text{ m/in.})(1000 \text{ kg/m})(0.83 \text{ m/s})}{5 \times 10^{-4} \text{ kg/m s}}$$
$$= 9784.$$

The tube-side Nusselt number calculated with the same correlation used for the double-pipe exchanger, Eq. (6.33), Table 6.5, is

$$Nu_1 = 0.023 Re^{0.8} Pr^{0.33}$$
$$= 54.$$

The shell-side Nusselt number can be estimated with a correlation suggested by Donohue (1949):

$$Nu_2 = 0.2 Re^{0.6} Pr^{0.33}$$
$$= 19.$$

The tube-side h value is

$$h_1 = \frac{54 \times 0.6}{0.0055} \frac{\text{W}}{\text{m}^2 \text{ K}}$$
$$= 5867.$$

The shell-side h value is estimated with the outside tube diameter $0.23 \times 2.54 \times 10^{-2}$ m $= 0.006$ m:

$$h_2 = \frac{19 \times 0.6}{0.006} \frac{\text{W}}{\text{m}^2 \text{ K}}$$
$$= 1900 \frac{\text{W}}{\text{m}^2 \text{ K}}.$$

The U value can be obtained with Equation (5.4.12):

$$U = \frac{1}{\dfrac{1}{1900} + \dfrac{0.23 \times 2.54 \times 10^{-2} \ln(0.23/0.218)}{16} + \dfrac{0.23}{5867 \times 0.218}}$$
$$= 1.4 \text{ kW/m}^2 \text{ K}.$$

This estimated value for U might be expected to vary by as much as a factor of two because of uncertainties in our shell-side calculations. If we use it to obtain a

Table 8.3. Technically feasible heat exchanger designs

Double-pipe exchanger	For both exchangers	Shell-and-tube exchanger
	$T_{1F} = 80\,°C$	
	$T_1 = 50\,°C$	
	$T_{2F} = 5\,°C$	
	$T_2 = 70\,°C$	
	$q_1 = 100$ L/min	
	$\quad = 1.67 \times 10^{-4}\ m^3/s$	
	$q_2 = 46$ L/min	
	$\quad = 7.6 \times 10^{-4}\ m^3/s$	
	$Pr_1 = 3.5$	
	$Pr_2 = 4.9$	
$D_1 = 2$ in.		$D_{in} = 0.25$ in.
$D_2 = 3$ in.		$N = 84$
		$D_2 = 4$ in.
$Re_1 = 80854$		$Re_1 = 9784$ (tube)
$Re_2 = 10087$		$Re_2 = 2328$ (shell)
$U = 1.0\ kW/m^2\ K$		$U = 1.4\ kW/m^2\ K$
$L = 20$ ft.		$L = 13$ ft.
$= 6\ m$		$= 4\ m$

design, we repeat Step 5 to estimate the area for heat transfer a, as:

$$\langle \Delta T_{21} \rangle = \frac{(5-50)-(70-80)}{\ln\left(\dfrac{-45}{-10}\right)}$$

$$= -23.3,$$

$$a = \frac{Q_{load}}{U\,\langle \Delta T_{21} \rangle}$$

$$= \frac{-2.1 \times 10^5\ W\,m^2}{-23.3 \times 1400\ W}$$

$$= 6.6\ m^2.$$

The shell-and-tube exchanger requires less area than the double-pipe exchanger, which needed 9.0 m². This could be achieved by use of shorter tubes, say 4 m (13 ft.).

The technically feasible designs are presented in Table 8.3. The logic to obtain such a design is presented in Figure 8.1.

The process stream is in the inner tube or tubes and the utility stream is in the outer pipe or shell. A sketch of the shell-and-tube exchanger is shown in Figure 8.2.

TUBE SHEET

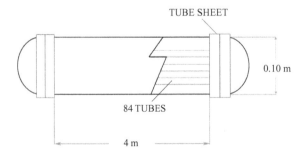

0.10 m   Figure 8.2. Technically feasible shell-and-tube exchanger.

84 TUBES

4 m

8.2 Technically Feasible Design of a Countercurrent Mass Contactor

As pointed out in Chapter 4, this type of contactor must be a vertical pipe or tower almost always with internals of trays or packing (strippers or scrubbers). Table 8.4 illustrates the types of tubular contactors most commonly employed. Table 8.5 and Table 8.6 show some less common tubular contactors. In this example we need to design a tower to remove oxygen from water by stripping the water stream with nitrogen. We follow the procedures outlined in Section 4.5:

Step 1. Determination of Objectives (Section 4.5). Water (50 L/min) free of dissolved oxygen is needed to reduce corrosion in the boiler tubes of power plants and is required for rinsing silicon wafers in semiconductor manufacturing. If oxygen is present in this rinse water, unwanted silicon oxide can be formed. There are a number of ways oxygen can be removed from water, but stripping by a nitrogen stream will illustrate the development of a technically feasible design for a stripper. The oxygen

Table 8.4. Most common tubular contactors

Contactor type	Packed tower	Tray tower
Operation	Continuous	Staged
Dispersed phase	Gas	Gas
Continuous phase	Liquid	Liquid
Diagram	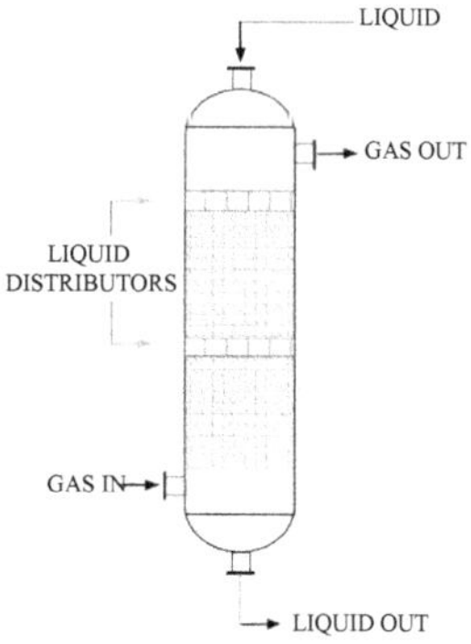 LIQUID GAS OUT LIQUID DISTRIBUTORS GAS IN LIQUID OUT	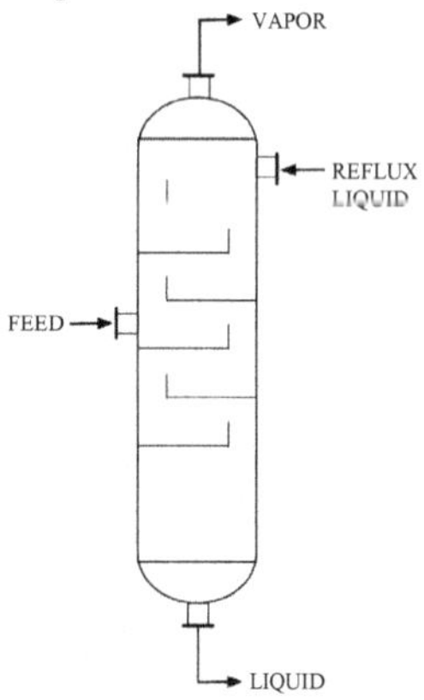 VAPOR REFLUX LIQUID FEED LIQUID
Fluid motions	Countercurrent	Countercurrent
Dispersed phase	Plug flow	Plug flow or well-mixed
Continuous phase	Plug flow	Well-mixed liquid holdup on each stage
Design issues	• Area determination (wetting of packing) • Poor distribution of fluid flow—may require staged operation with intermediate redistribution of flow	• Area determination per stage • Flooding of tower
Basic model classification	Continuous, countercurrent plug–plug contactor	Stage/tray: continuous mixed–mixed or plug–mixed contactors

Table 8.5. *Additional tubular contactors*

Contactor type	Wetted wall column	Spray tower
Operation	Continuous	Continuous
Dispersed phase	Gas	Liquid
Continuous phase	Liquid	Gas
Diagram		
Fluid motions	Countercurrent	Countercurrent
Dispersed phase	Plug flow	Ideally plug flow (can approach well mixed if entrainment exists)
Continuous phase	Plug flow	Ideally plug flow (more often well mixed)
Design issues	• Area determination and K_m complicated by ripples in falling film	• Coupled heat and mass transfer results from evaporation • Determination of drop velocity, entrainment
Basic model classification	Continuous, countercurrent plug–plug contactor	Continuous countercurrent plug–plug contactor *or* plug–mixed contactor

concentration in the inlet water, C_{AF}^I, is 8 mg/L. We need to reduce the oxygen concentration in the exit stream C_A^I to 8×10^{-3} mg/L. The tower will operate at 5 atm and 293 K. The proposed packed tower is shown in Figure 8.3. To design a packed tower we must specify its diameter, its height of packing, and its overall height. This can be done only after we have selected a nitrogen flow rate q^{II}.

Step 2. Calculation of the Mass Transfer Load. The m_{load} is obtained from Eq. (4.5.1). Water is designated as Phase I and the nitrogen as Phase II:

$m_{load} = 50 (0.008 - 8.0) = -400$ mg/min,

m_{load} is negative in accordance with our convention.

Step 3. Choose a Mass Transfer Agent. This has been designated in the problem statement as nitrogen.

Table 8.6. *Additional tubular contactors*

Contactor type	Sieve tray tower	Multistage agitator tower
Operation	Staged	Continuous or staged
Dispersed phase	Liquid	Liquid Gas
Continuous phase	Liquid	Liquid
Diagram		
Fluid motions	Countercurrent	Countercurrent
Dispersed phase	Plug flow through holdup on each tray	Plug flow
Continuous phase	Well-mixed liquid holdup on each tray	Limiting cases are plug and mixed fluid motions
Design issues	• Area determination • Flow patterns (require large difference in densities between phases)	• Area determination • Separation/isolation of stages
Basic model classification	Stage/tray; continuous, plug–mixed contactor	Continuous, countercurrent plug–plug contactor or plug–mixed contactor

Step 4. Flow Rate Determination. The proposed stripper is a tubular countercurrent mass contactor so Eq. (4.5.3) applies if we wish to determine minimum nitrogen flow rate:

$$q_{min}^{II} = q^I \left(\frac{C_{AF}^I - C_A^I}{C_{AF}^I M^{-1} - C_{AF}^{II}} \right). \tag{4.5.3}$$

To use this equation we need a value for M that we have defined as follows in Chapter 4:

$$C_A^I = M C_A^{II}. \tag{4.1.7}$$

To determine the distribution constant M, we assume that Henry's "Law" applies:

$$x_A = \frac{p_A}{H}, \tag{4.1.6}$$

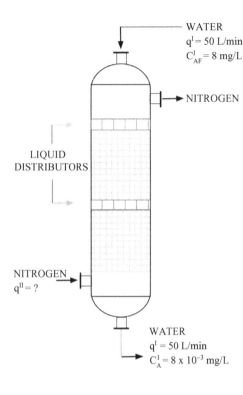

WATER
$q^I = 50$ L/min
$C^I_{AF} = 8$ mg/L

NITROGEN

LIQUID
DISTRIBUTORS

Figure 8.3. Packed tower.

NITROGEN
$q^{II} = ?$

WATER
$q^I = 50$ L/min
$C^I_A = 8 \times 10^{-3}$ mg/L

where x_A is the mole fraction of A in solution, H is the Henry's "Law" constant, and p_A is the partial pressure of A in the vapor phase. x_A can be expressed in terms of our C^I_A as follows:

$$x_A = \frac{C^I_A}{MW_A C_W}.$$

C_W is the concentration of the water in moles per liter $= 55$ mol/L.

The partial pressure p_A can be related to C^{II}_A by the ideal-gas law:

$$p_A = \frac{N_A}{V} RT = \frac{C^{II}_A}{MW_A} RT.$$

Substituting the expression for x_A and p_A into Eq. (4.1.6) yields

$$C^I_A = \frac{RT}{H} C_W C^{II}_A.$$

Equation (4.1.7) yields a value for M that we can obtain as follows by being careful with the units on R:

$$M = \frac{RT}{H} C_W.$$

Using
$$T = 293 \text{ K}; R = 8.314 \frac{J}{\text{mol K}},$$

and, from Table 4.3,

$$H = 4.01 \times 10^4 \text{ atm},$$

we find

$$M = \frac{(8.314\,\text{J/mol K})\,(293\,\text{K})(55\,\text{mol/L})(10^3\,\text{L/m}^3)}{(4.01 \times 10^4\,\text{atm})\left(1.01 \times 10^5 \dfrac{\text{J/m}^3}{\text{atm}}\right)}$$

$$M = 0.034.$$

The minimum nitrogen flow rate is

$$q_{min}^{II} = (0.83 \times 10^{-3}\,\text{m}^3/\text{s})\left[\frac{(8\,\text{mg/L}) - (8 \times 10^{-3}\,\text{mg/L})}{\left(\dfrac{8}{0.034} - 0\right)\text{mg/L}}\right]$$

$$q_{min}^{II} = 0.028 \times 10^{-3}\,\text{m}^3/\text{s}$$

$$= 1.68\,\text{L/min}.$$

The use of stage efficiency to calculate the nitrogen flow rate, as discussed in Section 4.5, is not used here because other operating constraints must be considered for this type of tower. Specifically, the gas flow rate must be selected to operate in a stable flow regime with good distribution of gas and liquid throughout the column.

To proceed we need to find a nitrogen flow rate q^{II} that will enable us to determine the tower diameter and pressure drop across the tower. This is a very complex problem in multiphase fluid mechanics. The liquid needs to be distributed uniformly through the packing, and the gas flow must be such that it does not carry liquid out of the column. If the liquid distribution devices are not properly designed, the downward-flowing liquid will not distribute evenly over the packing, and, in the most extreme cases, it may flow in separate channels from the flowing gas and there will be little contact for mass transfer. If the gas velocity is too high it will entrain liquid drops and carry them out of the tower in the overhead stream. Such issues have been investigated for many years experimentally in various commercial-scale towers with many different types of packing. The results of these efforts are summarized in Figure 8.4, which shows operability limits for packed towers. Each line on the graph corresponds to the pressure drop shown on the line. The uppermost line designates the flooding limit. Flooding is the term applied to the condition in which the tower may fill with liquid, causing the operation to go from the gas being the continuous fluid to the liquid being the continuous fluid. As the tower floods there is significant liquid carryover and there may be foaming. The y axis of Figure 8.4 contains G', the gas flux, and the x axis contains the ratio of the liquid-to-gas flux, L'/G'. The following steps allow us to estimate q^{II} and hence the tower diameter.

The symbols used in Figure 8.4, their units, and their equivalents in the nomenclature we use in this text are tabulated below:

$$L^I \frac{\text{kg}}{\text{m}^2\,\text{s}} = \frac{q^I \rho^I}{A_T \text{m}^2} \frac{\text{m}^3}{\text{s}} \frac{\text{kg}}{\text{m}^3},$$

A_T is the column cross-sectional area, in m^3,

$$G^I = \frac{q^{II} \rho^{II}}{A_T}$$

$$\rho_G = \rho^{II}, \text{ and } \rho_L = \rho^I. \tag{8.2.1}$$

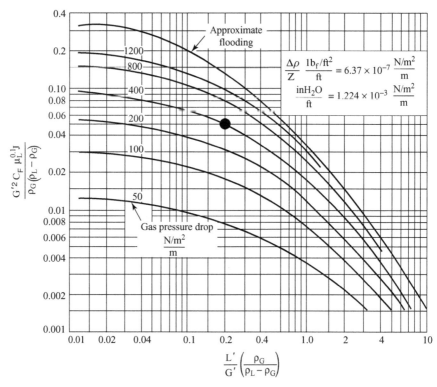

Figure 8.4. Flooding and pressure drop in random-packed towers. (Treybal RE. *Mass Transfer Operations*, 3rd ed., 1980. Reproduced with permission.)

One needs, as always, to be careful with the units. When SI units are used $J = 1$ and $g_c = 1$.

In the absence of any information specifying the allowable pressure drop or restrictions on L' or G' caused by other parts of the process, selection of a point somewhere in the middle of Figure 8.4 should be a reasonable choice to avoid flooding or other problems that could arise close to the boundaries of where the data were collected to construct Figure 8.4.

To develop a technically feasible design we select an x-axis value of 0.2 and a pressure drop of 400 N/m^2/m. The y-axis value is therefore 0.05. This is shown on Figure 8.4 as a dot.

$$\frac{L'}{G'}\left(\frac{\rho^{II}}{\rho^{I} - \rho^{II}}\right)^{1/2} = 0.2 \quad \text{(x-axis value from Figure 8.4)}, \quad (8.2.2)$$

$$\text{water density} = \rho^{I} = 1000\,\text{kg/m}^3,$$

$$\text{Nitrogen density} = \rho^{II} = \frac{N}{V}\,MW_{N_2}$$

$$= \frac{P}{RT}\,MW_{N_2}$$

$$= 5.8\,\text{kg/m}^3 \text{ at 293 K and 5 atm,}$$

where

$$MW_{N_2} = 28.$$

Table 8.7. Characteristics of random packings (*Chemical Engineer's Handbook*, 1997, Table 14-7b)

Name	Material	Nominal Size		Wall thickness (mm)	Bed weight (kg/m³)	Area (m²/m³)	% voids	Packing factor C_F (m⁻¹)	Dry packing factor C_{Fd} (m⁻¹)	Vendor
		mm	Number							
Raschig	Metal	19	—	1.6	1500	245	80	984	—	Various
rings		25	—	1.6	1140	185	86	472	492	
		50	—	1.6	590	95	92	187	223	
		75	—	1.6	400	66	95	105	—	

Then,

$$\frac{L'}{G'} = 0.2 \left(\frac{1000 - 5.8}{5.8} \right)^{1/2}$$

$$= 2.62.$$

With L'/G' known, we can obtain q^{II}:

$$\frac{L'}{G'} = \frac{q^I \rho^I}{q^{II} \rho^{II}},$$

$$q^{II} = \frac{8.3 \times 10^{-4} \times 1000}{5.8 \times 2.62}$$

$$= 0.055 \text{ m}^3/\text{s}$$

$$= 3300 \text{ L/min}.$$

With q^{II} known we can use Eq. (8.2.1) to obtain the tower cross-sectional area A_T if we compute G' by use of Figure 8.4:

$$\left[(G')^2 \frac{C_F \mu_L^{0.1}}{\rho^{II} (\rho^I - \rho^{II})} \right] = 0.05 \quad \text{(y-axis value from Figure 8.4)},$$

$$\mu_L = 10^{-3} \text{ kg/m s}. \tag{8.2.3}$$

To obtain G' we need C_F, and this can be obtained from Table 8.7. We pick 25 mm Raschig rings with $C_F = 472 \text{ m}^{-1}$:

$$G' = \left[\frac{0.05 \, (5.8) \, (1000 - 5.8)}{C_F \, (10^{-3})^{0.1}} \right]^{0.5}$$

$$= 1.1 \text{ kg/m}^2 \text{ s}.$$

Note that the units are fixed by the correlation as defined in Figure 8.4. We now have all the information we need to obtain the tower cross-sectional area A_T.

With G' known, the tower cross-sectional area A_T is obtained from Eq. (8.2.1):

$$A_T = \frac{q^{II} \rho^{II}}{G'}$$

$$= 0.29 \text{ m}^2.$$

The tower diameter D_T is

$$D_T = \left[\frac{0.29(4)}{\pi}\right]^{0.5}$$
$$= 0.6\,\text{m}.$$

With A_T determined, L', the liquid flow per cross-sectional area, is

$$L' = \frac{q^l\rho^l}{A_T}$$
$$= \frac{(8.3\times10^{-4}\text{m}^3\text{/s})\,(1000\,\text{kg/m}^3)}{0.285\,\text{m}^2}$$
$$= 2.9\,\frac{\text{kg}}{\text{m}^2\,\text{s}}.$$

One could select other sets of conditions and determine different tower diameters and pressure drops, but the flow rates previously selected will serve to illustrate our design process.

To find the tower height we need to estimate K_M and the interfacial area for mass transfer of the oxygen from the liquid into the nitrogen gas stream.

Step 5. Mass Transfer Coefficient Estimate. As pointed out in Subsection 6.4.2 the mass transfer of the oxygen in the liquid to the nitrogen will be controlled by the liquid-phase transfer, $K_m = k_m$. Equation (6.4.13), penetration theory model, can be used to obtain an estimate of k_m if we can estimate L and V_{max}. L is the distance over which the gas phase contacts the liquid, and V_{max} is the velocity of the liquid flowing over the packing. To obtain V_{max} we need to assume what portion of the column cross-sectional area is occupied by the flowing water. Let us assume that it will occupy about one quarter of A_T and see if we get reasonable results. We also need to estimate the length of the interface L. We approximate L as half the 25-mm dimension of the random-packed Raschig rings:

$$k_m = 2\sqrt{\frac{D_A V_{max}}{\pi L}}.$$

D_A can be found in Table 5.2. We need the diffusion coefficient for oxygen in water because the mass transfer is liquid-phase controlled:

$$D_A = 1.8\times10^{-9}\text{m}^2\text{/s},$$

$$V_{max} \simeq \frac{q^l}{0.25\,A_T}$$
$$\simeq \frac{8.3\times10^{-4}}{0.25\,(0.285)}$$
$$\simeq 0.01\ \text{m/s},$$

$$L \simeq \frac{25\,\text{mm}}{2\times1000\ \frac{\text{mm}}{\text{m}}}$$
$$\simeq 0.0125\,\text{m},$$

$$k_m = 2\sqrt{\frac{1.8\times10^{-9}\times0.01}{\pi\times0.0125}}$$
$$= 4.3\times10^{-5}\ \text{m/s}.$$

Step 6. Interfacial Area Determination. To estimate the interfacial area per unit volume a_v, we need to make use of an experimental correlation, of which there are as many as there are packing suppliers. We use the one reported in the *Chemical Engineer's Handbook* (1997), which is developed for 5-mm Rashig rings [Eq. (7.3.6)].

The interfacial area for mass transfer in a packed column is a function of the area of the dry packing a_w and the manner in which the gas and liquid flow rates interact with each other as they flow over the packing:

$$\frac{a_v}{a_w} = 0.54(G')^{0.31}(L')^{0.07},$$

$$G' = 1.1\frac{kg}{m^2\,s},$$

$$L' = 2.9\frac{kg}{m^2\,s},$$

$$\frac{a_v}{a_w} = 0.54(1.1)^{0.31}(2.91)^{0.07} = 0.6.$$

$a_w = 185\,m^2$ for the 25-mm Raschig ring packing from Table 8.7; so

$$a_v = 110\,m^2/m^3.$$

To determine the tower height we need to know what interfacial area is required for reducing the oxygen content in the water. Equation (4.5.5) applies with the countercurrent expression from Table 4.7:

$$\langle \Delta C^{I-II} \rangle = (-)\frac{\left(C_A^I - MC_{AF}^{II}\right) - \left(C_{AF}^I - MC_A^{II}\right)}{\ln\left(\dfrac{C_A^I - MC_{AF}^{II}}{C_{AF}^I - MC_A^{II}}\right)}.$$

The interfacial area a is calculated from Equation 4.5.5 and Table 4.7:

$$a = \frac{m_{load}}{k_m\langle \Delta C^{I-II}\rangle}.$$

To obtain $\langle \Delta C^{I-II}\rangle$ we need $C_A^I = 8 \times 10^{-3}$ mg/L, C_{AF}^{II}, which is 0, and C_A^{II}, which we obtain from the Level I analysis:

$$C_A^{II} = \frac{q^I}{q^{II}}\left(C_{AF}^I - C_A^I\right)$$

$$= \frac{8.3 \times 10^{-4}}{0.055}\left[(8\,mg/L) - (8 \times 10^{-3}mg/L)\right]$$

$$= 0.12\frac{mg}{L};$$

$$\langle \Delta C^{I-II}\rangle = -\frac{\left(8 \times 10^{-3}mg/L + 0\right) - (8\,mg/L - (0.034)0.12\,mg/L)}{\ln\left(\dfrac{8 \times 10^{-3}}{8 - (0.034)\,0.12}\right)}$$

$$= -1.16\,mg/L = -1.16 \times 10^3\,mg/m^3$$

$$a = (-400\,mg/min)\left(\frac{1}{60}\frac{min}{s}\right)\left(\frac{1}{-1.16 \times 10^3\,mg/m^3}\right)\left(\frac{1}{4.3 \times 10^{-5}\,m/s}\right)$$

$$= 134\,m^2.$$

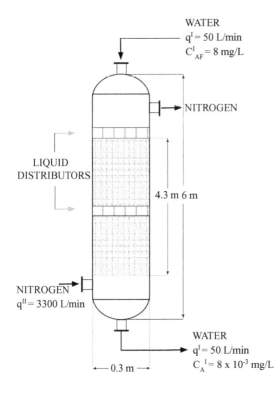

WATER
$q^I = 50$ L/min
$C^I_{AF} = 8$ mg/L

NITROGEN

LIQUID
DISTRIBUTORS

4.3 m 6 m

NITROGEN
$q^{II} = 3300$ L/min

WATER
$q^I = 50$ L/min
$C_A{}^1 = 8 \times 10^{-3}$ mg/L

0.3 m

Figure 8.5. Packed-tower design.

The tower height can be calculated because we know the required area and the area per unit volume of the packing.

$$\text{Tower packing volume} = \frac{a}{a_v}$$

$$= \frac{134}{110} = 1.2 \, \text{m}^3,$$

$$\text{Tower packing height} = \frac{1.2 \times 4}{\pi \, (0.6)^2}$$

$$= 4.3 \, \text{m}.$$

To obtain the tower height we need at least another 1–2 m to allow space for proper liquid draw-off, space for the nitrogen to separate from the liquid, and space for hardware installation such as liquid distributors. There are uncertainties in the estimation of k_m and the area per unit volume of the packing a_v that one needs to consider. To get better values, one needs to make a critical search of both the technical literature and the literature available from packing manufacturers. As with heat exchangers it is always possible to get bids from tower designers by going out on performance. The exercise we have just completed allows one to better evaluate bids from various vendors.

A sketch of our packed tower with the dimensions specified is shown in Figure 8.5. The logic to obtain a packed-tower diameter and height is summarized in Figure 8.6, which expands on the continuous-flow mass contactor design summary

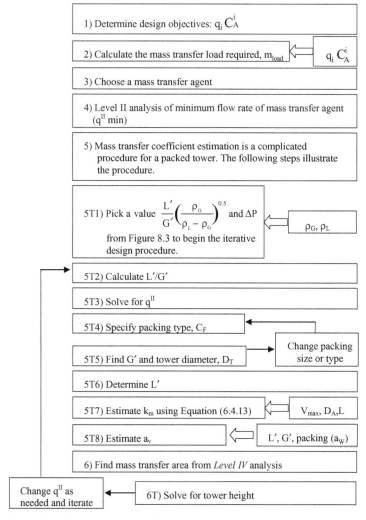

Figure 8.6. Design procedure for a packed tower. (Design summary in Section 4.5.)

in Section 4.5 to empirically account for the complex two-phase fluid motions in packed towers.

8.3 Analysis of a Pilot-Scale Bioreactor

In this case study we expand our analysis of the commercially available bioreactor or fermentor described in Chapter 2 to illustrate the importance of the mass transfer analysis in scaling up a pilot-scale device to commercial scale. Figure 8.7 is a sketch of a typical bioreactor.

The *Candida utilus* microorganism fermentor analysis discussed in Example 2.2 did not include a gas-phase component mass balance. Here we discuss a more realistic extension of this example, in which we consider mass transfer in this two-phase

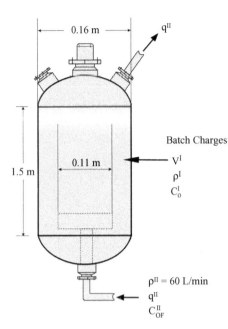

Figure 8.7. Semibatch bioreactor.

(mixed–plug) semibatch system. Recall that the model equations for the substrate glucose C_B and the biomass C_A are as follows:

$$\frac{dC_A}{dt} = \frac{k_1 C_A C_B}{K_B + C_B},$$

(2.2.15)

$$\frac{dC_B}{dt} = \frac{-k_2 C_A C_B}{K_B + C_B}.$$

(2.2.16)

The model equations that include the semibatch mass transfer for the oxygen are presented in Chapter 4 as Eqs. (4.3.13) and (4.3.14). To develop the model equation for the oxygen, we need to include an extra component, so we designate with a subscript O the concentration of oxygen in both phases.

We make the assumption that there is very little depletion of oxygen, C_O^{II}, from the gas phase. Thus $d/dz \, (C_O^{II} V^{II}) = 0$ in Eq. (4.3.14). This is a reasonable assumption, as the gas bubbles produced in this system rise through the liquid at a minimum velocity of the order of 0.3 m/s. Because the bubbles are in the liquid for only a few seconds, their oxygen content is not significantly depleted and C_O^{II} is essentially constant.

The discussion in subsection 4.3.2.2 for a mixed–plug mass contactor presents the model equation for the oxygen concentration in the liquid, C_O^I. In the absence of any chemical reaction consuming the oxygen, Eq. (4.3.17) applies:

$$\frac{dC_O^I}{dt} = \frac{-k_m a}{V} \left(C_O^I - C_{O,eq}^I \right).$$

(8.3.1)

An oxygen-depletion term needs to be added to this equation because of the cell consumption of oxygen, but we can determine the correct constitutive relationship only when we can compare model behavior with data. Application of the word statement, Figure 1.6, yields

The mass of oxygen in Phase I at time t+Δt	=	The mass of oxygen in Phase I at time t	+	The mass of oxygen entering Phase I by mass transfer	−	The mass of oxygen leaving Phase I by mass transfer
			+	The mass of oxygen entering Phase I by reaction	−	The mass of oxygen leaving Phase I by reaction.

Using r_{O+} and r_{O-} to denote the mass rate of production and consumption of oxygen by the biomass per volume, respectively (Section 2.1), and $\mathbf{r}_{O+}$ and $\mathbf{r}_{O-}$ to denote the rates of mass transfer (Section 4.2), we find

$$C_O^I\, V|_{t+\Delta t} = C_O^I\, V|_t + \left(\mathbf{r}_{O+}^I\, a - \mathbf{r}_{O-}^I\, a\right)\Delta t$$
$$+ \left(r_{O+}^I\, V - r_{O-}^I\, V\right)\Delta t.$$

Assuming that the volume is constant (i.e. that the amount of mass transfer of oxygen is small) and taking the limit as $\Delta t \to 0$ leads to

$$V\frac{dC_O^I}{dt} = \mathbf{r}_O^I\, a + r_O^I\, V.$$

Now we use the mass transfer rate expression, Eq. (4.2.11), which applies because Phase II has a significant excess of oxygen in the air. Our model equation becomes

$$\frac{dC_O^I}{dt} = -k_m a_v \left(C_O^I - C_{O,eq}^I\right) + r_O^I, \tag{8.3.2}$$

where $a_v = a/V$, the area per volume. Now, as was done in Problem 2.11, we need to examine the experimental data to determine an appropriate expression for the reaction rate. Recall that we have already solved the model equations for the biomass and substrate, (2.2.15) and (2.2.16), and determined the rate constants k_1, k_2, and K_B ($k_1 = 0.34$ L/g h, $k_2 = 0.64$ L/g h, $K_B = 0.48$ g/L).

In Figure 8.6 the experimental data for the oxygen concentration as a function of time are shown along with the biomass concentration data and the model fit from Chapter 2.

The solid circles show the biomass concentration and the triangles show the oxygen concentration in the broth, C_O^I. In Chapter 2 we were able to obtain a good fit to the data by assuming that the biomass had sufficient oxygen to grow properly. Thus we did not have to include the oxygen concentration in the constitutive relations nor did we need to consider the oxygen concentration balance equation. We retain the values that fit the substrate and biomass model equation and test our ability to fit the oxygen data given that information.

The total amount of oxygen that disappears from the control volume (in this case the Biostat UD 30) is due to the consumption by the biomass. Figure 8.8 shows that the batch operation begins with the bioreactor oxygen concentration at a value close to saturation at 7.2 mg/L. Equation (8.3.1) tells us that the oxygen

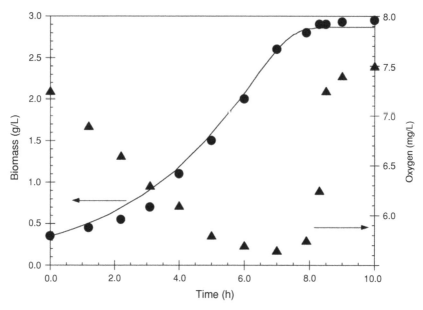

Figure 8.8. Experimental data (symbols) showing oxygen concentration and biomass concentration as functions of time in the bioreactor. [Line fit of biomass is a solution to model equations (2.2.15) and (2.2.16) (Tobajas et al., 2003).]

consumption is greater than the amount transferred from the sparged-air phase until between 6 and 8 h. At this point in time the amount transferred from the air is equal to that consumed by the biomass for cell growth. Beyond 8 h the amount transferred from the air is greater than that consumed. During the whole time that the experiment was performed there was more dissolved oxygen than that required by the biomass. The value of C_O^I never dropped below 5.3 mg/L such that there was sufficient oxygen for the cells. This is why we were able to carry out the analysis in Chapter 2 without including the oxygen component mass balance, Eq. (8.3.1).

To find the constitutive equation for the oxygen consumption we need to try various expressions to see if the model prediction will fit the data. We can make this process more effective if we first calculate K_m and the area by using the equations from Chapter 7.

From the data given by Tobajas et al. (2003), we know that the gas flow rate (q^{II}) in the fermentor is 60 L/min, or 10^{-3} m³/s. The bioreactor diameter, D, is 0.16 m and its height is 1.5 m.

Air is sparged into the bottom of the bioreactor via an 11-cm perforated ring with 24 orifices in the ring of 1-mm diameter each. Such a configuration will produce a single two-phase plume. The gas bubbles will entrain liquid in the plume of diameter d, and this liquid will then flow down in the annular space between the plume and the bioreactor wall. Although it is not necessary to confine the plume, the bioreactor of Tobajas et al. (2003) has a draft tube of 11-cm diameter. Figure 8.7 shows these dimensions on a standard dished-head vessel. Thus the superficial velocity (v_s) is defined as follows:

$$v_s = q^{II}/(\pi d^2/4) = \left(4 \times 10^{-3}\,\text{m}^3/\text{s}\right)/[\pi(0.11\,\text{m})]^2 = 0.1\,\text{m/s}. \qquad (8.3.3)$$

Using Eq. (7.2.6) for air–water systems, we can determine the bubble diameter at the orifice:

$$d_{B0} = 0.39(q^{II})^{0.32} = 1.5 \, cm, \qquad (8.3.4)$$

and the power input for the gas flow through the orifice driving the liquid circulation is determined from Eq. (7.2.8):

$$\begin{aligned} power &= q^{II}_{AVG} \, g \, \rho_L h_L \\ &= (1 \times 10^{-3} \, m^3/s)(9.8 \, m/s^2)(1000 \, kg/m^3)(1.5 \, m) \\ &= -14.7 \, W. \end{aligned} \qquad (8.3.5)$$

The plume mass is

$$mass = \rho^I \, \pi d^2/4 \, h_L = (1000 \, kg/m^3) \, 3.14(11/100 \, m)^2/4 \, (1.5 \, m) = 14.3 \, kg.$$
Thus, ε = power/mass = 1.03 W/kg.

The bubble diameter is then found from Eq. (7.2.11) for air–water systems:

$$d_B = (0.26 \times 10^{-2})\varepsilon^{-4} = 0.0026 \, m = 2.6 \, mm.$$

The liquid velocity in the plume, Eq. (7.2.13), becomes $v^I = 0.84$ m/s, and the rise velocity of the bubble is

$$u_r = [(1.45 \times 10^{-4}/d_B) + 4.9 \, d_B]^{1/2} = 0.42 \text{ m/s.}$$

The gas holdup ϕ, Eq. (7.2.15), ignoring contributions from static head, yields

$$\phi = \frac{q^{II}}{\dfrac{\pi d^2}{4}(v^I + u_r)} = \frac{0.10}{(0.84 + 0.42)}$$

$$\phi = 0.08,$$

telling us that the gas phase occupies 8% of the system volume.

Now, to determine the contact area for mass transfer, we utilize Eq. (7.2.12)

$$a_v = 6\phi/d_B = 6(0.08/0.0026) = 183 \, m^2/m^3 = a/V.$$

Because there are no bubbles in the downward-flowing liquid outside the plume, we multiply by the volume of the draft tube ($0.0095 \, m^2 \times 1.07$ m) to yield

$$a = 1.8 \, m^2.$$

We can calculate the value for K_m by using the parameters we have determined using Equation (7.2.16) modified for the air–water system:

$$\begin{aligned} k_m &= 4.79 \times 10^{-5}(u_r/d_B)^{1/2} = 1.93 \times 10^{-4} \text{ m/s,} \\ K_m a &= 1.93 \times 1.8 \times 10^{-4} \\ &= 3.47 \, m^3/s \times 10^{-4} \\ k_m a_v &= 1.93 \times 10^{-4} \times 183 \\ &= 3.53 \times 10^{-2} \, s^{-1} \\ &= 127 \, h^{-1}. \end{aligned}$$

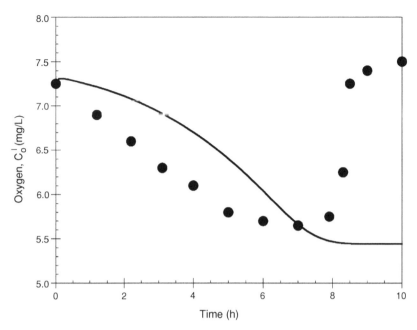

Figure 8.9. Fit of oxygen concentration data (Tobajas et al., 2003, symbols) to Eq. (8.3.6) (curve).

Tobajas et al. (2003) calculated $K_m a_v$ in their paper when studying oxygen transfer in this bioreactor, and they report values between 3.7×10^{-2} s^{-1} and 3.17×10^{-2} s^{-1}. Their experimentally determined values ranged from 2.63×10^{-2} s^{-1} to 4.95×10^{-2} s^{-1}, depending on the glucose concentration values.

Using the preceding value for $K_m a_v$ of 126 h^{-1}, we can now test various constitutive relations for the oxygen-depletion term in Eq. (8.3.2).

The total amount of oxygen that has disappeared from the control volume during Δt is due to the growth of the cells, and so the consumption of the oxygen is by the growing biomass. Thus the simplest constitutive relationship we can write is an expression that is first order in biomass, i.e., $r_A{}^I = - k_3 C_A$, and substitute into Eq. (8.3.2) to obtain

$$\frac{dC_O^I}{dt} = -k_m a_v \left(C_O^I - C_{O,eq}^I\right) - k_3 C_A. \tag{8.3.6}$$

Using the three model equations [(2.2.15), (2.2.16), and (8.3.6)] and our previous constants (k_1, k_2, and K_B) to solve for C_A, C_B, and C_O^I simultaneously, and fitting a value for k_3 to minimize the error between the model prediction and experiment gives us a value of $k_3 = 0.095$ h^{-1}. Figure 8.9 shows the model prediction as a solid curve.

This is not a satisfactory fit to the data, especially not for the later stages of growth where the substrate glucose is limiting. With that idea in mind, we add a dependency of the substrate into the constitutive equation, namely $r_A^I = - k_4 C_A C_B$:

$$\frac{dC_O^I}{dt} = -k_m a_v \left(C_O^I - C_{O,eq}^I\right) - k_4 C_A C_B. \tag{8.3.7}$$

The model fit to the experimental data shown in Figure 8.10 yields a value for $k_4 = 0.597$ L/g h.

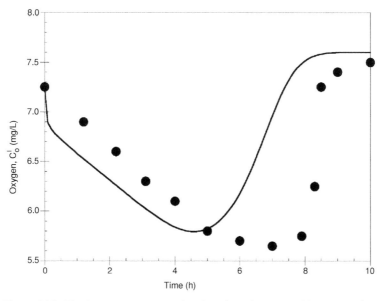

Figure 8.10. Fit of oxygen concentration data (Tobajas et al., 2003, symbols) to Eq. (8.3.7) (curve).

This fits the data better and leads us to believe that the dependence on substrate may be important. A final variation on the constitutive equation is to test the Monod form $r_A^I = -k_5 k_1 C_A C_B/(K_B + C_B)$. The use of $k_5 k_1$ assumes a one-to-one relationship between oxygen and cell mass growth, and fits best at early growth, during the most rapid growth phase. This yields the following model equation for the oxygen:

$$\frac{dC_O^I}{dt} = -k_m a_V \left(C_O^I - C_{O,eq}^I\right) \frac{k_5 k_1 C_A C_B}{K_B + C_B}. \tag{8.3.8}$$

Using this equation yields $k_5 = 0.41$ and the best fit of the three model equations (Figure 8.11).

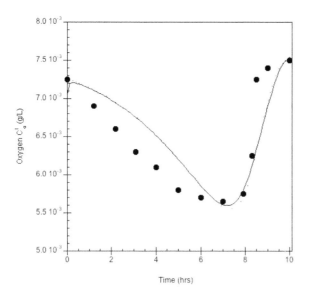

Figure 8.11. Fit of oxygen concentration data (Tobajas et al., 2003, symbols) to Eq. (8.3.8) (curve).

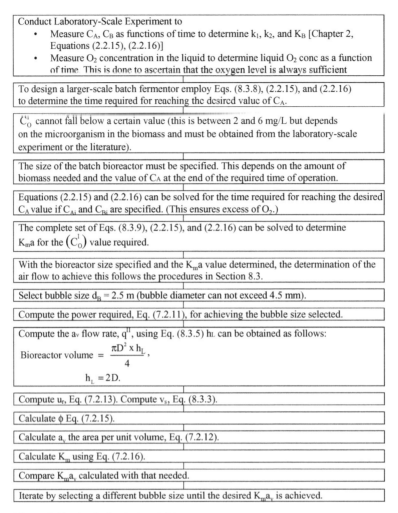

Conduct Laboratory-Scale Experiment to
- Measure C_A, C_B as functions of time to determine k_1, k_2, and K_B [Chapter 2, Equations (2.2.15), (2.2.16)]
- Measure O_2 concentration in the liquid to determine liquid O_2 conc as a function of time. This is done to ascertain that the oxygen level is always sufficient

---

To design a larger-scale batch fermentor employ Eqs. (8.3.8), (2.2.15), and (2.2.16) to determine the time required for reaching the desired value of C_A.

C_O^{si} cannot fall below a certain value (this is between 2 and 6 mg/L but depends on the microorganism in the biomass and must be obtained from the laboratory-scale experiment or the literature).

The size of the batch bioreactor must be specified. This depends on the amount of biomass needed and the value of C_A at the end of the required time of operation.

Equations (2.2.15) and (2.2.16) can be solved for the time required for reaching the desired C_A value if C_{Ai} and C_{Bi} are specified. (This ensures excess of O_2.)

The complete set of Eqs. (8.3.9), (2.2.15), and (2.2.16) can be solved to determine $K_m a$ for the $\left(C_O^l\right)$ value required.

With the bioreactor size specified and the $K_m a$ value determined, the determination of the air flow to achieve this follows the procedures in Section 8.3.

Select bubble size $d_B = 2.5$ m (bubble diameter can not exceed 4.5 mm).

Compute the power required, Eq. (7.2.11), for achieving the bubble size selected.

Compute the a$_v$ flow rate, q^{II}, using Eq. (8.3.5) h$_L$ can be obtained as follows:

$$\text{Bioreactor volume } = \frac{\pi D^2 \times h_L}{4},$$

$$h_L = 2 D.$$

Compute u_r, Eq. (7.2.13). Compute v_s, Eq. (8.3.3).

Calculate ϕ Eq. (7.2.15).

Calculate a$_v$ the area per unit volume, Eq. (7.2.12).

Calculate K_m using Eq. (7.2.16).

Compare $K_m a_v$ calculated with that needed.

Iterate by selecting a different bubble size until the desired $K_m a_v$ is achieved.

Figure 8.12. Analysis of a batch bioreactor.

We could also consider more complex models, such as coupling oxygen consumption to substrate utilization, bearing in mind the need of the model in relationship to our stated goals. That is, what are we using the model for — to determine parameters for scale-up of this system? To predict cellular behavior? To understand metabolism? Here we have achieved a good fit of the experimental data and obtained rate constants that would enable a consistent run under similar experimental growth conditions. Further, we have validated the use of the Chapter 7 equations for a complex air–liquid system for which there is experimental data for the determination of K_m.

To produce a desired amount of cells, $C_A V$, from a batch fermentor, a fermentor volume V must be specified. This requires one to decide how many batches can be accommodated in a given time period and how many fermentors can be employed.

Figure 8.12 shows how we would use the analysis to develop a technically feasible design for any batch bioreactor if the volume of the reactor is specified. Problem 8.1 shows how one can proceed for a continuously operating system.

Nomenclature

a	Area (m^2)
A	Cross-sectional area (m^2)
A_{cv}	Packing area/volume (m^2/m^3)
C	Concentration (kg/m^3)
C_F	Packing factor (m^{-1})
$\hat{C}_p$	Heat capacity (kJ/kg K)
D	Diameter (m)
D_H	Hydraulic diameter (m)
d_B	Bubble diameter (m)
G'	Gas flow rate (kg/m^2 s)
h	Local heat transfer coefficient (W/m^2 K)
H	Henry's "Law" constant
k	Rate constant
k	Thermal conductivity (W/m K)
L	Length (m)
L'	Liquid flow rate (kg/m^2s)
M	Distribution coefficient
m_{load}	Mass transfer load (kg/s)
MW	Molecular weight (g/mol)
N	Number of pipes
Nu	Nusselt number, hD/k
p	Partial pressure (Pa)
Pr	Prandtl number, $\hat{C}_p$m/k
P_W	Wetted perimeter (m)
q	Flow rate (m^3/s)
Re	Reynolds number
r	Rate of mass transfer
T	Temperature (K)
U	Overall heat transfer coefficient (W/m^2 K)
v_s	Superficial velocity (m/s)
v	Velocity (m/s)

Greek

Δ	Difference
μ	Viscosity (Pa-s)
ρ	Density (kg/m^3)

Subscripts

A	Species A
F	Feed
min	Minimum
G	Gas
L	Liquid
O	Oxygen

Superscripts

I	Phase I
II	Phase II

REFERENCES

Chemical Engineers' Handbook (7th ed.). New York: McGraw-Hill Professional, 1997.

Donohue D. Heat transfer and pressure drop in heat exchangers. Ind. Eng. Chem. 1949; 41(11): 490–511.

Tilton JN, Russell TWF. Designing gas sparged vessels for mass transfer. Chem Eng. November 29, 1982: 61–8.

Tobajas M, García-Calvo E, Wu X, Merchuk J. A simple mathematical model of the process of *Candida utilis growth* in a bioreactor. World J. Microbiol. Biotechnol. 2003; 19: 391–8.

Treybal RE. *Mass Transfer Operations*. New York: McGraw-Hill, 1980.

Seader JD, Henley EJ. *Separation Process Principles*. New York: Wiley, 1998.

PROBLEMS

8.1 Perform a technically feasible design of a multiplume gas-sparged contactor for oxygenating waste water in a treatment facility given the following process specifications. The process requires oxygenating a 0.5-m^3/s water stream (q^I) (initially oxygen free) to an oxygen content (C_A^I) of 2×10^{-3} kg/m^3 at a temperature of 20 °C. An existing on-site air line is to be employed as the oxygen source. A *multiplume contactor* [with a flow pattern (mixed–plug) identified in previous research, as shown in Figure P8.1] is to be used for the process. Critical design information can be found in Tilton and Russell (1982), which is summarized in the table that follows.

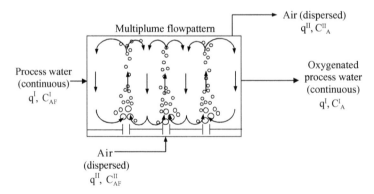

Figure P8.1.

a. Develop a list of any data you will require for this design.

b. As *preliminary* calculations, determine

   i. Minimum air flow rate required, q^{II}.

   ii. Design exit oxygen concentration of the air, C_A^{II}.

   iii. A design strategy (i.e., logic diagram with appropriate model equations).

Design Equations for Problem 8.1

1	T&R (5a)	*Liquid-side mass transfer coefficient*	$k_L = 4.79 \times 10^{-5} \left(u_r/d_B\right)^{1/2}$ [m/s]	Higbie penetration theory for oxygen transfer to water where d_B is the bubble diameter and u_r is the bubble rise velocity.
2	T&R (4a)	*Bubble rise velocity*	$u_r = \left[\left(\dfrac{1.45 \times 10^{-4}}{d_B} \right) + 4.9 d_B \right]^{1/2}$ [m/s]	Calculated based on Mendelson's relation for Re≫1 for air–water system where d_B is the bubble diameter.
3	T&R (7a)	*Liquid velocity in plume and annulus*	$v_L = \left[\dfrac{6.24 Q_M}{0.01D + \left(Kd^2/h_L\right)} \right]^{1/3}$ [m/s]	Q_M is the mean gas flow rate [m³/s], D is the diameter of the circulation element, K is the number of velocity heads lost (K = 1.3 is *suggested*), d is the plume diameter, and h_L is the liquid height.
4	T&R (20a)	*Stable bubble diameter for air–water system*	$d_B = (4.12 \times 10^{-2}) \left[P/V_p\right]^{-0.4}$ [m]	P/V_p is the power dissipation.
5	T&R (21a)	*Power dissipation*	$[P/V_p] = 12,480 \left(\dfrac{Q_M}{\pi t d^2} \right)$ [W/m³]	P/V_p is the power dissipation, d is the plume diameter [m], and Q_M is the mean volumetric gas flow rate [m³/s] per plume.
6	T&R (15)	*Plume diameter, d*	$d^2 = 0.5D^2$	D (typically 0.1–0.5 m) is the diameter of the circulation element.

(cont.)

(*Cont.*)

7	T&R (16)	*Tank diameter*	$D_T = N_s^{1/2} D$	D (typically $0.1 - 0.5$ m) is the diameter of the circulation element and N_s is the number of bubbling stations.
8	T&R (19)	*Fractional gas hold-up*	$\phi = \left(\dfrac{u_s}{u_r + v_L} \right) \left[\dfrac{p_o}{p_s + \rho_L g (h_L - z_L)} \right]$	u_s is the bubble rise velocity, v_L is the liquid velocity in the plume/annulus, p_o is the orifice pressure [N/m^2], p_s is the pressure at the liquid surface [N/m^2], ρ_L is the liquid density [kg/m^3], h_L is the liquid height, and z_L height corresponding to half the vessel depth.
9	T&R (8)	*Superficial gas velocity based on plume area*	$u_s = \dfrac{4 Q_{G_o}}{\pi d^2}$ [m/s] where $Q_{G_o} = \dfrac{Q_M (p_o - p_s)}{p_o \ln \left(\dfrac{p_o}{p_s} \right)}$ [m^3/s]	d is the plume diameter, Q_{G_o} is the gas flow rate per orifice, p_o is the orifice pressure [N/m^2], and p_s is the pressure at the liquid surface [N/m^2].
10	T&R (29)	*Tank height*	$h = \dfrac{h_L}{1 - \phi/2}$	h_L is the liquid height, ϕ is the fractional gas hold-up.
11	T&R (23)	*Interfacial area*	$a = 6\phi/d_B$ [m^{-1}]	area per unit volume in the two-phase plume

  c. Develop a technically feasible design, and specify

     i. The volumetric air flow rate, q^{II}, required.

     ii. The tank dimensions: diameter (D_T) and height (h).

     iii. The number (N_s) and spacing (D) of the bubbling stations.

     iv. The oxygen concentration in the outlet water (C_A^I) and airstreams (C_A^{II}).

     v. The mass transfer coefficient (K_L) and other critical information [e.g., area for mass transfer (a) and bubble diameter (d_B)].

     vi. Finally, identify three areas of *uncertainty* and how they may affect your design.

     Physical data:

     $T = 20\,°C = 293$ K,

     $P = 1$ atm,

     $\rho^I = 1000$ kg/m^3,

     $MW_{water} = 18$ g/mol,

     $MW_{oxygen} = 32$ g/mol,

     $R_{gas} = 0.082$ L atm/mol K.

     Henry's "Law" coefficient (O$_2$ in water, T = 20 °C):

     $H = 4.01 \times 10^4$ atm $\left(y_A^{II}P = Hx_A^I\right)$.

8.2 Perform a technically feasible design for the following process for removing acetone from water using trichloroethane (TCE). As TCE is valuable and also regulated, it is recycled after distillation. It removes essentially all of the acetone. The bottoms from the distillation (TCE) will be at a temperature of 55 °C and will be available for recycle for reuse in the acetone extraction.

  a. Process and design specifications:

    A 1000-L/min water stream at 25 °C containing 200 g/L of acetone must be treated to remove the acetone prior to waste discharge into a nearby river. A supply of 1,1,2-trichloreothane (TCE) is available to remove the acetone by liquid–liquid extraction.

  b. Mass contactor design:

    A liquid–liquid mass contactor(s) must be specified. The height, diameter, and the mixer power must be determined to complete your design. In addition, you must also specify exit stream flow rates and acetone concentrations. For this design, you will need to consider the experimental data for this acetone–water–TCE system presented in Figure 4.2.

  c. Separator design: The fluid–fluid mixture from the mass contactor must be separated into two streams, a TCE-free water stream, q^{II}, and a water-free TCE stream, q^I. This needs to be accomplished in a decanter or separator, which will allow the water droplets to separate from the TCE. The geometry and dimensions of the separator must be specified to complete the design.

  d. Heat exchanger design: The TCE stream from the separator is passed to a distillation column (*not to be designed*) to strip the acetone. The regenerated TCE is recycled to the beginning of the process at a temperature specified by the bottoms of the distillation column. Appropriate thermal management must be designed into the process so that the extraction previously discussed is carried out at 25 °C. Chilled water is available at 10 °C for heat exchange and can be returned to the utility company at no hotter than 60 °C. Any heat exchanger employed in the process will need to be specified in sufficient detail to send for a quote or fabrication.

   Provide a process flow diagram (showing *all* process equipment) with accompanying stream tables, technically feasible designs of the mass contactor(s), separator, and heat exchanger with associated equipment tables and supporting calculations. Comment on the amount of acetone that can be reasonably removed from the water and the suitability of discharging the effluent water into the environment, as well as on alternative strategies for efficiently removing the acetone.

   DATA: An experiment to determine the transient partitioning of acetone between an aqueous phase and an organic TCE phase has been performed. The experiment was performed in a batch system at 25 °C. 2 L of an acetone (A) – water (W) solution was contacted with 2 L of an initially acetone-free TCE phase in a batch mass contactor. This mass contactor consisted of a 5-L cylindrical tank ($D_{vessel} = 0.172$ m). A lab-scale impeller was employed to disperse the aqueous phase in the TCE phase. Relatively large bubbles (average diameter of 1 cm) of the aqueous phase resulted. Small samples of the TCE phase were taken throughout the experiment. The concentration of acetone in the TCE phase (I) was measured through gas chromatography analysis and is reported in the following table.

Acetone (A) concentration in the continuous TCE phase (I)

t (min)	C_A^I (g/L)
0.0	0.00
1.0	1.62
2.0	2.62
3.0	3.32
4.0	3.54
5.0	3.85
6.0	3.82
7.0	3.89

   The initial concentration of acetone in the dispersed aqueous phase was $C_A^{II} = 6.0$ g/L.

8.3   Methyl chloride (CH_3Cl) is used as a solvent in butyl rubber production. The monomers react in a packed-bed vessel catalyzed by an aluminum catalyst. After a certain amount of material has been processed, the catalyst has to be regenerated. This is done by sending a counterflow of nitrogen through the heated packed bed, stripping CH_3Cl and water from the bed. CH_3Cl is a toxic substance that can have serious effects on the nervous system with the possibility of coma or convulsions. Adverse effects on heart rate, liver, and kidneys have also been reported in humans following inhalation of methyl chloride. The EPA's reference concentration* (RfC) is 0.09 mg/m^3 and it is regulated under the Clean Air Act. The exhaust of any CH_3Cl into the atmosphere has to be limited. To avoid this, the nitrogen stream is internally recycled and reused during the regeneration processes. The CH_3Cl and water in this recycle stream have to be reduced to within acceptable limits to prevent accumulation of the CH_3Cl and water during the recurring regeneration cycles.

The rubber reaction and the following regeneration take 10 h each. If the CH_3Cl removal step is used for two reactors in parallel, the process can be operated continuously.

Nitrogen is used at a rate of 400 kg/h. Steam and CH_3Cl are released at a rate of 600 kg/h and 100 kg/h, respectively. The gas flow exiting the packed bed is at 150 °C, and 3 bars.

You are required to design a process to reduce the CH_3Cl and water in the nitrogen stream. The final exit streams have to be at 30 °C, and the process cannot take longer than 10 h. The maximum allowable amount of CH_3Cl in the nitrogen stream is 1 kg/h. The steam has to be removed from the gas stream to prevent accumulation during continuous recycling of the nitrogen. The stripped nitrogen is then compressed to 3.5 bars and sent back to the reactor, with a pressure drop of 0.5 bar. Selected physical and chemical data for CH_3Cl are given in the following table.

Physical and chemical properties of methyl chloride

Property	Value
Molecular mass	50.49 g/mol
Melting point	–97.7 °C
Boiling point	24.2 °C
Critical temperature	143 °C
Solubility in water at	mass%
0 °C	0.76
10 °C	0.676
20 °C	0.64–0.725
25 °C	0.48–0.70
30 °C	0.74
Vapor pressure at	10^2 Pa
–24.203 °C	1013
6.17 °C	2027
5.54 °C	3040
14.49 °C	4053
20 °C	5008
21.83 °C	5066
25 °C	5748
28.11 °C	6080
Henry's "Law" constant	$kPa\ m^3/mol$
Pure water	5
Seawater (salinity 30.4%)	0.1977

Develop a technically feasible design of a separation train as follows.

a. With 600 kg/h, steam is the largest component of the gas stream leaving the reactor. Water is also the highest boiling component of this gas stream, with a gap of more than 100 °C to the second highest boiler, methyl chloride (CH_3Cl). Thus it makes sense to condense the steam as a first step in the gas stream treatment. A condenser is constituted of a heat exchanger followed by a large tank. The heat exchanger is generally positioned so that the condensing fluid flows downward. Gravity will thus help the condensing liquid film to flow, without having to rely on the complex fluid dynamics inside the condenser. The tank is used to smooth out fluctuations. Rule of thumb usually dictates the tank to be large enough for a holdup time of 3–5 min when half full.

For the condenser design you are asked to calculate the heat load, surface area, outlet temperatures, and flow rate of the coolant. Perform calculations for two different cooling fluids: water and nitrogen (both coming into the heat exchanger at 25 °C). Because the preferred outlet temperature of the gas stream is not known yet (the second step in the gas stream treatment is yet to be designed), develop a spreadsheet solution so you can easily adjust your condenser design when designing the second part of the gas stream treatment.

You can use the following correlations to determine the heat transfer coefficient. Assume the resistance that is due to conductivity through the tube wall to be negligible, and the tube to have negligible thickness.

The correlation for the condensing fluid is

$$h = 0.729 \left[\frac{g \rho_l (\rho_l - \rho_g) k_l^3 H_v}{N \mu_l (T_{sat} - T_s) D} \right]$$

where

$g = 9.81$ m^2/s,
$\rho_i =$ density of the condensed liquid or gas respectively,
$k_l =$ conductivity of the condensed liquid,
$H_v =$ heat of vaporization,
$\mu_l =$ viscosity of the liquid,
$N =$ number of rows of tubes,
$T_{sat} =$ saturation temperature of the condensing vapor,
$T_s =$ surface temperature of the tube,
$D =$ tube diameter.

Note that the liquid properties have to be evaluated at the average liquid-film temperature, i.e., $(T_{sat} - T_s)/2$. Tube diameter and amount of tubes will be determined by the coolant that runs through the tubes.

For the coolant, use the following correlation from Chapter 6:

$$Nu = 0.027 \, (Re)^{0.8} \, Pr^{1/3} \left(\frac{\mu}{\mu_s} \right)^{0.14}.$$

Make sure the tube diameter is chosen so that $Re \geq 10000$ (turbulent flow). All coolant properties are taken at the average temperature of t.

b. After removing nearly all of the steam from the gas stream, there is essentially only methyl choride (CH_3Cl) and nitrogen left in the gas stream. The next step is a process to remove CH_3Cl from the gas stream so that the remaining, purified nitrogen can be reused. Neutralization of CH_3Cl with an aqueous solution of sodium hydroxide (NaOH) is a very common process. The resulting products, methanol (CH_3OH) and sodium chloride (NaCl) in water, are fairly harmless, and can easily be sent to a water treatment facility. You are asked to design a continuous two-phase reactor–contactor with an aqueous solution of NaOH, through which the nitrogen/CH_3Cl will be bubbled, operating at 25 °C. CH_3Cl will transfer from the gas into the aqueous phase and react with NaOH. An aqueous solution of 20 wt.% NaOH is available from within your chemical facility.

The reaction of CH_3Cl with NaOH can be described as follows:

$$NaOH + H_2O \longrightarrow Na^+ + OH^- + H_2O,$$

$$CH_3Cl + OH^- \longrightarrow CH_3OH + Cl^-,$$

$$\frac{dC_{CH_3Cl}}{dt} = -k_R C_{CH_3Cl} C_{OH^-},$$

with

$$k_R = 11.4 \, M^{-1} \, min^{-1} (\text{at } 25\,^{\circ}C).$$

The following operations are required:

1. Determine the amount of NaOH and aqueous NaOH solution needed per hour to reduce CH_3Cl to 1 kg/h in the gas stream (the inlet concentration of CH_3Cl is the concentration leaving the condenser in the gas stream).
2. Derive model equations for the concentration of all species in both phases.
3. Obtain estimates for the mass transfer coefficient, the interfacial area, and the bubble rise velocity.
4. Propose a technically feasible design.

NOTE

*RfC (inhalation reference concentration): An estimate (with uncertainty spanning perhaps an order of magnitude) of a continuous inhalation exposure of a chemical to the human population through inhalation (including sensitive subpopulations) that is likely to be without risk of deleterious noncancer effects during a lifetime. *Source*: www.EPA.org

Index

Lightning Source UK Ltd.
Milton Keynes UK
UKHW050648171118
332491UK00014B/251/P